CSR, Sustainability, Ethics & Governance

Series Editors

Samuel O. Idowu, London Metropolitan University, London, UK
René Schmidpeter, Cologne Business School, Cologne, Germany

In recent years the discussion concerning the relation between business and society has made immense strides. This has in turn led to a broad academic and practical discussion on innovative management concepts, such as Corporate Social Responsibility, Corporate Governance and Sustainability Management. This series offers a comprehensive overview of the latest theoretical and empirical research and provides sound concepts for sustainable business strategies. In order to do so, it combines the insights of leading researchers and thinkers in the fields of management theory and the social sciences – and from all over the world, thus contributing to the interdisciplinary and intercultural discussion on the role of business in society. The underlying intention of this series is to help solve the world's most challenging problems by developing new management concepts that create value for business and society alike. In order to support those managers, researchers and students who are pursuing sustainable business approaches for our common future, the series offers them access to cutting-edge management approaches.

CSR, Sustainability, Ethics & Governance is accepted by the Norwegian Register for Scientific Journals, Series and Publishers, maintained and operated by the Norwegian Social Science Data Services (NSD)

More information about this series at http://www.springer.com/series/11565

Stephen Vertigans • Samuel O. Idowu
Editors

Global Challenges to CSR and Sustainable Development

Root Causes and Evidence from Case Studies

Editors
Stephen Vertigans
School of Applied Social Studies
Robert Gordon University
Aberdeen, United Kingdom

Samuel O. Idowu
Guildhall School of Business and Law
London Metropolitan University
LONDON, United Kingdom

ISSN 2196-7075　　　　　　　ISSN 2196-7083　(electronic)
CSR, Sustainability, Ethics & Governance
ISBN 978-3-030-62500-9　　　ISBN 978-3-030-62501-6　(eBook)
https://doi.org/10.1007/978-3-030-62501-6

© Springer Nature Switzerland AG 2021
This work is subject to copyright. All rights are reserved by the Publisher, whether the whole or part of the material is concerned, specifically the rights of translation, reprinting, reuse of illustrations, recitation, broadcasting, reproduction on microfilms or in any other physical way, and transmission or information storage and retrieval, electronic adaptation, computer software, or by similar or dissimilar methodology now known or hereafter developed.
The use of general descriptive names, registered names, trademarks, service marks, etc. in this publication does not imply, even in the absence of a specific statement, that such names are exempt from the relevant protective laws and regulations and therefore free for general use.
The publisher, the authors, and the editors are safe to assume that the advice and information in this book are believed to be true and accurate at the date of publication. Neither the publisher nor the authors or the editors give a warranty, expressed or implied, with respect to the material contained herein or for any errors or omissions that may have been made. The publisher remains neutral with regard to jurisdictional claims in published maps and institutional affiliations.

This Springer imprint is published by the registered company Springer Nature Switzerland AG.
The registered company address is: Gewerbestrasse 11, 6330 Cham, Switzerland

Foreword

The global development of our economies as well as our common future will largely be influenced by our corporate and individual behaviours as well as the political measures taken by governments around the globe. It is already obvious today that the COVID-19 crisis has some major impact on our world economy. The main question seems to be: How are we going to be able to rebuild the global economy—and if so what will be the 'new normal' after the corona pandemic incident? We need a master plan to rebuild our economies and to simultaneously transform them towards issues relating to sustainability!

Is the current economic crisis the only driver for the necessary change, is in fact a relevant question to ask as things stand? I would personally say NO to this question for a number of reasons. Over the last decades, many of our business models have only survived due to the unsustainable exploitation of natural resources or social inequalities. We have seen a decline of 82% in the global population of mammals and a destruction of 47% of the ecosystems. Some even argue that this is one of the causes of pandemics like COVID-19 and more pandemics are probably ahead of us due to man's constant destruction of nature and some of its constituents.

Natural scientists have continuously argued that we live far above our planetary capacities, for example, CO_2 emissions, but even worse the loss of biodiversity which will inevitably lead to the breakdown of the world food production and the biosphere we live in if things do not change very soon. As the COVID-19 crisis can be overcome by hygiene measures and medicine, the measures we need in order to fight economic crises as well as the climate change will be even more far-reaching than these two actions.

Saying this is tantamount to arguing that the corona crisis is only a catalyst to much bigger impending challenges. It clearly shows the weaknesses of our current economic system and the need to change our economies in a sustainable way that enables us to solve our world's most challenging problems. The Sustainable Development Goals indicate what needs to be done, and how we need to realign our economic system to societal and environmental needs.

As John Elkington has already outlined before the corona pandemic, we are already in a situation of a high degree uncertainty and discomfort. We need to use the SDGs as a polar star to identify new business opportunities which outreach the current business models by a factor of 10 when it comes to impact and finally come up with an innovative economic and responsible political system. We are certainly in a real system change and not only in terms of the health crisis which some people might wrongly believe to be the case but also in terms of our other areas of sustainability and human existence.

The articles in this publication clearly outline what a positive future should look like and which measures have already been taken to transform our economic systems for a better future. It identifies the reasons why the old trade-off thinking between profitability and sustainability is no longer an appropriate mindset anymore, at least in the current global situation. The new thinking is based on 'AND' rather than on 'Either–Or'. Profit and positive impact for society are both integrated into the purpose of business, that is my own take of what we need and what we should all work towards in everything we do.

I would like to thank the two editors **Stephen Vertigans and Samuel O Idowu** as well as all the authors of the chapters that make up the book for their meaningful and impact-oriented work. My thanks also go to Springer which supports our academic global network on CSR, sustainability, ethics and governance. This book is certainly another milestone in the sustainable transformation of the global business world and makes a responsible impact to our common future! Everyone is invited to use the outlined experiences and good examples from around the world for their personal positive change. Let us promise that everyone would take care of each other and would ensure that no one is left behind in the crisis and even after the challenges that would follow the crisis. We need to change together—these might be the most important message to us all coming from the corona virus experiences!

Stay healthy, safe and well and let keep our promise to each other and to this our great planet!

Cologne Business School, Cologne,   René Schmidpeter
Germany
August 2020

Preface

Corporate social responsibility (CSR) and sustainable development (SD) have increasingly been positioned as solutions to widespread and deep-rooted environmental, social and economic problems across local, national and international boundaries. The concepts have grown in prominence following shifts in responsibilities that accompanied the rise of neoliberalism and deregulation initially in the global North before being rolled out within international financial packages to the global South. Alongside the concomitant civil demand for corporate engagement, growing awareness both of environmental disasters and their causes have led to widespread support for sustainable development in order to manage the world's finite resources. Consequently, the two concepts have become rallying calls through which to address corporate excess, social and geographic disadvantage, climate change and subsequent population vulnerability.

However, the impact of the development and implementation of CSR and sustainable development programmes has been mixed. Certainly, some local content, health and education initiatives are delivering longer-term benefits, community engagement can be more constructive and management of local resources can be directed towards the benefit of future generations. Yet there are also inconsistencies and contradictions often driven by self-interest, short-termism and fragmented ways of thinking that are underpinned by the continuing dominance of global northern ways of thinking in the application of global South solutions. This combination of factors has resulted in both concepts failing to achieve their potential at a time when confidence in leading global institutions and dominant nation-states to resolve problems continues to plunge. The need for CSR and sustainable development to therefore address fundamental social and environmental problems is arguably growing. In this book, we emphasize that if the concepts are to achieve their potential, then the problems that have bedevilled related strategies and programmes need to be acknowledged and addressed.

The multidisciplinary nature of those issues that revolve around CSR and SD makes them fertile areas for academics and practitioners from any discipline who are interested in the capabilities, challenges and potentials of CSR and sustainable development to continue to contribute to global knowledge about how corporate and individual citizens are faring in the attempt to ensure the survival of this planet. The intended readership is expected to be students and academics from across related disciplines, policymakers, NGOs and members of societies who study or have an interest in CSR and sustainable development. It is hoped that the chapters in the book will answer a number of related questions in CSR and SD.

Aberdeen, UK Stephen Vertigans
London, UK Samuel O. Idowu
Summer 2020

Global Challenges of Corporate Social Responsibility and Sustainable Development: An Introduction

Over recent years, the concepts, and associated practices, of corporate social responsibility and sustainable development have been subject to considerable challenges. These challenges range from basic questions concerning their purpose and organizational intent through to relevance to contemporary issues and suitability for both local and global applications. We can perhaps condense these challenges, into the questions—are corporate social responsibility and sustainable development meeting expectations, if not why not and what needs to change in order for their conceptual promises to be realized? These are questions that were formulated during the editing and as such were not set for the book contributors. Nevertheless, each author has in their own way explored these questions and in the process have provided us with some fascinating insights both into challenges and potential solutions.

A number of these challenges have been with us for many years and were instrumental in the concepts of CSR and sustainable development becoming prominent. For instance, the increasingly pervasive nature of globalization erasing nation-state autonomy, the diminishing capabilities of governments to regulate big business in environmental impact and fiscal matters such as taxation and the temporal loyalties of large corporation to shareholders rather than regions continue to resonate and cause alarm. Accompanying the international expansion of economic and technological processes has been the continuing rise in social inequalities between and within countries. These concerns stem in part from the frequent relocations of transnational corporations according to the best perceived short-term financial gain for their shareholders and competition between countries willing to waiver environmental, welfare and fiscal regulations in order to appeal to transnational companies. The loyalty of the companies both to the countries in which they operate and the staff who work on their behalf will often last only until a better opportunity appears. Consequences of these changing economic and political processes have included the growing numbers of vulnerable workers and reserve armies of labour who compete for diminishing wages and poorer conditions, while production and consumption of natural resources have increased alarmingly.

In the global North, the marginalization and national neglect of previously secure working class groups have fueled resentment and disillusionment with conventional politics. Their anger has also been misdirected towards perceived 'others' who are believed to have taken 'their' jobs. Yet companies responsible for the realignment of labour patterns are often distanced from the targeting of this protectionist nationalism. However, as the millennials mature and larger numbers start to experience the limited opportunities of the 'lost generation' who emerged following the 2007/8 financial crisis, we can expect different forms of protest. These protesters are less likely to follow populist explanations for their situations and, as recent protests have already indicated, are more likely to track the cause of their problems through political and economic processes. In other words, the critical attention being, and expected to be, directed at TNCs has similarities with the early 1980s. Hence 40 years after the neoliberal surge of deregulation and the shift of power from political institutions to economic organizations, many of the concerns that were raised then continue to provide fertile grounds for scrutiny and challenges to the nature of late capitalism and the global North hub.

Within the global South, the issues facing countries often differ from the global North. Although the migration of manufacturing and service operations has provided some job opportunities and economic trickle effects, the nature of these jobs is often insecure and low paid. Indeed, the relocation of business to these countries would not happen if such conditions did not exist. Alongside the mobility of transnational corporations are deep-rooted structural inequalities within both global and national processes which evolving TNC employment patterns do little to change. Moreover weaker regulatory frameworks, combined with the eagerness for TNCs to operate within their countries, have meant that countries in the global South face more immediate environmental problems concerning waste and pollution. With countries facing population growth and youth bulges the demand for jobs, education, food and pressures on education, welfare, housing and transportation will continue to grow. Further compounding difficulties, in the longer term, these locations are also more vulnerable to the impacts of climate change, rising sea levels, heavy rains, water shortages, food insecurity and desertification. Consequently, national and international failures to provide basic requirements and systemic expectations of increasing numbers of people within nation-states are becoming more noticeable.

These factors are contributing to global North and South locations sharing spiralling levels of disillusionment and cynicism about the ways in which national governments and corporate business operate and fail to deliver promises. Within the rising discontent, the concepts of CSR and sustainable development are at a crossroad. Both concepts have been significant in improving the quality of life within swathes of regions, benefitting large numbers of peoples within particular social, environmental and economic programmes. Nevertheless, despite decades of different levels of CSR and sustainable development activities, the threats outlined above remain and in some areas are growing. And while many of these problems cannot be directly attributed to business, there is much more that can be done. Many of these current challenges preceded COVID-19 and have been magnified by the pandemic. The extraordinary and uncertain times facing all societies and most businesses highlight a pressing need for TNCs and regional companies to reconsider their CSR and sustainable approaches in these evolving environments. Such reflection needs to take into consideration both many years of wasted investments and failed initiatives allied to some approaches that have delivered meaningful improvements to peoples' lives.

Hence while present COVID-19 crisis may be new, the requirement for CSR and sustainable development programmes to better concentrate resources towards meaningful solutions is not. Indeed, one of our previous publications (Idowu, Vertigans and Burlea 2017) explored how CSR had fared in the aftermath of the 2008 international financial meltdown. The banking led crisis shed light on various unethical principles and a need to tighten up regulations. Banks and related industries also became more conscious of their wider responsibilities or as a cynic would argue, they became aware of the need to be seen to be more responsible. Nevertheless, the wider ramifications of the crisis have been financial constraints that have restricted money being available for CSR across many organizations combined with further scepticism of corporate intentions. Our previous book picked up these arrangements across the world from small- and medium-sized enterprises through to national programmes. While this book also focuses on case studies within nation-states from around the world, the chapters take up a range of wider cross-cutting issues that underpin the challenges CSR and sustainable development face. One of the most noticeable differences has been how Sustainable Development Goals (SDGs) are closely associated with CSR programmes.

The application of CSR to the SDGs is symptomatic of both the realization about how global development needs constructive business engagement and that there are expectations and pressures on companies to deliver in ways they did not under the preceding uncritical forms of misdirected voluntarism. The book commences with a number of chapters exploring the impacts of SDGs on recent and future approaches to CSR and sustainable development. In the first chapter, Ásványi and Zsóka's explore CSR and corporate sustainability attachments to SDGs. Focusing on international and Hungarian examples enables the authors to illustrate the relevance of related programmes to achieving wider goals while also acknowledging how prioritising SDGs can have drawbacks to companies. Learning from these examples can collectively contribute towards greater integration of corporate and sustainable development goals.

Sarpong's requests in the Chapter 'Is Covid-19 Setting the Stage for UN Agenda 2030? In Pursuit of the Trajectory' for the COVID-19 pandemic to be a rallying call to learn lessons that will help progress the UN's Sustainable Development Agenda 2030. Hitherto the pandemic has highlighted weaknesses in global processes that have emphasized national and international inequalities and the iniquitous manner in which the most vulnerable peoples are facing the brunt of health, economic and social consequences. Conversely, the speed with which the world has been so adversely affected by a disease emphasizes the interconnections between countries and regions. By drawing upon these similarities, Sarpong points out the innovations and closer relationships that are at the forefront of the response to the pandemic and which can provide the basis to transform progress in achieving the SDGs.

Maphiri, Matasane and Mudimu focus, in the Chapter 'Challenges to the Effective Implementation of SDG 8 in Creating Decent Work and Economic Growth in the Southern African Hemisphere: Perspectives from South Africa, Lesotho and Zimbabwe', on SDG8, the promotion of inclusive economic growth, full and productive employment and decent work for all. Particular attention is placed on interlinking economic growth and decent work with the complementary objectives of the Southern African Development Community (SADC), namely to achieve economic development, peace and improve the quality of life for Southern Africans. Through specific reference to Lesotho, South Africa and Zimbabwe, the authors highlight both the opportunities there are to achieve decent work and the obstacles that are hindering developments.

The section on SDGs leads into chapters about CSR and sustainable development in the Global South. In the early days of CSR and sustainable development, much of the academic concentration was upon the global North. That related programmes were increasingly being directed in the South has belatedly been acknowledged both in the surge of publications and as the above chapters indicate, in the closer connections between business and SDGs which have a heavy Southern slant. Despite this growing emphasis, in the Chapter 'CSR in the Global South: The Continuing Impact of Postcolonial Power and Knowledge' Vertigans argues that evidence remains that CSR programmes continue to be dominated by the uncritical application of northern hemisphere derived ethics, governance and standards. Incorporating insights from postcolonial studies leads him to propose that colonial approaches and northern hemisphere sponsored postindependence programmes are being reproduced under the CSR rubric. To overcome this postcolonial continuation, he raises a number of proposals which can help overcome these deep-rooted relationships and processes.

Buckler takes forward some of the issues identified in the preceding chapter when exploring how companies navigate through local content regulations and the difficult balance between compliance and mutual benefit. By tracing roots and contemporary contexts, she evidences surrounding tensions for CSR and explores how shifts in power that accompany the implementation in local content policies are creating a backlash from within the global North. These shifts in dynamics and language raise questions for the future direction of the global economy.

In the Chapter 'Corporate Social Responsibility Practices and Its Implementation After the Legal Mandate: A Study of Selected Companies in India with Special Emphasis on the Mining Sector', Sumona undertakes an empirical study of mining companies operating in India. Studying companies' CSR practices from 2014–2015 to 2018–2019 enabled her to identify preferred activities such as education and health and corporate expenditure. The analysis highlights concerns about where companies direct their resources and in particular the relative neglect of energy, climate change, marine and terrestrial ecosystem that are so essential to sustainable development.

The next section focuses on Europe, highlighting growing recognition about the importance of the circular economy to sustainable development. The European Union (EU) has introduced numerous directives for waste that became part of the

2015 Circular Economy Package. The cornerstone of the approach is the encouragement of sustainable production and consumption behaviours. In the Chapter 'A Circular Economy Strategy for Sustainable Value Chains: A European Perspective', Camilleri provides a cost–benefit analysis of the EU's circular economy strategy and critically examines the most recent regulatory guidelines, instruments and principles. The findings indicate the circular economy approach is reducing industrial waste, emissions and energy leakages, leading to a number of suggested recommendations to policymakers and practitioners.

Alongside the shifting improvements within the EU strategy, Puiu outlines in the following chapter, a different picture of what is happening within Romania. She produces some comparative analysis of reuse/recycling of municipal waste within the European Union which raises concerns about Romania's levels of low recycling and high landfill. In order to find out about what is hindering progress, Puiu carried out fieldwork which enabled a better understanding of the obstacles and current levels of knowledge. This analysis has informed a number of proposed solutions to the identified difficulties facing reducing municipal waste in the country.

Underpinning a number of the above contributions are recent strategic approaches or often the lack of direction when such strategies are required. The following chapters directly address the importance of CSR and sustainable development strategies. Broomes positions four phases of CSR-related activities stemming from philanthropy through to strategic approaches whereby sustainability is embedded into core business. Integrating the triple sustainable development pillars of social, environmental and economic provides the basis for solving particular societal issues while becoming a catalyst towards SDGs. Applying a range of case studies allows Broomes to identify both strategic potential solutions and ongoing areas of concern such as health inequalities, energy inefficiencies, lack of sustainable livelihoods and human rights abuses in global supply chains. Through more targeted policies, these long-standing concerns can start to be more effectively addressed.

In the Chapter 'Anchoring Big Shifts and Aspirations in the Day-to-Day: A Case for Deeper Decision Making and Lasting Implementation Through Connecting Change-Makers with Their Values', Locher argues radical changes are required if there is to be greater social responsibility and sustainability. These changes need to integrate both big goals with personal drive and individual and organizational values. Placing values within decision-making processes will, Locher argues, provide a more holistic, systemic way of achieving sustainable better outcomes.

Amoako also considers in the Chapter 'The Role of Corporate Social Responsibility in Business Sustainability,' how CSR agendas can be taken forward in ways that benefit customers and companies. After identifying the benefits of strategic CSR, he argues that CSR needs to be embedded into the core business. He emphasizes the significance, and need for wider recognition, of human resource management and the interconnection with other corporate strategies that are central to the effective implementation of CSR activities.

The closer international interweaving of CSR with SDGs outlined above has led to the heightened requirement for accuracy and transparency which is noted within the drive behind this section arguing for improved reporting standards. Nwagbara and Kalagbor explore, in the Chapter 'Institutional Pressures and CSR Reporting Pattern: Focus on Nigeria's Oil Industry', reporting/disclosure in Nigeria and the role of institutions in these processes. The chapter focuses specifically on Nigeria's oil industry, which has a history of lacking accountability, legitimacy, responsibility and transparency. Locating the industry within politico-cultural, regulatory, governance and institutional contexts allows the authors to shed light on the level of CSR reporting. This reporting seeks to legitimize activities through social performance and philanthropy rather than accountability, business responsibility and wider stakeholders' interests.

Frederiksen, Pedersen, Nielsen and Idowu take forward reporting in their examination of the Global Reporting Initiative (GRI) Standards in the Chapter 'Corporate Social Indices: Refining the Global Reporting Initiative'. The authors' critical analysis shows how the lack of comparative social performance data for different organizations hinders informed decision-making. Drawing inspiration from the UN's Human Development Index and Sen's capability approach, they suggest GRI Standards can be improved by incorporating a set of cross-organization social indices that will aid comparison and decision-making.

In the final chapter, Neri provides a timely positioning of environmental, social and governance (ESG) factors within twenty-first century challenges. Drawing upon contemporary examples, she explains how separate financial reporting is ill-suited for the requirements of this period. Against the backdrop of the lack of global standards, Neri argues that the standardized of ESG will enable integrated reporting to better inform business managerial and investment decision-making across the broad spectrum of their risks and opportunities.

These chapters outline a range of challenges in different contexts that are facing CSR and sustainable development programmes. Building upon the authors' analysis, they also make a plethora of well-informed recommendations on how these obstacles can be overcome. Hence, while the challenges have grown during the COVID-19 pandemic, the contributors in this book are typical of the innovative solutions that can be found within experts in CSR and sustainable development. The next challenge is to find ways for those recommendations to be implemented ...

Robert Gordon University, Aberdeen, UK	Stephen Vertigans
London Metropolitan University, London, UK	Samuel O. Idowu

Acknowledgements

We are eternally grateful to a number of people who have made the publication of this book on Global Challenges of CSR and Sustainable Development possible. We express our thank you to all our contributors who are based in eight countries around the world and have worked tirelessly despite their heavy professional commitments to put together their chapters. Thank you to everyone. We are also grateful to Professor Rene Schmidpeter for writing a befitting foreword to the book.

We would like to thank our publishing team at Springer headed by the executive editor, Christian Rauscher, Barbara Bethke and other members of the publishing team who have supported this project and all our other projects.

We are also grateful to our respective family for bearing with us during those periods we should be with them but chose to spend time meeting our obligations towards the book.

Finally, we apologize for any errors or omissions that may appear anywhere in this book, please be assured that no harm was intended to anybody. Causing harm or discomfort to others is simply not the spirit of corporate social responsibility.

Contents

Part I Sustainable Development Goals

Directing CSR and Corporate Sustainability Towards the Most Pressing Issues 3
Katalin Ásványi and Ágnes Zsóka

Is Covid-19 Setting the Stage for UN Agenda 2030? In Pursuit of the Trajectory 21
Sam Sarpong

Challenges to the Effective Implementation of SDG 8 in Creating Decent Work and Economic Growth in the Southern African Hemisphere: Perspectives from South Africa, Lesotho and Zimbabwe ... 39
Mikovhe Maphiri, Matsietso Agnes Matasane, and Godknows Mudimu

Part II Global South

CSR in the Global South: The Continuing Impact of Postcolonial Power and Knowledge 67
Stephen Vertigans

CSR, Local Content and Taking Control: Do Shifts in Rhetoric Echo Shifts in Power from the Centre to the Periphery? 87
Sarah Buckler

Corporate Social Responsibility Practices and Its Implementation after the Legal Mandate: A Study of Selected Companies in India with Special Emphasis on the Mining Sector 105
Sumona Ghosh

Part III Europe

A Circular Economy Strategy for Sustainable Value Chains: A European Perspective 141
Mark Anthony Camilleri

Recycling Initiatives in Romania and Reluctance to Change 163
Silvia Puiu

Part IV Strategic

Catalyst, Not Hindrance: How Strategic Approaches to CSR and Sustainable Development Can Deliver Effective Solutions for Society's Most Pressing Issues 191
Veronica Broomes

Anchoring Big Shifts and Aspirations in the Day-to-Day: A Case for Deeper Decision Making and Lasting Implementation Through Connecting Change-Makers with Their Values 215
Christine Locher

The Role of Corporate Social Responsibility in Business Sustainability 229
George Kofi Amoako

Part V Reporting

Institutional Pressures and CSR Reporting Pattern: Focus on Nigeria's Oil Industry 249
Uzoechi Nwagbara and Anthony Kalagbor

Corporate Social Indices: Refining the Global Reporting Initiative 271
Claus Strue Frederiksen, David Budtz Pedersen, Morten Ebbe Juul Nielsen, and Samuel O. Idowu

Environmental, Social and Governance (ESG) and Integrated Reporting 293
Selina Neri

Index 303

Editors and Contributors

About the Editors

Stephen Vertigans is a sociologist who is currently Head of School of Applied Social Studies at Robert Gordon University, Aberdeen, UK. His research interests include corporate social responsibility, political violence and community development. At present, he is researching into processes of resilience in East African informal settlements during the COVID-19 pandemic. In his wide-ranging publications, he applies a sociological approach to try help levels of knowledge and understanding into social, political, economic and cultural processes that shape individual and community behaviours and activities. Positioning CSR within these wider processes can provide better insights into the successes and failures of CSR policies.

Samuel O. Idowu is a senior lecturer in accounting and corporate social responsibility at London Guildhall School of Business and Law, London Metropolitan University, UK. He researches in the fields of Corporate Social Responsibility (CSR), Corporate Governance, Business Ethics and Accounting and has published in both professional and academic journals since 1989. He is a freeman of the City of London and a Liveryman of the Worshipful Company of Chartered Secretaries and Administrators. He is the Deputy CEO and First Vice President of the Global Corporate Governance Institute. He has led several edited books in CSR, he is the editor-in-chief of three Springer's reference books—the Encyclopedia of Corporate Social Responsibility, the Dictionary of Corporate Social Responsibility and the Encyclopaedia of Sustainable Management (ESM). He is an editor-in-chief of the International Journal of Corporate Social Responsibility (IJCSR), the editor-in-chief of the American Journal of Economics and Business Administration (AJEBA) and an associate editor of the International Journal of Responsible Management in Emerging Economies (IJRMEE). He is also a series editor for Springer's books on CSR, Sustainability, Ethics and Governance. One of his edited books won the most

Outstanding Business Reference Book Award of the American Library Association (ALA) in 2016 and another was ranked 18th in the 2010 Top 40 Sustainability Books by, *Cambridge University, Sustainability Leadership Programme*. He is a member of the Committee of the Corporate Governance Special Interest Group of the British Academy of Management (BAM). He is on the editorial boards of the International Journal of Business Administration, Canada and Amfiteatru Economic Journal, Romania. He has delivered a number of keynote speeches at national and international conferences and workshops on CSR and has on two occasions 2008 and 2014 won Emerald's Highly Commended Literati Network Awards for Excellence. To date, he has edited several books in the field of CSR, Sustainability and Governance and has written seven forewords to CSR books. Samuel has served as an external examiner to the following UK universities—Sunderland, Ulster, Anglia Ruskin, Plymouth, Robert Gordon, Aberdeen, Teesside, Sheffield Hallam, Leicester De Montfort, Canterbury Christ Church and Brighton. He has examined PhD theses for universities in the UK, South Africa, Australia, The Netherlands and New Zealand.

Contributors

George Kofi Amoako is a senior lecturer and immediate past head of Marketing Department of Central Business School, Central University in Accra, Ghana. He also lectures as adjunct lecturer at Lancaster University Ghana Campus and Webster University Ghana Campus. He has taught Market Research, Digital Marketing, and Introduction to Marketing, Marketing Management Essentials and Fundamentals of Marketing at Lancaster University. He is an academic and a practising Chartered Marketer (CIM-UK) with specialization in Branding, CSR and Strategic Marketing. He has many years of industry experience. He was the regional sales manager for Fan Milk Ghana Ltd and The head of Lucas Desk at Mechanical Lloyd Ltd in Accra, Ghana. He has consulted for Decathlon Ghana Limited, Ghana Insurance Commission, Ghana Civil Aviation Authority, Bank of Ghana and many others. He was educated in Kwame Nkrumah University of Science and Technology in Kumasi, Ghana and at the University of Ghana and the London School of Marketing (UK). He obtained his PhD from London Metropolitan University, UK. He also received Postgraduate Certificate in Academic Practice (International) from Lancaster University, UK in November 2018. He has considerable research, teaching, consulting and practice experience in the application of Marketing Theory and principles to everyday marketing challenges and management and organizational issues. He is a Chartered Marketer with The Chartered Institute of Marketing, UK. He was a member of the University Council of the University of Ghana in 2000–2001. He has consulted for public sector and private organizations both in Ghana and UK. He has a strong passion for marketing strategy, branding, service quality and CSR issues in the corporate world. He has published extensively in internationally peer-reviewed academic journals and presented many papers at international conferences in Africa, Europe, America and Australia. He has contributed to the latest marketing book 'Break Out Strategies for Emerging Marketing', edited by Professor J. Sheth and recommended by Philip Kotler. This book was launched in the USA in June 2016 and was published by Pearson. He spoke at TEDx Accra and British Council organized seminar on Customer Profitability on 19 August 2016. He is currently a quality assurance consultant for British Council Jobs for the Youth Program and also Educational Consultant for the British Council Connecting Classroom Program in Ghana. He is a twice recipient of Academy of Management Scholarship for faculty development in Africa for 2012 and 2013; a recipient of Academy of Marketing Science J. Sheth Scholarship Award in 2013 and Africa CSR Leadership Award, Le Meridian (2015) in Mauritius.

Katalin Ásványi works as an associate professor at the Marketing Institute of Corvinus University of Budapest, Hungary. She got a PhD in CSR from Corvinus University of Budapest, Hungary in November 2013. Her research interests are CSR, sustainability, education for sustainability, tourism, cultural tourism and family-friendly tourism. Her teaching profile includes bachelor, master and postgraduate courses into corporate sustainability and CSR, CSR communication. Since 2018, she has been teaching as a guest professor at the University of Passau.

Veronica Broomes PhD, is UK-based consultant with expertise in Sustainability, Corporate Social Responsibility, Local Content in Supply Chains, Sustainable Development and Impact Assessment.

A dynamic, resourceful and experienced professional who has worked at the interface of research, consulting and training/lecturing, she served as a part-time lecturer to postgraduate students in the Human Resources Management master's degree programme at the University of Coventry in England. More recently, in late 2019, she was appointed as an associate lecturer in the School of Applied Social Studies at the Robert Gordon University in Aberdeen, Scotland. Her consultancy work has included research on sea-level rise and Commonwealth Island States conduct of sustainability audits for businesses, environmental and social impact assessments for agricultural, infrastructure and forestry projects evaluating grant funding applications for waste and food projects.

Having written widely on CSR, sustainability and associated topics for over 15 years, her publications include articles on CSR and the links with Governance, Human Resources and in leveraging inward investments. Her book 'Who Invests Wins: **Leveraging Corporate Social Responsibility for CEOs, Investors and Policymakers to Create 'Triple Wins"** provides insights into how CSR can be used strategically to create wins for CEOs, investors and policymakers. In 2013, she launched the first 'State of CSR' survey. It aimed to find out both what organizations across various industries were doing in CSR as well as inviting views on the future of CSR.

She has made meaningful contributions, including conference presentations and scholarly articles, to discussions about environment, social and governance (ESGs) in the decision-making process of investors and CSR as a catalyst for achieving development objectives in commonwealth countries. Her early career was in agricultural research in Guyana where she worked as a plant tissue culture specialist/biotechnologist, established the national facility for plant tissue culture. As a parttime lecturer at a specialist further education college she contributed to updating the botany curriculum as her lectures covered topics of agriculture biotechnology, biodiversity and climate change, including sea-level rise and implications of then contemporary issues for agriculture.

She is an affiliate member of the International Association for Impact Assessment (IAIA) and former head of its Guyana Chapter. She supports the start and growth of small and medium-sized enterprises (SMEs) through her contributions as a Volunteer Speaker with the Corporation of London's City Business Library, Portobello Business Centre in London and the Business and IP Centre at the Sheffield Central Library. In Sheffield, she volunteers also with the Sheffield Sustainability Network where she is an active member in its working groups on: (a) Sustainability Audits and (b) Diversity and Outreach.

She has worked in the UK and internationally holds a Masters of Law (LLM) degree from The Open University, PhD in Plant Science from University of Sheffield and postgraduate certificate in Environmental Impact Assessment from University of London.

Sarah Buckler PhD is a social anthropologist with an interest in the relationships between people and organizations, organizational cultures, human creativity and development. A particular interest in narrative and rhetoric and the ways these are used by individuals and organizations in order to achieve particular ends began with her PhD research with Gypsy Travellers and continued into her subsequent employment carrying out research for local and national government departments. More recently, this interest has expanded to focus on the ways different CSR and other development projects are delivered, especially in West Africa where she spent some time working on evaluation of a variety of CSR projects. In the light of Brexit and increasing populist rhetoric across the world, this interest has become a fascination with the ways power shifts are reflected in language and how this in turn influences national and international economic policies.

Mark Anthony Camilleri is an associate professor in the Department of Corporate Communication at the University of Malta. He successfully finalized his full-time PhD (Management) in three years' time at the University of Edinburgh in Scotland—where he was also nominated for his 'Excellence in Teaching'. He also holds an MBA from the University of Leicester and an MSc from the University of Portsmouth, among other qualifications. During the past years, he taught business subjects at undergraduate, vocational and postgraduate levels in Hong Kong, Malta, UAE and the UK.

He is a member in the Global Reporting Initiative (GRI)'s Stakeholder Council, where he is representing the European civil society. He is a scientific expert in research for the Ministero dell' Istruzione, dell' Universita e della Ricerca (in Italy) and a reviewer for the Austrian Science Fund (FWF). He is an editorial board member in a number of academic journals, conferences and committees. He has authored and edited 7 books for Emerald, IGI Global Springer, among others. He has published more than 100 contributions in high-impact, peer-reviewed journals, chapters and conferences.

Claus Strue Frederiksen Ph.D., is a trained philosopher and has been working with CSR for more than a decade. He has published in a variety of scientific journals and books, including in *Journal of Business Ethics*, *International Journal of Applied Philosophy* and in Springer's *Encyclopedia of Corporate Social Responsibility*. Currently, he is working as a sustainability consultant at the Danish biotech company Gubra, where he has developed and implemented ethical approach to CSR.

Sumona Ghosh has been associated with St. Xavier's College, Kolkata since 2002. She was the head of the Department of Law from 2003 to 2018. Presently, she is the Joint Coordinator of the Foundation Course. After completing her post-graduation in Commerce with rare distinction, she has been conferred with the Degree of Philosophy in Business Management by the University of Calcutta on 31st of July 2014. Her area of research was on Corporate Social Responsibility (CSR). The title of her doctoral dissertation was 'Pattern of Participation of Public and Private Sector Companies in Corporate Social Responsibility Activities'.

She has published in journals of national and international repute. She has been highly acclaimed for her guest lectures on CSR in premier institutes of higher learning including the Indian Institute of Management (Calcutta) and Indian Institute of Management (Shillong). She has taken sessions in Management Development Programmes conducted by premier institutes on CSR. She has presented papers on CSR at various national and international conferences. Her research interest lies in Corporate Social Responsibility, Sustainable Development, Integrated Reporting and Philosophies of Management. She is also a Certified Assessor for Sustainable Organizations (CASO), certification conferred upon her by UBB GmBH Germany.

She is also the recipient of the 'Bharat Jyoti Award' for meritorious services, outstanding performance and remarkable role given by India International Friendship Society for the year 2012 given by Dr Bhishma Narain Singh, Former Governor of Tamil Nadu and Assam, on 20th of December 2012.

Anthony Kalagbor holds BSc and MBA, respectively, from Plymouth University and University of Sunderland in marketing and international human resource management, respectively. He is currently lecturing at Cumbria University London Campus. He also lectured at Greenwich School of Management (GSM) London and was a tutorial lecturer at Rivers State University of Science and Technology, Nigeria as well as worked at Mercantile Bank of Nigeria plc. for upwards of six years before moving to Europe for further studies. While in Europe, he worked as sales manager in Drettmann Yachts, a reputable company in Europe (Germany). He also has been awarded Chartered Management Institute (CMI) Level 7 Diploma in Strategic Management and Leadership. He is a prospective doctoral candidate and interested in CSR research and consults in the areas of business responsibility and sustainability.

Christine Locher M.A. FRSA, FLPI is a consultant, coach and author, working with leaders looking to put their vision into practice based on their core values, following a first career in newspaper and radio journalism and a global career in learning and development. She has degrees in Communication, Intercultural Communication and Psychology (M.A. from Ludwig-Maximilian-University Munich, Germany with studies at Kyushu University Fukuoka, Japan) and postgrads in Systems Thinking in Practice (Open University Milton Keynes, UK) and Conflict Resolution (Fernuniversitaet Hagen, Germany) and a professional certificate in Solution Focused Business Practice (University of Wisconsin-Milwaukee, USA). Her book: Values-based: Career and Life Changes that make Sense just launched in 2019, her article 'Stepping into Your Values – Literally' will launch in November 2019 in the Book of the House of Beautiful Business and her book on decision-making based on values will launch in 2020. She is a fellow of the RSA (Royal Society for the encouragement of Arts, Manufactures and Commerce) and a fellow of the Learning and Performance Institute (LPI). She is an active member of the Institute of Directors' Entrepreneurship network. She is a certified coach since 2007 (first ICF-accredited training in Germany). She is a trained yoga teacher studying yoga, mindfulness and various body-centred approaches with teachers in Germany, India and the USA. She is also licenced for psychotherapy in Germany (Heilpraktiker fuer Psychotherapie) trained in Client-Centred Therapy, Gestalt and Psychosynthesis/Transpersonal Psychotherapy. She coaches for WYSE (World Youth Service and Enterprise, a charity affiliated with the UN Department of Public information that serves emerging leaders all around the world). She runs her own coaching and leadership company.

Mikovhe Maphiri is attorney of the High Court, South Africa. She is an academic (lecturer) at the Department of Commercial Law, University of Cape Town, South Africa. She holds LLB from the University of Limpopo and LLM from the University of Cape Town, where she is also currently a PhD candidate. She is an emerging researcher whose research interests are in corporate governance, CSR and sustainability in South Africa. She is a lecturer in the Department of Commercial law at the University of Cape Town where she teaches law to both law and non-law students.

Matsietso Agnes Matasane holds LLB from the University of Venda, South Africa, LLM Mercantile Law from the University of Pretoria, South Africa, she is currently LLD Candidate at the University of Pretoria. She is an emerging researcher from the University of Pretoria. She is a PhD candidate in Banking Law at the University of Pretoria. Her area of research is in deposit guarantee system in the South African banking sector in which she interrogates the envisaged introduction of explicit deposit insurance system in South Africa. She is the recipient of the Absa Barclays Africa Chair in Banking Law in Africa scholarship which is awarded to PhD candidates in the field of Banking Law.

Godknows Mudimu is a Research Associate Mineral Law in Africa, a University of Cape Town PhD Candidate. He holds LLM from the UCT, LLB and BSocSc from Rhodes University, South Africa. He is a PhD candidate at the University of Cape Town. His thesis assesses the effectiveness of self-regulation on labour standards in selected South African mining sectors. His areas of interest include regulation, labour law, mineral law, occupational safety and health at mines, and transparency and accountability in the extractive industry.

Selina Neri is professor of management and corporate governance at HULT International Business School (Dubai and London campuses), where she designs and teaches postgraduate courses in global leadership, talent management, business ethics and luxury marketing. In 2020, she was awarded Best Professor of the Year for the HULT London campus. Her most recent research focuses on corporate purpose, the role of boards of directors and investors in creating sustainable value and fostering sustainable development. Her research has been published in the *Academy of Management Proceedings (2020),* in the *Encyclopaedia of Sustainable Management (2020)* edited by Springer, and in peer-reviewed journals including *Business Ethics: A European Review (2019)*, the *Journal of Studies in International Higher Education (2019),* the *Journal of Higher Education Policy and Management (2018)* and *The International Journal of Management Education (2018).*

For over 26 years, she has worked in executive management in the technology and luxury industries in global roles in Sales, Customer Service and Marketing. She has served as non-executive director in the travel industry and is a senior industry advisor in the area of corporate governance. In recognition for her services in governance, she was included in 'The Female FTSE Board Report 2016: 100 Women to Watch', published by *Cranfield School of Management* for the UK Government, Department for Business, Innovation and Skills (Lord Davies Review).

She holds a B.A. in Economics from the University of Parma (Italy), an M.B.A. from the University of Clemson (USA) and a Ph.D. in Business Management from The British University in Dubai (UAE). She is fluent in four languages and a resident of Dubai.

Morten Ebbe Juul Nielsen Ph.D., associate professor in philosophy, Institute for communication, Copenhagen University. He works in practical philosophy, with an emphasis on political and applied philosophy, including business ethics and CSR. He has published extensively internationally.

Uzoechi Nwagbara holds both BA and MA in English as well as MSc and PhD in human resource management and management, respectively, from University of Wales, UK. He has been teaching in higher education for upwards of 19 years. He is a published academic, management consultant and researcher having published

6 books and over 100 publications in peer-reviewed international journals including ABS rated journals (1, 2 and 3 star journals). He is also on the editorial boards of some international, peer-reviewed journals including *Journal of Sustainable Development in Africa* and *Economic Insights: Trends and Challenges*. He is ad hoc reviewer for some ABS rated journals including *Accounting, Auditing and Accountability Journal (AAAJ)*, *International Journal of Human Resource Management* and *Employee Relations*, among others. He is a visiting professor at Coal City University, Nigeria and Director of Studies (DoS) at Cardiff Metropolitan University and University of the West of Scotland as well as teaches at University of Sunderland. Her research interests are eclectic involving corporate social responsibility, corporate social responsibility reporting, sustainable development, human resource management, leadership, management, corporate governance and postcolonial studies.

David Budtz Pedersen (b. 1980) is professor of science communication and impact studies and director of the Humanomics Research Centre at Aalborg University, Denmark. His research focuses on responsible research and innovation, research impact and science and technology policy. He has published and edited about 150 scientific papers, monographs, chapters, policy reports and op-eds.

Silvia Puiu is a PhD lecturer in the Department of Management, Marketing and Business Administration at the Faculty of Economics and Business Administration, University of Craiova, Romania. She has a PhD in management since 2012 and she teaches Management, Ethics Management in Business, Marketing, Public Marketing, Creative Writing in Marketing, Marketing in NGOs and Marketing of SMEs. In 2015, she graduated in postdoctoral studies after a 16-month period in which she made a research on *Ethics Management in the Public Sector of Romania*. During the scholarship, she spent two months in Italy at University of Milano-Bicocca conducting a comparative research between Romania and Italy regarding Corruption Perception Index, ethics management in the public sector and regulations on whistleblowing. During the last years, she published more than 40 articles in national and international journals and the proceedings of international conferences. Her research covers topics from corporate social responsibility, strategic management, ethics management, public marketing and management. She is a reviewer for journals indexed in DOAJ, Cabell's, REPEC, EBSCO, Copernicus or Web of Science (Sustainability, The Young Economists Journal and Annals of Eftimie Murgu University). Since 2014, she is a member of Eurasia Business and Economics Society. Some of the relevant works: *Strategies on Education about Standardization in Romania*, chapter in the book S*ustainable Development*, Springer, 2019; *NEETs: A Human Resource with a High Potential for the Sustainable Development of the European Union*, chapter in the book *Future of the UN Sustainable Development Goals*, Springer, 2019; *Entrepreneurial Initiatives of Immigrants in European Union*, The Young Economists Journal, no. 31, 2018.

Sam Sarpong PhD is associate professor, School of Economics and Management, at Xiamen University Malaysia. He is also a senior research fellow at the Department of Business, Competition, Ecological Law Institute, South Ural State University. He holds a PhD from Cardiff University and an MBA from University of South Wales. His research interests lie in the relationship between society, economy, institutions and markets. He tends to explore the nature of and ethical implications of socio-economic problems and has authored numerous scholarly publications in management and marketing journals. He is also a reviewer for many leading journals and also serves as an associate editor for the *International Journal of Corporate Social Responsibility* (Springer).

Ágnes Zsóka works as full professor at the Department Marketing Management of Corvinus University of Budapest, Hungary. Since 1997, her key research areas have been closely related to environmental and social sustainability, including responsible corporate behaviour, sustainable consumption and education for sustainability. Her main interest lies in exploring the gaps within sustainable organizational and individual behaviour, with a special focus on how to close those gaps, for the sake of consistent and conscious behaviour. Her teaching profile includes international (English and German) as well as native master courses into corporate sustainability and CSR, corporate environmental management, global climate strategy and responsible consumer behaviour. Since 2015, she has been teaching as guest professor at the University of St. Gallen and the Vienna University of Economics and Business, and she has a regular international CSR master course at the University of Passau.

Part I
Sustainable Development Goals

Directing CSR and Corporate Sustainability Towards the Most Pressing Issues

Katalin Ásványi and Ágnes Zsóka

Abstract Several pressing issues of the world today are strongly related to the failures of sustainability as well as corporate and individual responsibility. Obviously, we need a wide range of toolset to tackle those problems, including regulatory and voluntary measures, according to the nature and severity of the issue. The awareness of the necessity to incorporate sustainability and responsibility into the decision making also at the micro level became common. We can witness how specific companies committed themselves to manage their CSR activities at a strategic level and how they mobilise their core business expertise to tackle social and environmental problems – through their operation, volunteering, in-kind assistance, financial assistance, and further CSR tools. However, we can also witness several other companies obviously missing the point which areas and problems they really need to focus on when it comes to responsible business practices. In 2015, the United Nations set up 17 Sustainable Development Goals (UN SDGs) which provide an applicable framework for companies to find the right direction how they can contribute best to solve the most pressing sustainability issues. By 2030, several companies committed themselves to end poverty, manage inequalities, handle the problem of climate change, etc. thus creating a better future for the earth and humanity. This chapter highlights perspectives and approaches of CSR and provide case studies upon how economic, social, and ecological problems are addressed by the corporate world. Those examples can serve either as role models or as lessons for companies to direct their interests to solving the most pressing issues and to properly develop their own CSR strategies.

K. Ásványi (✉) · Á. Zsóka
Corvinus University of Budapest, Budapest, Hungary
e-mail: katalin.asvanyi@uni-corvinus.hu; agnes.zsoka@uni-corvinus.hu

© Springer Nature Switzerland AG 2021
S. Vertigans, S. O. Idowu (eds.), *Global Challenges to CSR and Sustainable Development*, CSR, Sustainability, Ethics & Governance,
https://doi.org/10.1007/978-3-030-62501-6_1

1 Introduction

Challenges of responsible corporate behaviour became widely known and discussed topics, both in academia and the society. Pressing social and environmental issues are various and have grown in number and severity in the last decades, however, companies still tend to discretionally select among the issues they address. Quite a few of them look at CSR as an opportunity to solve the ecological crisis, while others undertake social commitments, help the local community, or serve a good case. Unfortunately, a lot of them simply use CSR as compensation for the damage they cause or as a marketing tool.

This Chapter aims to provide an overview how the meaning of CSR has changed over time, and which are the main approaches to CSR orientation and corporate philanthropy, illustrated by positive and negative examples of corporate practice. Going beyond isolated actions and the philanthropy focus, value-creating features and potential of CSR will be discussed. Finally, the opportunities of corporate contribution to the UN Sustainable Development Goals are examined, as a common field of CSR and corporate sustainability. The Chapter follows a discussing manner, contrasting some relevant CSR theories with corporate practice, by focusing on the most pressing issues humanity is facing today.

2 CSR Perspectives: A Changing Phenomenon

Recognition and meaning of corporate responsibility has significantly changed over the last century. The term Corporate Social Responsibility and its origin are unclear. According to Windsor (2001), it can be traced back to the 1920s, when corporate leaders were already talking about responsibility and accountability practices. However, according to Post (2003) and Turner (2006), the concept of CSR appeared among managers in the 1930s.

In the judgment of Carroll (1999), Howard Bowen was the "Father of Corporate Social Responsibility" who opened a modern period in the CSR literature. In the light of this approach, the starting point for CSR as a concept is Bowen's (1953) book, The Social Responsibility of the Businessman. The author emphasizes the responsibility of the businessman in terms of social values and goals, and already points to corporate responsibility.

Corporate social responsibility as interpreted by Friedman (1970) is a sign of the principal-agent problem. In his view, corporate resources are misused by companies when they engage in CSR activities, which would require a much greater need to focus on value-added internal projects or to return returns to shareholders. In Friedman's judgment, managers use CSR to advance their own careers or achieve other personal goals and not for business purposes.

In 1984, Drucker, the father of management theories, went beyond the previous interpretation of the concept of CSR, stating that the management of social problems should be transformed into an economic opportunity. Corporates should act responsibly, taking into account social and business interests (Drucker 1984), so responsibility must be integrated into the operation of the company.

In Porter and Kramer's formulation from 2006, the practice of CSR is defensive in nature and is intended to mitigate or compensate for the harm the company has caused to society and the environment (Porter and Kramer 2006). At the same time, they draw attention to the fact that it is worthwhile for companies to move in the direction of value creation, by which social and economic goals together provide the company a competitive advantage. Porter and Kramer (2006), Kerekes and Wetzker (2007) argue that "Companies are neither responsible for all the problems in the world, nor they have the resources to solve them all. Every company needs to identify the range of social problems that it can most effectively participate in solving while at the same time strengthening its competitiveness the most" (Porter and Kramer 2006, p. 14). Porter and Kramer also reinforce the idea that a company should engage in CSR activities that can be incorporated into a business strategy, as this is the only way to become a successful and responsible company. The authors further developed their view to the CSV (creating shared value) concept (Porter and Kramer 2011) which reflects the double dividend opportunities of.

The definition of the European Commission (2011) highlights the responsibility of companies for their impact on society, in which the priority of responsibility for a wide range of stakeholders, beyond owners and shareholders became clear.

3 Approaches to CSR Orientation and Corporate Philanthropy

Lantos (2001) differentiates three types of CSR: ethical, altruistic, and strategic CSR. Ethical CSR means complying with economic and legal obligations and avoiding negative measures that may cause harm to a stakeholder of the company, it includes morally, ethically required CSR activities (Simon et al. 1983). Altruistic CSR, also known as humanitarian or philanthropic CSR, means contributing to the common good. It refers to activities that are not ethically required but can be beneficial to the company. Altruistic CSR is a very noble and virtuous responsibility, but precisely for that reason it is also rare, as it is outside the scope of the company's core activities (Smith and Quelch 1993). Strategic CSR addresses such issues of social welfare which result in a win-win situation. The company focuses primarily on the nature of the problem (Lantos 2001). Thus, there is an obvious trend in CSR that inspires companies to focus on problem solving within the framework of their responsibilities, which can be implemented most effectively on a strategic basis, in a way that benefits both the company and the society.

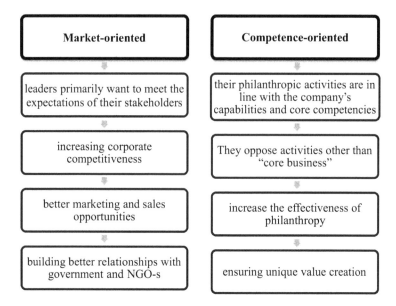

Fig. 1 Market-oriented and competence-oriented philanthropy. Source: based on Bruch and Walter (2005)

> **Box 1 Strategic CSR at National Australia Bank (NAB)**
> From 2008, the slogan of NAB is "give more, take less". The bank has realized to be able to give back much more to society if they volunteer in an area, they are also familiar with. Hence, they have stopped planting trees and painting fences and switched completely to skill-based volunteering and providing consultancy towards other companies on how they can implement skills-based volunteering. (NAB 2020).

Carroll (1991) argues that social responsibility can be divided into four groups that fully cover the concept of CSR: economic, legal, ethical, and philanthropic. He articulates philanthropic responsibility: that it is a corporate activity that meets social desires. It includes programs and activities that promote social well-being and goodwill. (Carroll 1991) Although philanthropy seems to be a less important tool, it is at the same time the most visible form of helping society, which, although not materially profitable, in any case provides an opportunity to discriminate against competitors.

According to Bruch and Walter (2005), philanthropic activities stem either from market-oriented or from competence-oriented perspectives (Fig. 1).

Based on a market-oriented approach, managers primarily want to meet the expectations of their stakeholders, shape their corporate philanthropic activities according to external needs, from which they expect increased corporate competitiveness, better marketing and sales opportunities, and better relationships with government and non-profit organisations. These companies are more interested in influencing stakeholder attitudes than in increasing social outcomes. At the same time, societal benefits can also be achieved with market-oriented philanthropy as it contributes to meeting basic needs and the needs of stakeholders.

> **Box 2 Market-oriented Approach at Wells Fargo**
> Wells Fargo donates up to 1.5% of its revenue to charitable causes each year, to more than 14,500 non-profit organisations through philanthropy, in form of food banks and incubators to hasten the speed to market for start-ups. The company is also engaging its employees to volunteer and give back to the community, as charity of their own choice. (Businesswire 2018).

According to the competence-oriented approach, corporate leaders strive to align their philanthropic activities with the company's capabilities and core competencies. They oppose activities other than core business, aiming to increase the efficiency of philanthropy and ensure unique value creation.

However, the strong internal orientation of the competence-oriented approach makes the reconciliation with the interests of external stakeholders difficult, and it may result in a situation where beneficiaries do not receive the greatest value. On the other hand, competence-oriented corporate philanthropy can also generate unique benefits because it focuses on individual expertise rather than financial support, based on the core business of a company. Some companies combine an external and an internal approach, while others focus specifically on one perspective, and some companies do not accept a strategic approach to philanthropy at all. Based on the degree of internal and external orientation, Bruch and Walter (2005) define four types of corporate philanthropy: peripheral, dispersed, constricted, and strategic philanthropy (Fig. 2).

Dispersed philanthropy is characterized by a complete lack of strategic direction. Initiatives are uncoordinated, managers and employees do not have a comprehensive picture of the company's activities, there are no clear decision criteria as to why they even support a charity project. As a result, many smaller projects are supported without a guideline, and funding is arbitrary towards institutions operating in different fields. Dispersed philanthropic activities include the negative effects of peripheral and constricted types. The background of these initiatives is mostly the personal interests of the members of the management. However, in special cases, dispersed philanthropy can also be useful. In times of severe crises where immediate action is needed, this type of activity can be extremely valuable and achieve more significant social impacts and benefits than using a strategic approach.

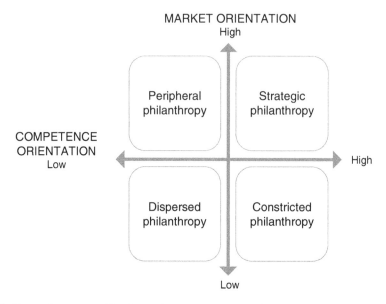

Fig. 2 Tpyes of corporate philanthropy. Source: Bruch and Walter (2005, p. 51)

> **Box 3 Dispersed Philanthropy in Time of COVID-19**
> Major consumer brands have recognized that in addition to the inevitable limitation of store opening hours, the most obvious way of dealing with the virus is to launch campaigns that take advantage of their vast scope and outreach. McDonalds, Nike and Coca-Cola have come up with revamped, themed logos and slogans, highlighting the importance of social distancing to avoid physical contact with each other, cited as the most effective way to deal with the epidemic. In addition, these companies also play a significant role in crisis management through their various charities, including the Coca-Cola Foundation which has raised some $ 20.5 million to help economies in particularly affected countries like China, Italy, the United States and Canada. (CNN Business 2020).

In the context of peripheral philanthropy, companies engage in charitable initiatives that are based primarily on external demand and stakeholder expectations. Most of these companies expect a better competitive position from corporate philanthropy. Their activities are independent of the company's core mission. The strategic consequences of peripheral philanthropy are mixed: it helps stimulate demand for products and services, retain skilled workers, and reduce state and regulatory controls. However, there are also companies that, in the context of peripheral philanthropy, divert both financial and managerial resources from their core business, leading to ambiguity in strategy. Overall, these activities are both ethically and economically relevant and bring benefits to companies but are generally unsustainable in the long run.

> **Box 4 Misguided Philanthropic Action at Wal-Mart**
> A photo that revealed great publicity and crawled the Internet revealed an ambiguous action of Walmart employees of a store in Canton, Ohio, organizing food gathering for their needy colleagues. This philanthropic action left the impression as if Walmart did not pay enough wages to their employees to live, thus the company tried to explain it by stating they aim to help employees who suddenly got into a serious situation (such as getting divorced). (Cleveland 2013).

Constricted philanthropy embodies the synergy between the company's core and charitable activities. They use their core competencies for social purposes, largely ignoring the perspectives of external stakeholders. The strategic effects of constricted philanthropy are also mixed. Its benefits include the fact that existing expertise, resources and opportunities increase the efficiency of activities, and managers strive for this philanthropic spirit to permeate the entire company, applied as a kind of innovation. However, its disadvantage is that it neglects the needs and expectations of stakeholders due to its internal focus. Managers only want to benefit through their own products and specialized employee expertise and help resolve emergencies, thereby losing the opportunity to increase reputation and contribute to the strategy of the activity. While constricted philanthropy does not systematically target key stakeholders, it lacks a strategic approach that limits a company to be the best in its competitive environment.

> **Box 5 Constricted philanthropy at Casper**
> Casper is an online mattress sales company that provides a 100-day warranty to its customers, opting for in-kind donations as one of the tools of its responsibility activities. If, within 100 days, the consumer finds the mattress not suitable, the company takes it back, provides full refund, and in addition, the returned good-quality mattress will be donated to further use. (Casper 2020).

Strategic philanthropy is combining a strong external (market) and internal (competence) orientation in a highly effective way. It considers the business strategy of the company and the interests of the stakeholders at the same time, which also promotes the sustainability of the activity. It provides an opportunity to apply key competencies in new business areas, increase employee intrinsic motivation and labour market attractiveness, stimulate consumer demand, and strengthen company identity with company-wide, coordinated philanthropy.

> **Box 6 Strategic Philanthropy at Pfizer**
> For Pfizer, CSR initiatives are an important part of the company's operations. To help the society through core business is essential, by considering market needs. Consequently, women and children in need are provided with health care and a significantly reduced price of Prevenar 13 vaccinations. (Pfizer 2015).

There is also a special version of strategic philanthropy, place-based strategic philanthropy, the essence of which is to focus on a specific geographical area, in this case the company wants to support not a single issue, but rather a special community (Backer et al. 2004). In this approach, societal challenges can be well linked, community involvement and collaboration with other organizations can be achieved, collaboration with other donors can be more cost-effective, revitalization of a given "cluster" can take place, and results can be achieved. Easier to observe. However, it also has difficulties, as it is not easy to start and maintain relationships with partners along divergent interests, and many want to avoid thinking at a "too strategic" level. (Murdock 2007).

> **Box 7 Place Based Strategic Philanthropy at AUDI Gyor**
> Audi Hungaria has been an important factor in the development of the Hungarian economy and the city of Gyor (site of the company), since 1993. Audi Hungaria is an integral part of the city's life, by playing a crucial role in improving the quality of life of citizens, via supporting cultural and sports events. The company also supports high school education as well as higher education in Gyor and is committed to improving the quality of healthcare system in the region. In addition, employees deliver volunteering activities specifically in and around the city. (Audi 2020).

4 Value Creation with CSR

Beyond philanthropic CSR, Chen et al. (2018) also distinguishes promotional and value-creating CSR. Promotional CSR can be linked to short-term sales goals, so it includes cause-related marketing and sponsorship activities. Cause-related marketing is realized through direct product sales. This goal of support can also be attractive to customers, as through the purchase of a product or service, they also contribute to the support of a particular cause.

> **Box 8 Cause-Related Marketing at TOMS**
> The TOMS brand used the One-for-One concept in its business model. For every pair of shoes sold, it donated a pair of shoes to those in need, donating a total of 60 million pairs of shoes. The concept was suitable for both generating profit and building up good relationship with the community, which later inspired other companies such as Solo sunglasses to launch similar campaigns. (Daniels Fund Ethics Initiative 2020).

Value-creating CSR is a tool for creating common value (Porter and Kramer 2006, 2011), the possible forms of which are social alliances and value chain CSR. The research of Chen et al. (2018) shows that consumers respond more favourably to value-creating CSR than to philanthropic CSR or promotional CSR. Promotional CSR is more accepted from lower-competency companies, while value-creating CSR is expected from high-competency companies. With respect to philanthropic CSR, there is no difference according to competence.

Elements of a company's competence are the company's resources, tools, skills, and knowledge. Several theoretical approaches support that CSR activity and corporate competence can reinforce and improve each other's outcomes. Based on resource-based theory, CSR can provide a company resources to improve its competencies, most notably in the areas of HR, innovation, and corporate culture. (Grant 1991; Srivastava et al. 2001; Surroca et al. 2010; Wernerfelt 1995). The theory of the corporate perspective argues that a company's CSR attributes are largely determined by what competencies the company has, however, the introduction of CSR elements can improve those competencies. (McWilliams and Siegel 2001). The theory of social networks (Dyer and Singh 1998; Gulati 1998) sees the potential for growth in company competence in long-term and multifaceted collaboration with NGOs. All three theoretical approaches support that a company's competence can greatly influence the effectiveness of a CSR activity, however, an effective CSR activity can also greatly improve a company's competence. (Chen et al. 2018).

> **Box 9 SME Example for Value Creation at Hatos Language School**
> In Hatos Language School in Hungary, in-service training for own teachers is also open to teachers working in public education. Additionally, in collaboration with surrounding schools, talented, needy children are provided the opportunity to attend free language courses organized by the company. (Hatos Language School 2020).

Expectations towards high-competency multinational companies may differ from low-competency SMEs, and the same is true for long-established, traditional companies versus new, innovative companies. Companies with low competencies typically have limited corporate resources, expertise, and capabilities, giving consumers the impression that they may not necessarily be qualified to implement value-creating CSR programs. In their case, they consider CSR activity much more as a compensation for lower quality products. However, in the case of high competency companies, value-creating CSR activities are much more expected than charitable and promotional CSR actions. (Chen et al. 2018).

> **Box 10 Multinational Example for Value creation at Symantec**
> The IT security company has been offering education to disadvantaged young people without a college degree since August 2014, where, in addition to obtaining a cybersecurity certificate, further employee competencies are developed (such as resume writing or communication). Thus, the company creates skilled and loyal employees in a segment where there are currently 60,000 people in America in unskilled positions. Meanwhile, it also helps reduce the labour market exposure of the unskilled. (Triplepundit 2020).

Sen and Bhattacharya (2001) call the amount of consumer perception of overlaps between CSR campaigns and companies' core business congruence. Consumer expectations can set companies to differentiate themselves from their competitors through CSR, reduce uncertainty for the company, and increase buying intent.

> **Box 11 Failure of Congruence at KFC**
> Susan G. Komen, owner of a foundation for breast cancer research, partnered with KFC in a nationwide campaign aimed at fund-raising and awareness-raising, by selling pink-branded fried and grilled chicken. KFC donated 50 cents for every bucket sold in a month, for a total of $ 4.2 million. Financially, the campaign was successful, but in most other ways, it was a public relations nightmare. Experts have accused both Komen and KFC to encourage people to buy unhealthy food which, according to research results, leads to an increased risk of breast cancer for women by 25 percent. Komen was criticised for the partnership, while KFC for using cause-related marketing for the sake of women's health with the help of an unhealthy product. The product clearly did not fit the cause the company wanted to support. (abcNEWS 2010).

Webb and Mohr (1998) and Barone et al. (2000) support that consumers value altruistic CSR, which is related to the well-being of society, more than CSR, which is related to the main activity of companies, because in this case companies are responsible not for themselves but for society. Nan and Heo (2007) also found that CSR campaigns more closely related to the core business of firms are not more effective compared to CSR campaigns that are less tied to the core business. Various researches support that consistency between corporate CSR and a company's core business:

increases the credibility of the company (Lucke and Heinze 2015) increases brand loyalty with CSR brand fit (Cha et al. 2016)

leads to a more positive WOM when collaborating with nonprofits (Rim et al. 2016)

increases the possibility of partnership (Till and Nowak 2000)

increases the willingness to buy due to a more positive perception of products (Gupta and Pirsh 2008)

Bower's (1981) research shows that if the consistency of a company and the sponsored event is high, it improves the perception of the company. In the case of sponsorship, the company gets a better perception in the eyes of consumers if the company profile and the sponsored cause are linked. Based on this research, it can be concluded that in order to better perceive consumers, it is more worthwhile for companies to coordinate their CSR actions with their core activities, while taking care not to reflect selfish behaviour.

> **Box 12 Sponsorship at Nicholson International**
> Nicholson International sponsors five conferences of Economist Intelligence Unit at the regional headquarters, with an aim of making itself visible for professionals participating at those conferences. This is of strategic importance to the HR consultancy firm. Through the sponsorship, Nicholson appears together with the well-known Economist (co-branding), collaborates professionally with its staff, benefits from its press relations, and reaches important regional and national decision-makers who appear at those events.(Nicholson International 2020).

5 Corporate Contribution to UN Sustainable Development Goals

The United Nations' Sustainable Development Goals (UN SDGs, 2020) is a useful framework which provides a transparent and understandable categorisation of the most pressing sustainability-related issues which need to be addressed, on all levels of decision-making. Companies were welcoming this formulation of sustainability goals because it enables them to clearly express their commitment in contributing to the achievement of those goals by 2030. These goals often serve as an expression of corporate social responsibility (KPMG survey 2017).

However, sustainability goals seem not to be equally important to companies; they tend to select which of the 17 SDG goals to highlight in their objectives (PwC Survey 2015). The extent to which a company can contribute to each goal and how much risk and opportunity each goal holds depends on several factors. For a strategic approach to SDGs, the first step is to assess the current and potential, positive and negative impacts of the company's business on the Sustainable Development Goals. This helps identify where positive effects can be realised or negative effects could be reduced or avoided. Companies determine their priorities accordingly. As a result of a PwC survey (2015), the following 5 goals were reported by companies to be most important for them:

Goal 8: Decent work and economic growth
Goal 13: Climate action
Goal 9: Industrial innovation and infrastructure
Goal 4: Quality education
Goal 3: Good health and well-being

Although focusing on just a few, specific goals appears as attractive and simple for companies, they are often criticized by citizens for "cherry-picking" and being superficial, for using CSR efforts more as a self-serving PR campaign instead of expressing real corporate value (PwC Survey 2015). PWC proposes a holistic approach for companies, where implementation should be treated as a single framework, covering all 17 objectives. However, for most companies, it may be unrealistic to expect the initial implementation of the SDGs to include all the 17 goals. The short-term strategy is more likely to focus on specific goals, probably those that are already aligned with their existing CSR activities. The SDG Compass (2020) proposes a process which is based on prioritization of 4–6 goals, setting concrete commitments to them, implementing them together with the stakeholders, and finally, reviewing those priorities and make progress every year. Commitment to progress can result in a transition process to achieve all 17 goals in the long run. Sector-specific best practices, included in an SDG Industry Matrix also make the understanding and application of SDG-related corporate action easier. (KPMG 2020).

Experts are not unified in their opinion about the prioritization of SDGs. Some argue with the importance of Goal 13: Climate action, as the achievement of this goal positively influences several sustainability goals. Setting climate goals is popular in corporate practice.

> **Box 13 Innovation at Johnson and Johnson**
> The firm Johnson and Johnson has been addressing the issue of climate change and environment protection for decades. They purchased a privately-owned energy supplier which made possible for the company to produce green energy. It is a double dividend solution as it results in less greenhouse gas emission, while it is also an economically profitable alternative of electricity production. (Johnson and Johnson 2017).

Other experts argue that Goal 4: Quality education should be a top priority as it can break the cycle of poverty and change systems in other urgent areas of global need. (Pierce 2018).

Bretton Woods II and the OECD asked 85 experts from around the world to identify and prioritize the SDG goals. The five main SDG goals from 169 goals for experts are almost evenly distributed between institutional and individual welfare issues (NewAmerica 2020):

Goal 10: Reduced inequality - Promoting the rule of law and access to justice.

Goal 1: No poverty - Eliminate the most extreme poverty.

Goal 3: Good health and wellbeing - Ensure access to safe, effective and affordable health care, medicines and vaccines.

Goal 5: Gender equality - Ensure women's rights to economic opportunities, property rights and inheritance.

Goal 16: Peace and Justice Strong Institutions - Government accountability and transparency must be ensured.

Goal 1, 5, 10 and 16 are obviously more difficult for companies to relate to in a short run, however, they are crucial for sustainable societies, so companies most probably cannot avoid addressing them in their longer-term goal setting.

> **Box 14 Negative Example for Ending Poverty**
> Mastercard started the Football World Cup with the slogan that every goal of Messi or Neymar will feed 10,000 children in Latin America and the Caribbean, donated by the company, through specific foundations. Consumers reacted by asking: what if they don't shoot the goals? What happens if they drop out? At the World Cup, Neymar scored 2, Messi 1, so it was a huge mistake instead of sending a positive message to the public. (Lepitak 2018).

On the other hand Goal 3 often overlaps with the priorities of the surveyed companies.

> **Box 15 Positive Example in the Covid-19 Situation for Good health and Wellbeing**
> As the Covid-19 situation has shown, companies with different profiles may effectively contribute to handling issue and ensure access to the necessary health care measures.
>
> universities and research centres work with pharmaceutical companies on finding the vaccine to the virus, taxi companies take doctors and nurses from home to the hospital every day, hotels provide free accommodation for quarantine purposes, sport equipment companies provide specific products for free to the hospitals, etc. The pandemic made it obvious that, with some creativity, every company can find a way to have a positive impact on society through its own core business. (Mfor.hu, 2020).

6 Conclusion

The chapter focused on the focal approaches of CSR and corporate sustainability which are driving company practice today, in order to generate value and positive impact for the society. The chapter aimed to highlight the huge variety of relevant types of corporate social responsibility, their applicability and limitations. Positive and negative examples - mainly well-known transnational companies and some Hungarian ones - illustrate the importance of understanding the purpose and the possible impacts of the different types of CSR activities when selecting and implementing them.

Today, corporate sustainability and CSR are becoming more and more overlapping fields of action, which reflects in the strong attachment of companies to the UN Sustainable Development Goals and the ways how companies make use of these goals as an expression of their corporate social and environmental responsibility. Prioritization of SDGs has both its benefits and drawbacks, but companies obviously face the challenge of increasing expectations from the society which they need to respond to in a consistent, congruent way. Both good and ambiguous practical examples help navigate among the different approaches and goals, to enhance responsible behaviour at the corporate level.

References

abcNEWS. (2010). Fried chicken for the cure? Retrieved from https://abcnews.go.com/Health/Wellness/kfc-fights-breast-cancer-fried-chicken/story?id=10458830
AUDI. (2020). Audi Hungary and the society. Retrieved from https://audi.hu/en/corporate-responsibility/audi-hungaria-and-the-society/
Backer, T. E., Miller, A. N., & Bleeg, J. E. (2004). *Donor perspectives on place-based philanthropy*. Encino CA: Human Interaction Research Institute.
Barone, M. J., Miyazaki, A. D., & Taylor, K. A. (2000). The influence of cause-related marketing on consumer choice: Does one good turn deserve another? *Journal of the Academy of Marketing Science, 28*(2), 248–262.
Bowen, H. P. (1953). *Social responsibilities of the businessman*. New York: Harper.
Bower, G. H. (1981). Mood and memory. *American Psychologist, 36*(2), 129–148.
Bruch, H., & Walter, F. (2005). The keys to rethinking corporate philanthropy. *MIT Sloan Management Review, 47*(1), 49–55.
Businesswire. (2018). Wells Fargo's Corporate Philanthropy Totals $286,5 Million for 2017. Retrieved from https://www.businesswire.com/news/home/20180221005972/en/Wells-Fargo%E2%80%99s-Corporate-Philanthropy-Totals-286.5-Million
Carroll, A. B. (1991). The pyramid of corporate social responsibility: Toward the moral management of organizational stakeholders. *Business Horizons, 34*, 39–48.
Carroll, A. B. (1999). Corporate social responsibility: Evolution of a definitional construct. *Business and Society, 38*(3), 268–295.
Casper (2020). Mattresses. Retrieved from https://casper.com/mattresses/
Cha, M.-K., Yi, Y., & Bagozzi, R. P. (2016). Effects of customer participation in corporate social responsibility (CSR) programs on the CSR-brand fit and brand loyalty. *Cornell Hospitality Quarterly, 57*(3), 235–249.
Chen, X., Huang, R., Yang, Z., & Dube, L. (2018). CSR types and the moderating role of corporate competence. *European Journal of Marketing, 52*(7/8), 1358–1386.
Cleveland. (2013). Is Walmart's request of associates to help provide Thanksgiving dinner for co-woekers proof of low wages? Retrieved from https://www.cleveland.com/business/2013/11/is_walmarts_request_of_associa.html

CNN Business. (2020). McDonald's and other brands are making 'social distancing' logos, Retrieved from https://edition.cnn.com/2020/03/26/business/social-distancing-brand-logos-coronavirus/index.html

Daniels Fund Ethics Initiative. (2020). TOMS: One for One Movement, Retrieved from https://danielsethics.mgt.unm.edu/pdf/TOMS%20Case.pdf

Drucker, P. F. (1984). The new meaning of corporate social responsibility. *California Management Review, 26*, 53–63.

Dyer, J. H., & Singh, H. (1998). The relational view: Cooperative strategy and sources of interorganizational competitive advantage. *Academy of Management Review, 23*(4), 660–679.

European Commission. (2011). A renewed EU strategy 2011–14 for Corporate Social Responsibility. A renewed EU strategy 2011–2014 for corporate social responsibility. Retrieved from http://eurlex.europa.eu/LexUriServ/LexUriServ.do?uri=COM:2011:0681:FIN:EN:PDF

Friedman, M. (1970). The social responsibility of business is to increase its profits. *New York Times, 13*(1970), 122–126.

Grant, R. M. (1991). The resource-based theory of competitive advantage: Implications for strategy formulation. *California Management Review, 33*(3), 114–135.

Gulati, R. (1998). Alliances and networks. *Strategic Management Journal, 19*(4), 293–317.

Gupta, S., & Pirsh, J. (2008). The influence of retailer's corporate social responsibility program on re-conceptualizing store image. *Journal of Retailing and Consumer Services, 15*(6), 516–526.

Hatos Language School. (2020). About us. Retrieved from https://hatos.hu/

Johnson & Johnson (2017). 4 ways Johson & Johnson is going above and beyond to meet its citizenship & sustainability goals. Retrieved from https://www.jnj.in/corporate-citizenship/4-ways-johnson-johnson-is-going-above-and-beyond-to-meet-its-citizenship-sustainability-goals

Kerekes, S. & Wetzker, K. (2007). Keletre tart a "társadalmilag felelős vállalat" koncepció. Műhelytanulmány (working paper). Budapesti Corvinus Egyetem. Retrieved from http://unipub.lib.uni-corvinus.hu/1349/1/Kerekes_Wetzker_2007.pdf

KPMG. (2020). Sustainable Development Goals Insutry Matrix. Retrieved from https://home.kpmg/xx/en/home/about/our-role-in-the-world/citizenship/sdgindustrymatrix.html

KPMG Survey. (2017). The road ahead. Retrieved from https://assets.kpmg/content/dam/kpmg/xx/pdf/2017/10/kpmg-survey-of-corporate-responsibility-reporting-2017.pdf

Lantos, G. P. (2001). The boundaries of strategic corporate social responsibility. *Journal of Consumer Marketing, 18*(7), 595–632.

Lepitak, S. (2018). The worst marketing I've ever seen' – Mastercard's World Cup children's meals campaign stirs debate, Retrieved from https://www.thedrum.com/news/2018/06/02/the-worst-marketing-ive-ever-seen-mastercards-world-cup-childrens-meals-campaign

Lucke, S., & Heinze, J. (2015). The role of choice in cause-related-marketing – Investigating the underlying mechanisms of cause and product involvement. *Procedia – Social and Behavioral Sciences, 213*, 647–653.

McWilliams, A., & Siegel, D. (2001). Corporate social responsibility: A theory of the firm perspective. *Academy of Management Review, 26*(1), 117–127.

Mfor (Menedzsment Fórum). (2020). Új mutató indult: hétmilliárdnyi felajánlás érkezett a járvány ellen Magyarországon (A new indicator was launched: seven billion offers were received against the epidemic in Hungary), Retrieved from https://mfor.hu/cikkek/befektetes/uj-mutato-indult-hetmilliardnyi-felajanlas-erkezett-a-jarvany-ellen-magyarorszagon.html

Murdock, J. (2007). The place- based strategic philanthropy model. In *Centre for Urban Economics, June.*

Nan, X., & Heo, K. (2007). Consumer responses to corporate social responsibility (CSR) initiatives: Examining the role of brand/cause fit in cause-related marketing. *Journal of Advertising, 36*(2), 63–74.

National Australien Bank (NAB). (2020). Social impact approach, Retrieved from https://www.nab.com.au/about-us/social-impact/shareholders/corporate-responsibility-approach

NewAmerica. (2020). SDG orders. Retrieved from https://www.newamerica.org/digital-impact-governance-inititiative/sdgs-order/

Nicholson International. (2020). About Us, Retrieved from https://www.nicholsoninternational.com.tr/en/pages/about-us.html

Pfizer (2015). Pfizer commits to further reduce priced for prevenar 13 in the world's poorest countries through 2025. Retrieved from https://www.pfizer.com/news/press-release/press-release-detail/pfizer_commits_to_further_reduce_price_for_prevenar_13_in_the_world_s_poorest_countries_through_2025

Pierce, A. (2018). Why SDG 17 is the most important UN SDG? Retrieved from https://www.sopact.com/perspectives/sdg17-most-important-sdg

Porter, M. E. & Kramer, M. R. (2006). Strategy and Society. The Link Between Competitive Advantage and Corporate Social Responsibility. Harvard Business Review, December, 1–15.

Porter, M. E., & Kramer, M. R. (2011). Creating shared value. *Harvard Business Review*, 62–77.

Post, F. R. (2003). A response to the social responsibility of corporate management: A classical critique. *Mid - American Journal of Business, 18*(1), 25–35.

PwC Survey. (2015). Make it your business:Engaging with the Sustainable Development Goals. Retrieved from https://www.pwc.com/gx/en/sustainability/SDG/SDG%20Research_FINAL.pdf

Rim, H., Yang, S. U., & Lee, J. (2016). Strategic partnerships with nonprofits in corporate social responsibility (CSR): The mediating role of perceived altruism and organizational identification. *Journal of Business Research, 69*(9), 3213–3219.

SDG Compass. (2020). SDG Compass. Retrieved from https://sdgcompass.org/

Sen, S., & Bhattacharya, C. B. (2001). Does doing good always lead to doing better? Consumer reactions to corporate social responsibility. *Journal of Marketing Research, 38*(2), 225–243.

Simon, J. G., Powers, C. W., & Gunnemann, J. P. (1983). The responsibilities of corporations and their owners. In T. L. Beachamp & N. E. Bowie (Eds.), *Ethical theory and business* (2nd ed.). Englewood Cliffs, NJ: Prentice-Hall.

Smith, N. C., & Quelch, J. A. (1993). *Ethics in marketing*. Homewood, IL: Irwin.

Srivastava, R. K., Fahey, L., & Christensen, H. K. (2001). The resource-based view and marketing: The role of market-based assets in gaining competitive advantage. *Journal of Management, 27*(6), 777–802.

Surroca, J., Tribo, J. A., & Waddock, S. (2010). Corporate social responsibility and financial performance. *Strategic Management Journal, 31*(5), 463–490.

Till, B. D., & Nowak, L. I. (2000). Toward effective use of cause-related marketing alliances. *Journal of Product and Brand Management, 9*(7), 474–484.

Triplepundit. (2020). Symantec corporate responsibility, Retrieved from https://www.triplepundit.com/author/symantec-corporate-responsibility/1721

Turner, R. J. (2006). Corporate Social Responsibility: Should disclosure of social considerations be mandatory? Submission to the Parliamentary Joint Committee on Corporations and Financial Services Inquiry.

UN SDGs. (2020). About the Sustainable Development Goals. Retrieved from https://www.un.org/sustainabledevelopment/sustainable-development-goals/

Webb, D. J., & Mohr, L. A. (1998). A typology of consumer responses to cause-related marketing: From skeptics to socially concerned. *Journal of Public Policy & Marketing, 17*(2), 226–238.

Wernerfelt, B. (1995). The resource-based view of the firm: Ten years after. *Strategic Management Journal, 16*(3), 171–174.

Windsor, D. (2001). The future of corporate responsibility. *International Journal of Organizational Analysis, 9*(3), 225–256.

Katalin Ásványi works as an associate professor at the Marketing Institute of Corvinus University of Budapest, Hungary. She got a PhD in CSR from Corvinus University of Budapest, Hungary in November 2013. Her research interests are CSR, sustainability, education for sustainability, tourism, cultural tourism and family-friendly tourism. Her teaching profile includes bachelor, master and postgraduate courses into corporate sustainability and CSR, CSR communication. Since 2018, she has been teaching as a guest professor at the University of Passau.

Ágnes Zsóka works as full professor at the Department Marketing Management of Corvinus University of Budapest, Hungary. Since 1997, her key research areas have been closely related to environmental and social sustainability, including responsible corporate behaviour, sustainable consumption and education for sustainability. Her main interest lies in exploring the gaps within sustainable organizational and individual behaviour, with a special focus on how to close those gaps, for the sake of consistent and conscious behaviour. Her teaching profile includes international (English and German) as well as native master courses into corporate sustainability and CSR, corporate environmental management, global climate strategy and responsible consumer behaviour. Since 2015, she has been teaching as guest professor at the University of St. Gallen and the Vienna University of Economics and Business, and she has a regular international CSR master course at the University of Passau.

Is Covid-19 Setting the Stage for UN Agenda 2030? In Pursuit of the Trajectory

Sam Sarpong

Abstract The coronavirus (Covid-19) pandemic has caused much havoc across the world. It has upended lives and livelihoods and created a state of despair. Many countries are still reeling from the impact of the pandemic. Notwithstanding this, the pandemic has also exposed mankind to a startling reality - specifically the fundamental weaknesses in our global system, the prevalence of poverty, the lack of social-protection mechanisms to protect the most vulnerable, weak health systems and above all the need for global cooperation. The on-going crisis has also re-enforced the interdependence of our world and has brought to the fore the urgent need for global action to meet people's basic needs, to save our planet and to build a fairer and more secure world, all of which have been canvassed within the framework of the UN Sustainable Development Goals (SDGs).

The chapter explores issues relating to the pandemic and their implications for sustainable development. It seeks to draw parallels between the two by arguing that the crisis, despite its ferocity, has proved an opportunity for mankind to ramp up actions necessary to achieve the SDGs. This is evidenced by the spirit of solidarity, the global network of innovations, and the synchronised efforts and the shared vision being displayed by civil society, private sector, and governments.

1 Introduction

The chapter explores issues relating to the coronavirus (Covid-19) pandemic and their implications for sustainable development. It does this through the lens of the UN Sustainable Development Agenda 2030 which calls for action to end poverty, the protection of the planet and a desire for people to enjoy peace and prosperity by the year 2030. The issues the chapter engages in are manifold: Firstly, it explores the

---

S. Sarpong (✉)
School of Economics and Management, Xiamen University, Malaysia, Sepang, Selangor, Malaysia
e-mail: samsarpong@xmu.edu.my

nature of the pandemic and examines the sort of arrangement it is fostering on the world today. It also provides an insight into how the pandemic has spiked an uncertainty, a situation which has since led to the strengthening of global cooperation, the embrace of humanity and the desire to forge ahead with capabilities that can strengthen innovative ideas.

In the light of this, the chapter advances some questions: Is Covid-19 shining a light on our underlying mind-sets and behaviour? Will we come out of this as a changed world, and, a world for the better? How can we get our communities and economies back on track and in what condition can we have them? Would be able to meet the measures of sustainability as a result of what is currently happening?

The questions listed are provided as a framework for the analysis rather than as research questions that are explicitly answered. The issues raised by them are threaded into the analysis that follows, which explores the advent of the pandemic and the impact it has had on people, governments, businesses and the environment as a whole. It also looks at the resilience shown by people, the innovative spirit being inculcated and displayed by people and the deep sense of humanistic tendencies that have evolved since the pandemic began.

The chapter approaches this inquiry through the following themes: Resilience, Leadership, Innovation and Humanity. These themes were arbitrarily chosen because they seem to be standalone pillars through which the current crisis can be seen. The chapter begins with an introduction to the pandemic. It then assesses the SDG goals in order to gauge the parallels between the current happenings and the fulfilment of SDGs. After that, the chapter evaluates the thematic strands to see the development in the wake of the pandemic. The stage is then set for a discussion before the conclusion comes in.

2 The Coronavirus Pandemic

The coronavirus (COVID-19) was identified in Wuhan, China, in December 2019 and recognised as a pandemic on 11 March 2020 (WHO 2020). The Covid-19 pandemic has triggered a massive spike in uncertainty (Baker et al. 2020) Travelling has virtually ground to a halt, sporting activities have been cancelled and many parts of the world have experienced lockdowns all in an attempt to stem the tide of the pandemic (Johnson and Boone 2020; LePan 2020). Since the outbreak, numerous communities have been devastated, services industries continue to suffer, jobs have been lost, families have been torn apart, economic and social activities have slowed down as the world battles to find a remedy for the pandemic (Johnson and Boone 2020). There has also been a huge economic impact of the coronavirus on financial markets and vulnerable industries such as manufacturing, tourism, hospitality and aviation (Baker et al. 2020).

Although pandemics have occurred over the years, covid-19 seems to have taken may countries by surprise in view of its scale and the sheer havoc it is causing (Solberg and Akufo-Addo 2020). It has left everyone in no doubt that it is no respecter of person or country as developed countries especially, and many important personalities have contracted and died from the disease (Berg 2020; Brown 2020). Major uncertainties surround almost every aspect of this pandemic. The issue around this pandemic ranges from its prevalence and lethality; the incapacity of healthcare systems to meet this extraordinary challenge; the ultimate size of the mortality shock; the duration and ineffectiveness of social distancing, market lockdowns, and the inadequacy of other mitigation and containment strategies; the near-term economic impact of the pandemic and policy responses; the long stretch of recovery as the pandemic recedes, among others (Baker et al. 2020).

The pandemic has also fomented significant social and political consequences. In the US, some American-Asians are bearing the brunt. They are reportedly being discriminated against on the grounds that they are carriers of the virus (Zhou et al. 2020). The current situation has also revealed racial disparities in the U.S. healthcare system where the disease is reportedly killing many African-Americans, many of whom are deemed to lack medical adequate medical care (Connley 2020). In China, some Africans were maligned and subjected to harassment because some Chinese people felt they could be carrying the virus (HRW 2020). Meanwhile, social distancing and the closure of schools and childcare to combat the spread of coronavirus also created additional pressures on working parents and students alike.

The challenging and uncertain time has generated many debates. These have centred on: the source of the crisis; humanity and what it pertains; how to navigate the crisis and build a better, more sustainable world; the role of government and the needed economic approaches that can spearhead development and keep economies afloat after the pandemic.

3 Pandemic Preparedness

The Global Health Security (GHS) Index 2019 has indicated that it is likely that the world would continue to face health outbreaks since most countries are ill-positioned to combat epidemics in view of their fundamentally weak national health security systems. The report fundamentally makes a huge impression of what is happening today. One significant aspect of this report is the supposition that it makes - that the same scientific advances that help fight epidemic disease also have allowed pathogens to be engineered or recreated in laboratories and that disparities in capacity and inattention to biological threats among some leaders have exacerbated preparedness gaps.

Several conspiracy theories have circled around covid-19 and how it came about. In some instances, covid-19 has been described as a hoax (Sullivan 2020). Unlike most other natural disasters, pandemics do not remain geographically contained, and damages can be mitigated significantly through prompt intervention (LePan 2020). The pandemics have increased over the past years because of increased global travel and integration, urbanisation, changes in land use, and greater exploitation of the natural environment (Morse 1995). Indeed, the same pathways of global commerce and other aspects of the world's integration become the transmission vectors for disease (LePan 2020).

4 What is the Current Situation?

Despite the huge concerns over covid-19 as a threat to humanity, it has seemingly recreated a new sense of humanity by rekindling the human spirit. In many parts of the world, a real sense of camaraderie is showing. People have also shown a great sense of gratitude to health workers and other front-line workers who have been helping to battle this disease. Vulnerable people are also being assisted in the course of the pandemic.

The current phenomenon also seems to be subjecting the role of the state to a lot of scrutiny. The vulnerabilities occasioned by the pandemic has alerted and tasked governments to do more for their citizens, albeit, on a larger scale than what currently prevails. But nowhere has this been played out much more than in developing countries where structural inequality, extreme poverty, lack of amenities, poor governance issues and social disharmony, among others, often tend to have a foothold. In many of these places, with their extremely large informal economy, most workers survive on daily wages and have no savings or stockpiles of food. They also lack social safety nets like unemployment insurance. These vulnerable people, often left out of any contributory schemes, work as traders, labourers or craftsmen in the informal sector, and in Africa, they account for as much as 85 percent of employment across the African continent, (George and Houreld 2020).

Many governments with an eye for reviving their economies are now racing to implement economic stimulus and support packages to keep individuals, businesses and economies afloat.

4.1 Business

Whilst almost every business is faced with navigating the current landscape for their long-term-survival, there are also short-term challenges and opportunities that ought to be addressed. For now, the intentions of business are becoming clearer by the day - there are those who have furloughed their staff, others have laid off their workers without any compensation, whilst some companies have ramped up their efforts to support covid-19 relief and also support their workers. With keenness, people now keep watch over which companies are stepping up at this time and also those ones failing society. In all likelihood, it seems, the way companies act now will eventually affect the way people perceive them in the near future. Hence, the decisions businesses make today will define them well after this pandemic has passed, which only reinforces the importance of leading with integrity, honesty, compassion and, ultimately pursuing objectives that lie much in societal interest. The support and cooperation of the private sector, as a driving force for development, is much more than needed now to improve and maintain the living conditions of people and societies altogether.

The envisaged corporate contributions are most often discussed in terms of corporate social responsibility (CSR) (Sarpong 2017). Concerns about and the need for CSR have grown significantly during the past decades and the current situation, to some extent makes a case for that. One of the core components of CSR is about putting a human face on business entities by communicating empathy, having an understanding and support for societal concerns. Such values have a place in this time of uncertainty and anxiety. For top business personnel and companies that have seized the moments, playing more pro-active roles by getting involved in environmental and societal strategies, collaborating with governmental and civil institutions as well as with international organisations represents a duty unsurpassed and a good corporate citizenship.

It is, therefore, heartening to see major corporations, including those beginning to make their mark in the business community, assisting in diverse ways. Companies like Gap and Zara are now producing masks in their sourced factories, with General Motors and Ford building ventilators. Meanwhile, leaders from some of the companies most affected by the pandemic, particularly those working with airlines, are forfeiting their pay as the pandemic worsens. These leaders include the co-founders of Lyft, executives at Airbnb, and the CEO of Marriott. Disney's Executive Chairman Bob Iger is forgoing his salary for 2020, while the top five Comcast executives are donating theirs to charity (Brandt 2020).

These happenings have reinforced the notion that businesses should be a meaningful partner of the community in which they operate in and, to the extent their finances would allow, contribute to the wellbeing of their stakeholders. The world of business indeed possesses unique responsibilities and offers much potential for achieving just and compassionate care relations in the broadest sense (Sarpong 2018a, 2018b) and as Gaylin (1976) indicates, caring is biologically embedded in the human species as a mechanism for its survival, continuation and an on-going development.

4.2 Funds

In many countries, funds have been set up to seek for donations to fight the pandemic.

These funds have attracted donations from both private and public sector institutions, individuals and other organisations. In India, Prime Minister Narendra Modi Modi launched the 'PM Cares' fund to provide relief to those affected by the coronavirus and to help millions of poor workers caught up by a nationwide lockdown (Kalra and Ulmer 2020). In March 2020, U.S. lawmakers also agreed on the passage of a $2 trillion stimulus bill called the CARES (Coronavirus Aid, Relief, and Economic Security) Act to lessen the impact of an economic downturn set in motion by the global coronavirus pandemic. Other countries have followed suit all in a bid to provide relief to vulnerable people within their countries and localities.

4.3 Solidarity

The solidarity and generosity of people and organisations everywhere has been quite remarkable. China donated ventilators, masks and sent medics to a number of countries, whilst some countries have also come to the aid of others. Hundreds of researchers around the world have also been collaborating with other research centres to develop vaccines. Private individuals have also shown a high level of philanthropy by donating various items from food to Personal Protection Equipment (PPE) to less privileged people and even to government agencies.

Such care and support can be framed within the ethics of care. The ethics of care has become a distinct moral theory or normative approach, relevant to global and political matters as well as to the personal relations that can most clearly exemplify care (Held 2005). It deals with the relational aspect of human life (Sarpong 2018a, b). Ethics of care starts from the fundamental position of the relationality of all humans with each other and the environment and the interweaving of multiple relationships as the basic human condition throughout life (Hawk 2011). Schuman (2001) has argued that aside from our own needs, we must also be of help to those people in our web of relationships, which include the people with whom we have close relationships with as well as those in the larger communities in which we live in. Proponents of ethics of care have emphasised the roles of mutual interdependence and emotional response that play an important part in our moral lives (Sarpong 2018a, b).

4.4 Pollution

Since the onset of the pandemic and the subsequent shutdown of many economic activities, emissions from fossil fuel combustion have dropped substantially in many countries (Stone 2020). India, for instance, experienced a significant decline in some pollutants following the lockdown. The skyline in Venice, Italy also became more visible since the lockdown. In China, emissions reportedly fell by 25% as people were instructed to stay at home (Henriques 2020). The visible cloud of toxic gas hanging over industrial powerhouses almost disappeared in major Chinese cities between January and February 2020 (Wright 2020). In Europe, satellite images showed nitrogen dioxide (NO_2) emissions fading away over northern Italy following the lockdown. A similar story played out in Spain and the UK.

High levels of air pollution is currently seen as one of the most important contributors to deaths from Covid-19, according to research conducted by Martin

Luther University Halle-Wittenberg in Germany (Carrington 2020). The analysis showed that of the coronavirus deaths across 66 administrative regions in Italy, Spain, France and Germany, 78% of them occurred in just five regions, and these were the most polluted. The results indicate that long-term exposure to NO_2 may be one of the most important contributors to fatality caused by the Covid-19 virus in these regions and maybe across the whole world.

The next section examines how the recent issues play in the SDGs. The chapter, therefore, delves into the requirements of the 17 goals as against the present happenings. SDGs target the 5Ps (people, planet, prosperity (originally profit), peace and partnership) and encompass seventeen goals which are of importance to everyone in the world. The United Nations Sustainable Development Goals (SDGs) focuses on sustainable agenda and action plan for people, planet and prosperity (UN 2015, 2018).

The UN recognises the important role stakeholders play in implementing sustainable development and in shifting the world on to a sustainable and resilient path (UN 2015, 2018).

5 The consequences of Covid-19 and SDGs

The presence of the pandemic has strengthened the course being charted by sustainable development goals which call for improvements in the way we live (Solberg and Akufo-Addo 2020. Many calls have been made in recent times for a conscious attempt to be made in taking the bold and transformative steps needed to shift the world onto a sustainable and resilient path (UN 2015). Besides, the need for governments and societies at large to the provision of support and encouragement to people, the need to tackle vulnerability, the determination of increased health care and community engagement, progress in environmental conditions and the realisation that the world should act in collaborative partnership, it can be inferred that the role of the SDGs has been greatly heightened by the pandemic as a trajectory to shape the future through its goals (Solberg and Akufo-Addo 2020).

To some extent, the coronavirus seems to be the 'transformational' agent, in that, it disrupts just about everything that humans are used to and offers an opportunity for us to fall in line with the goals of sustainable development. The UN proposes 17 Sustainable Development Goals and 169 targets, mainly within the economic, social and environmental dimensions. Some actions from individuals, governments and businesses since the pandemic began, have to a large extent, helped in attaining some of the objectives of the SDGs (see Table 1).

The responses to the issues raised in Table 1 may not be deemed as absolute or adequate responses to the requirements of the SDGs. Despite that, they mark a considerable shift from the pre-covid-19 era. Whilst not all the goals can be fulfilled for now, the shift towards their accomplishment, in a sense, is quite remarkable. This is in view of the need to adhere to the goals as a means to ensure a better world. What the chapter seeks to point out are the changes that have occurred following the

Table 1 17 SDGs and the role of the pandemic

SDG	Outcome
SDG1: No poverty End poverty in all its forms everywhere	This goal seems a tough call, however, many individuals and organisations are helping to ease the issues regarding poverty. Much more efforts are now being put in the light of the observed vulnerabilities arising from the pandemic. Governments have also prepared economic stimulus packages to aid sectors affected by the pandemic. Although poverty cannot been ended outright, the fact that many governments are now much more opened to the idea of ending poverty is worth noting
SDG2: Zero hunger End hunger, achieve food security and improved nutrition and promote sustainable agriculture	The need to ensure the entire food ecosystem is safe, nutritious, and sustainable has become even more critical. The low inventory in some grocery shops during the pandemic points to the fact that the entire food system cannot pause. Covid-19 has exposed how vulnerable our global food systems are. Many countries are now realising the benefits of having food security and sustainable agriculture as a major backbone of their economies. The pandemic has accelerated the risk of famine as lockdown measures affected incomes and trade. This calls for more investment in agricultural research to improve yields, develop drought-resistant crops, early warning systems, and promote sustainable farming methods It is important more than ever to make food systems more resilient so that our most vulnerable people are better placed to cope with the next drought, flood or plague
SDG3: Good Health and Well-being Ensure healthy lives and promote Well-being for all at all ages	The pandemic provides a watershed moment for health emergency preparedness and for investment in critical twenty-first century public services. It has also become obvious to many governments that comprehensive medical care needs to be provided for their citizens. The advent of the pandemic has helped to widen the scope of actors and efforts, in order to ensure that no one is left behind due to lack of access to health care and healthy lifestyle options.
SDG4: Quality education Ensure inclusive and equitable quality education	The full ramifications of COVID-19 on education and learning are still unfolding. There is now a shift to online teaching among major educational establishments. We are at a point where half of the global population doesn't have internet access. To mitigate this, the global Education coalition was founded in the early days of the COVID-19 crisis to find and implement distance learning strategies that

(continued)

Table 1 (continued)

SDG	Outcome
	enabled students around the world to continue their education, including remote learning through TV and radio for those without access to the internet Major technology businesses have offered free services. Whilst this does not in any way fulfil absolutely the accomplishment of the goal, it has opened up another avenue for learning. The whole community approach and strengthened partnerships - with governments, non-governmental organisations, civil society, teachers, parents, youth and the private sector – Seems to be the only way forward for sustainable quality education solutions
SDG5: Gender equality Achieve gender equality and empower all women and girls	Though there is limited development in terms of how the pandemic impacts on this goal, it can be recognised that the pandemic gives an opportunity for radical positive action to redress longstanding inequalities in multiple areas of women's lives and to build a more just and resilient world. SDG5 is a lofty goal with many unique complexities. The COVID-19 pandemic underscores society's reliance on women both on the front line and at home, while simultaneously exposing structural inequalities across every sphere, from health to the economy, security to social protection. Responding to the pandemic is not just about rectifying long-standing inequalities, but also about building a resilient world in the interest of everyone with women at the centre of recovery
SDG6: Clean water and sanitation; Ensure availability and sustainable management of water and sanitation for all	SDG 6 focuses on clean, accessible water for all. The goal aims to address water scarcity, poor water quality and inadequate sanitation globally. The essence of using clean water has been a major concern to health authorities, especially in many developing countries, where hygiene levels are less followed. Since the pandemic, many of these countries have offered free treated water to their citizenry as a means to prevent the occurrence of the pandemic and also to lessen the burden with the cost of paying for such a service. Fumigation has also been taking place in many areas like markets and social places in order to keep such places free from any disease
SDG7: Affordable and clean energy Ensure access to affordable, reliable, sustainable and modern energy for all	There is the need for international cooperation to bridge the energy access gap and the placement of sustainable energy at the heart of

(continued)

Table 1 (continued)

SDG	Outcome
	economic stimulus and recovery measures. Renewable energy is key to achieving SDG 7 and building resilient, equitable and sustainable economies in a post COVID-19 world. Since the pandemic set in, electricity costs in households have been absorbed in some countries by their governments. This is meant to cushion off the burden many people are facing now
SDG8: Decent Work and economic growth Promote sustained, inclusive and sustainable economic growth	Disadvantaged groups are already suffering disproportionately from the adverse effects of the pandemic. The pandemic could also hamper people's income generating activities as many are now out of work. For the millions of vulnerable people whose livelihoods hang in the balance, an ambitious commitment by the state to confront these challenges will be decisive Now is a good time to restructure and rebuild the systems we have in place. Implicit in the framings of the economic crisis due to COVID-19 is an expectation of the future where we await economic growth
SDG9: Industry, Innovation and infrastructure Build resilient infrastructure	The COVID-19 pandemic is primarily a health issue in which scientific advancement and research breakthroughs play a central role. The development of new, affordable, and effective vaccines and treatments have become key priorities. Smart technological breakthroughs and tracing and tracking systems are being developed, as are technologies that improve analytical and decision-making processes. The research and development that underpins scientific advance are leading to international collaboration between the private sector, the scientific community and research universities, and governments. Innovations and production of items to fight off the pandemic have intensified of late
SDG10: Reduced inequalities Reduce inequality within and among countries	The pandemic has highlighted the deep inequalities around the world in terms of access to certain services. Communities have realised the enormity of this and have rallied together to support vulnerable people by providing them with their basic needs. Meanwhile, governments are setting up funds and also providing economic stimulus packages to help in addressing the issue. Interventions seeking to cushion the gap between the haves and have-nots are now being given some consideration

(continued)

Table 1 (continued)

SDG	Outcome
SDG11: Sustainable cities and communities Making cities and human settlements safe, resilient and sustainable	Stories are being highlighted about everyday acts of kindness. From small gestures to big projects, ordinary people are making a lasting difference in the lives of those around them. Besides there has been improved resilience in communities
SDG12: Responsible consumption and production Ensure sustainable consumption and production patterns	Excessive production has dwindled. Major retail outlets have suspended their operations. Consumption seems to be more responsible than ever. The pandemic offers countries an opportunity to build recovery plans that will reverse current trends of irresponsible consumption and change our production patterns towards a more sustainable future
SDG13: Climate action Take urgent action to combat climate change and its impacts	There has been a massive reduction in emissions and pollutants owing to less industrial activities and travels of late. Clean and abundant fresh water, healthy oceans and a stable climate are arguably the foundation of all the other socio-economic goals which, to some extent, the pandemic has brought about
SDG14: Life below water Conserve and sustainably use the marine resources	The lockdown has led to less pollution. What prevails now are productive land, clean and abundant fresh water, healthy oceans and a stable climate, which are arguably the foundation of all the other socio-economic goals
SDG15: Life on land Protect, restore and promote sustainable use of terrestrial ecosystems	Less industrial activities are leading to less environmental problems. Pollution is at its barest in heavily polluted places like India and Italy.
SDG16: Peace, justice and strong institutions	The world is pulling together. Ceasefires have been declared through some major confrontations during the pandemic era. Saudi Arabia announced a ceasefire in Yemen with the Houthi rebels. Shia militants also announced a ceasefire with the Americans. The pandemic has been able to put armed conflict on lockdown and ensured people rather focus on it
SDG17: Partnerships for the goals	Interventions are bolstering togetherness. Confronting the pandemic has required coordinated global humanitarian and socio-economic responses. Addressing this crisis requires worldwide collaboration and partnerships, across all aspects of the response

Source: Author

outbreak of the pandemic, albeit temporarily. It does not seek to indicate that these shifts that could remain permanent.

6 Outcome of Pandemic

The chapter recognises four strands that can be developed from the responses to the covid-19 pandemic - Resilience, Innovation, Leadership and Humanity- as strands from the residues of the pandemic. The constructs of resilience, innovation, leadership and humanity are examined in order to see their defining roles and what they hold for society.

6.1 Resilience

The pandemic has provided a lot of lessons for mankind. It has revealed our state of preparedness in the face of a case of this magnitude. It has helped us to appreciate our coping skills and to find resilience in terms of what we do. The literature advocates a number of variables that denote resilience. According to Henderson and Milstein (1996), resilience is the ability to recover from negative life experiences and to become stronger whilst overcoming adversity. This is the position that a lot of people can identify with as at now. Survival, recovery, and thriving are concepts associated with resilience and often describe the stage at which a person may be, during or after facing adversity (Ledesma 2014). These attributes have become the refrain in recent times as people strive for the best despite the glaring hardships they face in the wake of lockdowns and failings within the social system.

Magis (2010, p. 401) describes the resilience displayed by communities as 'existence, development and engagement of community resources by community members to thrive in an environment characterised by change, uncertainty, unpredictability, and surprise.' It is also seen by Berkes and Ross (2013) as a social system's capacity to unite and collaborate toward a shared goal or objective. Resilience and thriving provide positive self-esteem, hardiness, strong coping skills, a sense of coherence, self-efficacy, optimism, strong social resources, adaptability, risk-taking, low fear of failure, determination, perseverance and a high tolerance of uncertainty (Bonanno 2004; Ungar 2004; Masten 2005).

The concept of 'thriving' refers to a person's ability to go beyond his or her original level of level of functioning and to grow and function despite repeated exposure to stressful experiences. Recent studies in resilience have started to look at the concept of 'thriving.' Thriving is grounded on an individual's positive transformation resulting from the experience of adversity (Nishikawa 2006). The belief that 'people are capable of transmuting traumatic experiences to gain wisdom, personal growth, positive personality changes, or more meaningful and productive lives has been a central theme in centuries of literature, poetry, and personal narratives.' (Saakvitne et al. 1998, p. 281).

As already alluded to, the construct of resilience argues that it is not the nature of adversity that is most important, but how we deal with it (Moore 2020). When we face adversity, misfortune, or frustration, resilience helps us to retrace our steps in a bid to recover quickly from the said difficulties. It helps us to survive, recover, and even work out a recovery in the face of misfortune. The resilience to the Covid-19 pandemic as at now reflects the massive human attempt to find solutions to the huge difficulties being experienced.

6.2 Leadership

The pandemic has also brought to the fore questions regarding leadership in both governance and industry. It created leadership challenges for leaders at all levels of governments and industries (CCL 2020). Throughout the COVID-19 pandemic, organisations around the globe demonstrated remarkable agility, changing business models literally overnight: setting up remote-work arrangements; offshoring entire business processes to less-affected geographies; initiating multi-company cooperation to redeploy furloughed employees across sectors (Renjen 2020). In each situation, the urgency for results prevailed over traditional bureaucratic responses.

Within the political arena, there have leaders who felt the need to take immediate action when the issue came up, others, meanwhile, were a bit slow in terms of how to react. Some Americans were embittered by the slow response and inconsistencies in the responses from the erratic leadership of the Trump Administration, which devalued and often refused to follow the advice of his scientific advisers (Bennhold 2020). Other countries have won praise for their swiftness in tackling the pandemic. What has emerged from the current situation is that good leadership is invaluable to the sustainability and progress of a country. The action that some took yielded good result, whilst others had difficulties stemming the tide of the pandemic because of their inaction earlier on.

6.3 Innovation

A lot of ingenuity has been on display since the coronavirus pandemic occurred. Many people have worked through various means to help find solutions to the nagging problems associated with the pandemic. It has been a moment in time for people to stretch the limits of their abilities and the boundaries of their creative capacities. The pandemic has also led to a situation where people had to improvise to produce extraordinary things to help in fighting the impact of Covid-19.

In places where personal protective equipment (PPE) were in short supply, factories sprang up overnight to produce these items. Health robes, medical scrubs and medical gowns were produced locally by some garment manufacturing companies in many countries. For instance, a consultant anaesthetist at Carmarthen, South Wales, UK was so 'desperately concerned' about the lack of ventilators to treat Covid-19 patients that he went on to design his own. Initiatives to help improve people's conditions during the pandemic also gained momentum. Jamaican student, Rayvon Stewart developed a device to sanitise door knobs which uses ultraviolet light to kill harmful micro-organisms, thereby reducing the risk of contagion. Meanwhile, Apple and Google, two of the richest and most successful competitors also teamed up on a project to help health care experts track patterns of exposure to the coronavirus using bluetooth. Tesla, a car manufacturer, also produced ventilators to aid covid-19 victims.

6.4 Humanity

The crisis has revealed the competence and generosity of those who are in frontline care and are quite keen to do their best for the afflicted in society. We have also seen the generosity of people, organisations and governments working hand-in-hand with each other. Hence, community engagement and solidarity exemplified by the relationality of mankind has been commonplace. People checked on each other and also helped in feeding the less privileged ones in society during the pandemic. The amazing initiatives that took place around the world, from the very small acts of kindness that people showed towards each other to big donations from say, private organisations, churches and hotels supplying empty rooms for quarantined people during this tragedy, are all testimonies of what humanity brought into people. Many people ramped up their responses to the pandemic though philanthropic efforts to fulfil a role that should mostly be played by government. The moments of crisis highlighted how important people's priorities are to serve humanity.

7 Forging Ahead

But what are the deeper lessons from the coronavirus pandemic that can help companies, society and governments to be more sustainable going forward? The rapid global spread of a novel coronavirus has taught us that we are all interconnected in terms of our health and well-being and that our collective well-being is fundamental. If there was any doubt that our world faces common challenges, this pandemic should categorically put that to rest. The crisis has re-enforced the interdependence of our world and also brought to the fore the urgent need for global action to meet people's basic needs, to save our planet and to build a fairer and resilient world (Solberg and Akufo-Addo 2020). It has also provided a basis for us to take the bold and transformative steps which are urgently needed to shift the world onto a sustainable and resilient path. This is what the SDGs, the global blueprint to end poverty, protect our planet and ensure prosperity, are all about.

From the earlier discussions, we are left in no doubt that covid-19 has ensured a commonality among people. It has shown us how fragile the world is (Sarpong 2018b) and the vulnerabilities of our interconnected and global economy (Baubion 2013). At the same time, its challenges have helped us to pursue advancements in science and technology as well as medicine by seeking remedies for the virus. It has also brought to the fore the importance of establishing and improving upon the social safety net mechanisms especially in developing countries. Many countries have fallen out on this and would have to improve the conditions of their people.

More particularly, development agencies, national governments, civil society and the private sector need to come together in a global effort to create a conducive situation where people can enjoy fulfilling lives. How we do this will be very crucial for mankind because the SDGs have a powerful call: that no one should be left behind mankind's development. That is why we need to act now taking into consideration the lines of action that can lead us towards the universal agenda for people, planet and prosperity that the SDGs advocate for. Through such activities, we can strengthen universal peace in larger freedom.

8 Conclusion

The rapid spread of Covid-19 has generated a global public health crisis which is being addressed at various local and global scales through various measures. The chapter explored issues relating to the pandemic and their implications for sustainable development. It did this through the lens of the UN Sustainable Development Agenda 2030 which calls for action to end poverty, the protection of the planet and a desire for people to enjoy peace and prosperity by the year 2030. The issues the chapter explored were meant to provide a basis for us to know what the current situation holds for our future endeavours. It was also to bring to the fore the impetus that this covid-19 has led to, especially the embrace of humanity and the desire to forge ahead with capabilities that can strengthen innovative ideas.

References

Baker, S., Bloom, N., Davis, S., & Terry, S. (2020). COVID-induced economic uncertainty and its consequences. *CEPR Policy Portal.* Retrieved April 20, 2020, from https://voxeu.org/article/covid-induced-economic-uncertainty-and-its-consequences

Baubion, C. (2013) 'OECD risk management: Strategic crisis management' , OECD working papers on public governance 23, OECD Publishing. Retrieved May 12, 2020, from https://www.oecd-ilibrary.org/governance/oecd-risk-management_5k41rbd1lzr7-en

Bennhold, K. (2020). *Sadness' and disbelief from a world missing American leadership*. NY Times, April 23. Retrieved June 5, 2010, from https://www.nytimes.com/2020/04/23/world/europe/coronavirus-american-exceptionalism.html

Berg, S. (2020). Q&A with Oprah shines light on COVID-19 impact on minorities. *American Medical Association*. 15 April. Retrieved May 20, 2020, from https://www.ama-assn.org/delivering-care/health-equity/qa-oprah-shines-light-covid-19-impact-minorities?gclid=EAIaIQobChMImIG40p3g6QIVlqqWCh0seAAyEAAYASAAEgK9j_D_BwE

Berkes, F., & Ross, H. (2013). Community resilience: Toward an integrated approach. *Society & Natural Resources: An International Journal, 26*(1), 5–20.

Bonanno, G. A. (2004). Loss, trauma and human resilience. *American Psychologist, 59*, 20–28.

Brandt, L. (2020). 13 business leaders who have cut their salaries to $0 to help struggling workers as the coronavirus wreaks havoc on their industries, 10 April. Retrieved April 22, 2020, from https://www.businessinsider.my/list-of-business-leaders-giving-up-salaries-during-the-pandemic-2020-3?r=US&IR=T

Brown, A. (2020). 'Celebrity' or 'Notable' Covid-19 deaths call into question rates of. *Real Clear Markets*. Retrieved May 20, 2020, from https://www.realclearmarkets.com/articles/2020/04/07/celebrity_or_notable_covid-19_deaths_call_into_question_rates_of_104095.html

Carrington, D. (2020). The Guardian (2020) 'Air pollution may be 'key contributor' to Covid-19 deaths – study. 20 April. Retrieved April 15, 2020, from https://www.theguardian.com/environment/2020/apr/20/air-pollution-may-be-key-contributor-to-covid-19-deaths-study

CCL. (2020). Leading in Times of Crisis and beyond. *Centre for Creative Leadership*. Retrieved June 5, 2020, from https://www.ccl.org/coronavirus-resources/

Connley, C. (2020). Racial health disparities already existed in America— the coronavirus just exacerbated them. *CNBC*, 15 May. Retrieved May 30, 2020, from https://www.cnbc.com/2020/05/14/how-covid-19-exacerbated-americas-racial-health-disparities.html

Gaylin, W. (1976). *Caring*. New York: Knopf.

George. L. & Houreld, K. (2020) Millions face hunger as African cities impose coronavirus lockdowns. *Reuters*, 16 April. Retrieved April 28, 2020, from https://www.reuters.com/article/us-health-coronavirus-hunger-africa/millions-face-hunger-as-african-cities-impose-coronavirus-lockdowns-idUSKCN21Y14E

Hawk, T. F. (2011). An ethic of care: A relational ethic for the relational characteristics of organizations. In M. Hammington & M. Sander-Staudt (Eds.), *Applying care ethics to business. Issues in Business Ethics* (pp. 3–34). Dordrecht: Springer.

Henriques, M. (2020). Pollution and greenhouse gas emissions have fallen across continents as countries try to contain the spread of the new coronavirus. Is this just a fleeting change, or could it lead to longer-lasting falls in emissions? *BBC Future*. 27 March. Retrieved April 4, 2020, from https://www.bbc.com/future/article/20200326-covid-19-the-impact-of-coronavirus-on-the-environment

Held, V. (2005). *The ethics of care: Personal, political and global*. Oxford: Oxford University Press.

Henderson, N., & Milstein, M. M. (1996). *Resiliency in schools: Making it happen for students and educators*. Thousand Oaks, CA: Corwin Press.

HRW (2020) China: Covid-19 Discrimination against Africans – forced quarantines, evictions, refused services in Guanzhou, Human rights watch. Retrieved June 12, 2020, from https://www.hrw.org/news/2020/05/05/china-covid-19-discrimination-against-africans

Johnson, S. & Boone, P. (2020). From lockdown to locked in, here's what post-pandemic travel could look like. *World Economic Forum*. Retrieved May 20, 2020, from https://www.weforum.org/agenda/2020/05/coronavirus-lockdown-travel-tourism

Kalra, A. & Ulmer, A. (2020). Donations pour in but India's 'PM CARES' coronavirus fund faces criticism. *The Star*. 8 April. Retrieved April 2, 2020, from https://www.thestar.com.my/news/world/2020/04/08/donations-pour-in-but-india039s-039pm-cares039-coronavirus-fund-faces-criticism

Ledesma, J. (2014) 'Conceptual frameworks and research models on resilience in leadership, SAGE Open. Retrieved April 10, 2020, from https://doi.org/10.1177/2158244014545464.

LePan, N. (2020). Ranked: Global Pandemic Preparedness by Country. *Visual Capitalist.* March 20. Retrieved April 25, 2020, from https://www.visualcapitalist.com/global-pandemic-preparedness-ranked/

Magis, K. (2010). Community resilience: An indicator of social sustainability. *Society & Natural Resources: An International Journal, 23*(5), 402–426.

Masten, A. S. (2005). Ordinary magic: Resilience processes in development,' *Annual Progress in Child Psychiatry and. Child Development, 56,* 227–238.

Morse, S. S. (1995). Factors in the emergence of infectious diseases. *Emerging Infectious Diseases, 1*(1), 7–15.

Moore. C. (2020) 'Resilience theory: What research articles in psychology teach us', Positive psychology. Retrieved April 11, 2020, from https://positivepsychology.com/resilience-theory/

Nishikawa, Y. (2006). *Thriving in the face of adversity: Perceptions of elementary-school principals.* La Verne, CA: University of La Verne.

Renjen, P. (2020). COVID-19: How leaders can create a new and better normal. WEF, 11 May. Retrieved June 5, 2020, from https://www.weforum.org/agenda/2020/05/leaders-should-use-their-vision-and-trust-to-create-a-new-and-better-normal/

Sarpong, S. (2017). Corporate social responsibility in Ghana: Issues and concerns. In S. Idowu, S. Vertigans, & A. B. Schiopoiu (Eds.), *Corporate social responsibility in times of crisis, CSR, sustainability, ethics and governance* (pp. 191–205). Cham: Springer.

Sarpong, S. (2018a). Sweatshops and a duty of care – To what extent? The case of Bangladesh. In S. Seifi & D. Crowther (Eds.), *Stakeholders, governance and responsibility* (pp. 229–247). Bingley: Emerald.

Sarpong, S. (2018b). Crisis management and marketing. In D. Gursoy & C. G. Chi (Eds.), *The Routledge handbook of destination marketing* (pp. 101–109). New York: Routledge.

Saakvitne, K. W., Tennen, H., & Affleck, G. (1998). Exploring thriving in the context of clinical trauma theory: Constructivist self-development theory. *Journal of Social Issues, 54,* 279–299.

Schuman, P. L. (2001). A moral principles framework for human resource management ethics. *Human Resource Management Review, 11,* 93–111.

Solberg, E. & Akufo-Addo, N. A. D. (2020). Why we cannot lose sight of the Sustainable Development Goals during coronavirus. *World Economic Forum,* April. Retrieved April 25, 2020, from https://www.weforum.org/agenda/2020/04/coronavirus-pandemic-effect-sdg-un-progress/

Stone, M. (2020). Carbon emissions have falling sharply due to coronavirus. But not for long,' National Geographic. 3 April. Retrieved April 27, 2020, from https://www.nationalgeographic.com/science/2020/04/coronavirus-causing-carbon-emissions-to-fall-but-not-for-long/

Sullivan, B. (2020). Fox news faces lawsuit for calling COVID-19 a hoax. *Forbes.com,* 10 April. Retrieved April 27, 2020, from https://www.forbes.com/sites/legalentertainment/2020/04/10/covid-19-lawsuit-against-fox-news/#674bbd715739

Ungar, M. (2004). A constructionist discourse on resilience. *Youth & Society, 35,* 341–365.

UN. (2015). *Transforming our world: The 2030 agenda for sustainable development.* United Nations A/RES/70/01. Retrieved April 21, 2020, from http://www.un.org/ga/search/view_doc.asp?symbol=A/RES/70/1&Lang=E

UN. (2018). Sustainable development knowledge platform. Progress of Goal 16 in 2018. Retrieved April 10, 2020, from https://sustainabledevelopment.un.org/sdg1

Wright, R. (2020) There's an unlikely beneficiary of coronavirus: The planet. CNN World. Retrieved April 15, 2020, from https://edition.cnn.com/2020/03/16/asia/china-pollution-coronavirus-hnk-intl/index.html

WHO (2020) WHO Director-General's opening remarks at the media briefing on COVID-19 - 11 March 2020. *World Health Organization, 11 March.* Retrieved April 5, 2020, from https://www.who.int/dg/speeches/detail/who-director-general-s-opening-remarks-at-the-media-briefing-on-covid-19%2D%2D-11-march-2020

Zhou, M., Yu, Y., & Fang, A. (2020) We are not COVID-19: Asian Americans speak out on racism. *Nikkei Asian Review*. Retrieved May 21, 2020, from https://asia.nikkei.com/Spotlight/Coronavirus/We-are-not-COVID-19-Asian-Americans-speak-out-on-racism

Sam Sarpong PhD is associate professor, School of Economics and Management, at Xiamen University Malaysia. He is also a senior research fellow at the Department of Business, Competition, Ecological Law Institute, South Ural State University. He holds a PhD from Cardiff University and an MBA from University of South Wales. His research interests lie in the relationship between society, economy, institutions and markets. He tends to explore the nature of and ethical implications of socio-economic problems and has authored numerous scholarly publications in management and marketing journals. He is also a reviewer for many leading journals and also serves as an associate editor for the *International Journal of Corporate Social Responsibility* (Springer).

Challenges to the Effective Implementation of SDG 8 in Creating Decent Work and Economic Growth in the Southern African Hemisphere: Perspectives from South Africa, Lesotho and Zimbabwe

Mikovhe Maphiri, Matsietso Agnes Matasane, and Godknows Mudimu

Abstract In September 2015, the UN Generally Assembly embraced the International Labour Organisations Agenda on decent work for sustainable development. Formally known as the 2030 Agenda for Sustainable Development, the agenda is made up of 17 sustainable goals and targets. SDG8, focuses on promoting inclusive economic growth, full and productive employment and decent work for all. The decent work component of SDG8 is anchored on four pillars namely job creation, rights at work, social protection and social dialogue. Notably, the objectives of the SDGs are interlinked with the objectives of the Southern African Development Community (SADC) which are to achieve economic development, peace and improve the quality of life for Southern Africans. Decent work and economic growth within the SADC region are critical for the respective governments and for regional integration. The SADC region has aligned the SDGs to its existing policies and to other regional and international normative instruments and policy documents in a bid to improve the life and welfare of citizens in the region. This paper seeks to investigate the challenges and opportunities in achieving decent work within the SADC region, with specific reference to South Africa, Lesotho, and Zimbabwe.

1 Introduction

In 2015, 195 nations came together for a unified purpose of changing the world for the better by bringing together their respective governments, business, media, local non-profit organizations and institute of higher learning to achieve unified goals by

M. Maphiri (✉) · G. Mudimu
University of Cape Town, Cape Town, South Africa
e-mail: mikovhe.maphiri@uct.ac.za

M. A. Matasane
University of Pretoria, Pretoria, South Africa

the year 2030 (Chowdhury and Koya 2017). These goals are ambitious and inclusive in nature, with a too good to be true undertone as the aspirations seek to achieve what most developing African economies would describe as an elusive African dream.

The SDGs require member states to end extreme poverty in all forms by 2030, end world hunger and achieve food security through a commitment to end malnutrition as well as commit to good health and wellbeing for all at all ages (Pekmezovic 2019). This includes, among other things: reducing the number of infant mortality and the death of children through HIV and Aids (Cluver et al. 2018); promoting inclusivity by ensuring quality in education; championing the economic growth and decent work agenda (Moodley and Cohen 2012). The SDGs encourage member states to build sustainable communities and promote justice for all, by monitoring climate change and ensuring environmental justice through global partnerships and innovation (United Nations Sustainable Development Goals, [APA] n.d.).

The intrinsic feature of the SDGs is to achieve common objectives that make people human. These objectives are interlinked with the objectives of the Southern African Development Community (SADC) which are to achieve economic development, peace and improve the quality of life for Southern Africans by supporting the socially disadvantaged through a commitment to development, peace and security, alleviate poverty and enhance the standard and quality of life for Southern Africans (SADC 2013).

This article examines SDG8, focusing on the decent work component and economic growth for all. The subject of decent work within the SADC region remains a prickly issue as challenges to the effective provision for decent work remained an intangible goal for decades (Southern African Development Community Decent Work Programme 2013–2019, 2013). Challenges of colonization in Lesotho (Gocking 1997) and Zimbabwe (Chigwedere 2001) as well as apartheid and colonization in South Africa (Connolly 2001) created deeper levels of inequality and uneven economic growth across these countries (Southern African Development Community (SADC) Decent Work Programme 2013–2019, 2013). The century long struggle for decent work and sustainable economic growth continues with efforts put in place by members of the SADC community to achieve this goal and build societies free from the plagues of poverty and unemployment.

The approach of the SADC community has been to promote sustainable and equitable economic growth that ensures poverty alleviation and improve the quality of life for the socially disadvantaged through regional integration, promoting and maximising productive employment and using existing resources by member states. In addition, the SADC approach seeks to ensure that the region is self-sustaining and that there is a collective self-reliance and interdependence of member states (SADC Consolidated Treaty 2011).

This paper argues that the SDG8 of decent work and economic growth is not a new agenda within the SADC community. As such, pre-existing limitations and challenges for decent work and economic growth cannot be realised outside the partnership of countries within the SADC region and the investment made by the international community. The goals set out by the UN General Assembly should be commended for their recognition of the key elements needed to sustain humanity which include an emphasis on innovation, collaboration and partnership. SDG8 as well as the other SDGs will struggle in the face of local governance challenges of corruption, political instability, unemployment and inequality as well as disunity among the countries in the SADC community.

This paper provides an overview of the challenges and opportunities in achieving SDG8 within the SADC community, with specific reference to South Africa, Lesotho and Zimbabwe. The review highlights the gap between the global aspirations of the UN and the local realities within which SDG8 is being implemented in South Africa, Lesotho and Zimbabwe. In what follows, we provide an overview of the challenges embedded in implementing decent work and economic growth within the three countries before evaluating the structures for improvement. Each section of this paper draws on scholarship from law, industrial sociology and economics to consider the barriers to implementation as well as the local challenges encountered in achieving SDG8 within the remaining 10 years of the 2030 deadline. The paper also recommends the robust and decisive actions that need to be taken to attain SDG8. Finally, we review the synergy and contradictions between SDG8 and the SADC agenda.

2 Decent work and SADC Initiatives

The significance of achieving decent work has formed part of crucial debates in realising sustainable development in both developing and developed economies (Decent work, ILO.org n.d). Thus, the concept of decent work is not a novel concept as its origins are traced to the International Labour Organisation (ILO) (Theron 2012). According to the ILO, decent work is a very broad and multifaceted agenda. It is about the aspirations of people in their working lives which involves the process of ensuring fair income, freedom of expression and participation in organisational and or corporate decisions that affect their working lives (Decent work, ILO.org n.d) Decent work can be summarised in terms of the four pillars of the ILO's decent work agenda, namely employment creation, social protection, rights at work and social dialogue.

As Theron correctly puts it, decent work is not all about work in the literal sense, it is about sustainable job creation which embodies, inter alia: the notion of creating sustainable livelihoods; guaranteeing rights at work by giving workers a collective voice through labour laws; and supporting structures conducive for collective bargaining and worker participation by advancing structures for social protection and welfare (Theron 2012). To attain this objective, work must be non-discriminatory, safe and provide structures for collective bargaining and social security (Moodley and Cohen 2012).

One of the key contributions of the decent work agenda is that it applies broadly to all workers, including those operating in formal and informal sectors, unregulated wage workers, home workers and to self-employed workers (Ghai 2003). This is significant in the SADC region where a lot of work is being created in the informal economy, then in the formal sector (Smit and Mpedi 2010). It is the quality of such jobs that the decent work agenda aims to improve, across all economic divides including in the formal and informal sector. More importantly, the decent work agenda recognises the critical role of various stakeholders, particularly corporations in the creation of jobs. It is through this recognition of stakeholders that strong social dialogue platforms between employers and employees can contribute towards the advancement and fostering of sustainable solutions (Rider 2015).

Furthermore, the concept of decent work presupposes that such work is pre-existing or rather structures for work are in place or have the potential to be put in place in a sustainable manner. The challenge for decent work in the SADC region is not only embedded in the decency aspect of it which include amongst other things workplace safety, fair wages and structures for workers to organise through collective bargaining and workplace forums (Bendix 2001). It is the lack of work which forms the biggest challenge to achieving decent work. This is created by the sluggish and reclining economic growth conundrum companied by wide gaps of inequality and poverty together with poor and ineffective corporate and public governance structures. This is an area that requires the immediate attention of the SADC community since it is grappling with high unemployment rates and high levels of poverty (SADC 2013). As will be demonstrated in the South African, Lesotho and Zimbabwean context, decent work is interconnected to other SDGs and basic human rights. Thus, the success of SDG8 is interlinked to other rights, particularly participation rights.

3 Perspectives on Decent work from South Africa: An Overview

South Africa is a diverse country with a unique socio-political history. The country possesses a great wealth of metallic and non-metallic mineral resources that are located across the nine provinces (Davenport 2013). The country's wealth in mineral resources led to the colonization of the country as well as the apartheid regime which ruled in the country from the late 1940s until the early 1990s (Allen 2003). The exploration of mineral resources and the mining industry formed the foundation to South Africa's economic development and industrialisation process. Apartheid in South Africa was a system of oppression put in place to ensure the marginalisation and discrimination of black persons through laws, policies and systems oppression (Saul 2012).

Work in South Africa has continuously been categorised from the lenses of race and gender (Hans 1935). This has been largely influenced by the history of apartheid and colonisation which not only fostered inhumane working conditions especially in the mining sector by employing cheap black labour (Allen 2003), but ensured that the SADC region supplied labour to South African mines through the migrant labour system. This in turn created systems of inequality that are still plaguing society through wide gaps of inequality and poverty (Tabata 1973). The history of South Africa has played a key role in the development of the economy. However, despite the country achieving its democracy in 1994, and putting systems in place to redress the injustices and legacies of the past (Buhlungu 2006), South Africa remains one of the most unequal societies in the world. The country is grappling with high rates of unemployment and low job security (Collier 2015). Since democracy, South Africa has been on a quest to redress the patent inequality in society, this is done through laws that promote human dignity, equality and freedoms of all persons underscored by the spirit of Ubuntu (Wenckebach et al. 2019).

The South African decent work agenda is pursued under the umbrella of labour law with over 16 legislative instruments promoting the rights and freedom of workers. The Constitution of the Republic of South Africa, Act 108 of 1996 which is the highest law in the land forms the basis of the rights and freedoms of workers (Du Toit 2003). In particular, section 23 of the Constitution, read together with sect. 5 of the Labour Relations Act 66 of 1995, aim to promote decent work by ensuring equality at the workplace through the creation of forums where workers can organise and have a collective voice (Bendix 2001). The pursuit for the attainment of decent work in South Africa led to the promulgation of legislation supporting decent work, which inter alia include legislation such as the Occupational Health and Safety Act 85 of 1993, the Employment Equity Act 55 of 1998 which is derived from the equality clause entrenched in sect. 9 of the South African Constitution as well as the Basic Conditions of Employment Act 75 of 1997 which seeks to ensure that the agenda of decent work is pursued and regulated within existing labour laws.

Opportunities to achieve decent work in South Africa cannot be described outside the economic and socio-political setting of the South African society. The agenda for decent work has formed part of the history of the country during which the forces of apartheid and colonization were against the provision of decent work for historically disadvantaged racial groups under the apartheid regime (Stares 1977). Workers in the mining industry have been fighting for the enforcement of decent work within the mines for centuries (Allen 2003). The demand for sustainable wages, the quest to improve working conditions, the recognition of workers' rights to organise have all formed part of the fight against apartheid and the promotion of decent work in the mines (Budeli 2007). Twenty-six years post democracy, the position has remained largely unchanged as the challenges for many workers in achieving decent work have remained unsolved. This was witnessed in the 2012 Marikana massacre where over 34 mine workers and police officers were killed by policemen in their fight for decent work and improved working conditions (Alexandra et al. 2013). In 2012 the mining sector underwent intense waves of labour unrest (Chinguno 2013). On 16 August 2012, the mining industry in South Africa received the world's attention, when 34 mineworkers who took part in unprotected strike were shot and killed by the officers of the South African Police Services (SAPS) at the hills of Rookopies also known as Marikana in the Rustenburg (Farlam et al. 2015). This was the culmination of the preceding waves of unrest and violence by mine workers in post-apartheid South Africa (Luiz 2002). The Marikana Massacre was essentially a dispute about living conditions and the salary by the rock drill operators, who demanded a wage increase to R12 500 per month, and to support these demands they embarked on a wildcat strike (Marikana Commission of Inquiry 2015). These events left so many questions on the state of the country's democratic system, the question of decent work and the accountability of corporates through corporate social responsibility and accountability, the government's role in promoting structures of decent work and labour governance in South Africa. On 23 of November 2018, the president of the Republic of South Africa, President Cyril Ramaphosa assented the National Minimum Wage Act 9 of 2018. The purpose of the Act is to advance economic development and social justice by raising wages of the lowest paid workers, protecting workers from unreasonably low wages, preserving the value of the national minimum wage as well as promoting collective bargaining by supporting structures for economic transformation. Although the promulgation of a Minimum Wage Act can be viewed as a commendable step towards alleviating the disparities in income in the national labour market, the functionality of the Act in promoting decent work is still a subject of much debate as trade unions argue that the South African minimum wage is nowhere near promoting platforms for eradicating poverty and inequality as it is below the inflation rate.

Decent work includes the payment of decent wages and decent working hours. In South Africa the Basic Conditions of Employment Act sets out the ordinary working hours for work at 45 hours a week and hours beyond the 48 hours indicate that the workers are working overtime. In 2018, South Africa adopted the National Minimum Wage Act and in March 2020, the new national minimum wages were pegged at R20.76 an hour. This rate is almost equivalent to $1 United States dollar per hour. This minimum wage remains a subject of much debate as trade union representatives argue that the minimum wage cap is below the inflation rate, thus undermining efforts to achieve decent work.

4 Opportunities to Achieving Decent work and Economic Growth in South Africa

4.1 *Opportunities for Decent Work Through Law and Policy*

The agenda for decent work for all in South Africa can be traced back to the South African socio-economic policy framework implemented by the African National Congress (the ANC) (Cameron 1996). The Reconstruction Development Program (RDP) was introduced in South Africa after months of deliberations between the ANC government and the Congress of South African Trade Unions (COSATU) and the South African Communist Party (SACP) (Cameron 1996). Trade unions in South Africa play a unique role in that they have not only assumed power by fighting for decent work in South Africa prior to and post democracy but have been key players in the socio-political movement against apartheid by acting as key drivers for socio-political and economic transformation in South Africa

(Webster 1987). The RDP programme sought to oversee the transformation of the South African economy by embarking on a macro-economic stabilisation program formally known as the Growth, Employment and Redistribution (GEAR) which was aimed at driving the stabilisation of the economy and the settlement of the debts owed by the apartheid government which now passed on to the new democratic government (Michie and Padayachee 1998).

The creation of decent work has, since the dawn of the new democracy, been at the centre of new economic policies championed to promote decent work by the ANC government (SPPII, 30). The New Growth Path which was launched by the ANC government in 2010 envisaged the creation of five million new jobs by the year 2020. The vision was that by the year 2020 half of all the working-age would be in paid employment and that unemployment would have dropped by 15 percent. The New Growth Path focuses on identifying areas where large scale employment creation is possible through a series of partnerships between the government and the private sector (New Growth Path 2011). However, the objectives of the New Growth Path (Fine 2012) are yet to be realised as the expanded unemployment rate currently stands 35,313 percent (StatsSA 2020).

In a bid to promote stability and security of work in South Africa, the South African government has put in place structures for decent work through the Commission for Conciliation, Mediation and Arbitration (CCMA). The CCMA is a dispute resolution body established in terms of the Labour Relations Act 66 of 1995 (Bhorat et al. 2009). The overarching goal of the CCMA is to give effect to the constitutional right to fair labour practices by providing platforms for conciliating workplace disputes, arbitrating certain categories of work disputes and conducting inquiries by arbitrators to name a few of the broad services offered by the CCMA. The provision of platforms for dispute resolution through conciliation and mediation forms part of the decent work agenda. During the 2018 to 2019 financial year, the CCMA received a total of 193,732 cases, making it the largest labour dispute resolution agency in the world by volume of referrals (Tshabalala 2019).

In addition, other legislative interventions like the Employment Equity Act are key drivers towards pursuing the decent work Agenda in South Africa. The Act has been hailed as a key instrument in promoting equality in the workplace (Du Toit 2003). The Act has two key objectives namely: to ensure that employees receive equal opportunity and fair treatment through the prohibition of unfair discrimination against all employees; and to achieve equitable representation of black people, including women and people with disability by means of affirmative action. Affirmative action's procedures coincide with the country's plight to promote transformation and redress areas of socio-economic inequality through black economic empowerment programmes such as South Africa's Black Economic Empowerment (BEE) policy which later developed into the Broad-Based Black Economic Empowerment Act 53 of 2003 (BBBEEA).

The BBBEEE Act can be described as an advancement to the BEE policy as it seeks to encompass a wider range of historically disadvantaged persons and not to exclusively empower black persons. In other words, the difference in these policies is that while the BEEE policy sought to address the wrongs of the past through exclusive Black economic empowerment, BBBEEE aims to redistribute the wealth of the nation to all races and genders including people with disability. The BBBEE Act seeks to promote affirmative action by promoting a preferential procurement process that aims to ensure the effective participation of historically disadvantaged persons into the economy. Historically disadvantaged persons being the Blacks, Indians and Coloured persons, women as well as people with disability through socio-economic strategies (Esser and Dekker 2008) into the mainstream economy which was previously exclusively white and inaccessible (Horwitz and Jain 2011). However, opponents of the BBEEA policy argue that the BEE policy has led to the enrichment of a small elite from the previously disadvantaged groups and its value in promoting decent work and addressing the challenges of inequality are yet to be seen. This is also the case as its implementation has been plagued by corruption with BEE contracts benefiting political elites (Simkins 2011).

4.2 The Barriers to Effective Implementation to SDG8 in South Africa

Despite the provision of legal structures and institutions to support decent work in South Africa, several problems persist. Key challenges to the attainment of decent work in South Africa include, among other things: high levels of unemployment; skills shortages; inequality; and; corruption. A quick glimpse at the Quarterly Labour Force Survey (QLFS) for the fourth quarter of 2019 paints a gloomy picture. The QLFS is a household-based sample survey conducted by Statistics South Africa. It collects data on the labour market activities of individuals aged 15 years and older who live in South Africa. The official unemployment rate remained unchanged at 29.1% between Quarter 3: 2019 and Quater4: 2019. All provinces recorded increases in the expanded unemployment rate. The largest increase was recorded in Limpopo up by 5.2% followed by Free State up by 3.0% points Mpumalanga up by 2.7%. While it is too soon to depict the impact of the COVID-19 pandemic on South Africa's plight to achieving decent work and the results of the 10 years New Growth Path put in place by the ANC government in 2010 to promote decent work and overall employment, the efforts made by the New Growth Path have substantially been destroyed by the COVID −19 pandemic which led to the lockdown of the country for more than 5 months.

The lockdown in South Africa started from 26 March 2020 leading to a shutdown of large parts of the economy for 5 weeks and some smaller sectors remained in lockdown for 5 months post the initial date of the national lockdown. Due to the COVID 19 pandemic, South Africa has experienced a massive decline in production and sales which then led to a vast number of job losses as many industries closed indefinitely due to the lockdown. Therefore, the gains made in 2019 in terms of employment prior to the pandemic continue to drop amid the COVID 19 pandemic.

A second barrier to effective implementation to decent work relates to high levels of corruption. Corruption is a plague that has been eroding national resources for decades. The fight against corruption also ties in with South Africa's priority in achieving effective service delivery which is crucial in promoting decent work and improving the quality of life for all South Africans. In the last quarter of 2019, South Africa was recorded to have lost close to 27 billion Rands annually, leading to thousands of job losses as a result (Shuma 2018). Corruption allegations in State owned enterprises and some international companies doing business in South Africa has led to serious financial losses. This, in turn, led to massive job losses. The report published by the office of the public protector State of Capture in 2016 records in detail the extent of improper and unethical conduct by the office of the President and other state functionaries (State Capture 2016). The state of corruption amid COVID times has increased as South Africa has lost close to five hundred billion Rands set aside to assist financially distressed companies and unemployed South Africans during the pandemic. The prospects of achieving decent work in South Africa within the remaining 10 years of the 2030 goal remain improbable as the present state of the economy and the growing pangs of corruption amid the COVID 19 pandemic places unprecedented strains in the economy.

5 Perspectives from Lesotho Opportunities and Challenges

Lesotho was a British colony from 1884 until it gained its independence in 1966 (ILO EESEL 2014). Since its independence, Lesotho has sought to promote national development planning through the implementation of five-year development plans which were carried out until the year 2000 (Mashinini 2019). In 2000, the country formulated Vision 2020 to provide a long-term perspective within which national short to medium-term plans could be articulated (Lesotho Vision 2020). The Vision 2020 visualized Lesotho as a country full of employment opportunities and a magnet retaining its people to decent jobs by the year 2020 (Lesotho Vision 2020).

Alongside the Vision 2020, Lesotho also adopted the Millennium Development Goals (MDGs) in 2000, as well as the Poverty Reduction Strategy Plan (PRSP) to serve as the country's development frameworks (Mashinini 2019). The MDGs included, among other things, plummeting poverty and hunger through the creation of employment (Lesotho MDGs Status Report 2015). The Poverty Reduction Strategy Plan, on the other hand, was anchored on three inter-connected approaches: rapid job creation through the establishment of a beneficial operating environment that facilitates private sector-led economic growth; delivery of poverty-targeted programmes that empower the poor by empowering them to gain access to income opportunities; and establishment of policies and legal frameworks that are beneficial to the full implementation of priorities (Lesotho PRS 2004/2005).

Lesotho adopted SDGs in 2015 to carry the baton from the MDGs and as a reflection of its commitment to the UN Agenda 2030. The implementation of SDGs in Lesotho contributes to a larger pool of policy frameworks put in place for the development of Lesotho and Basotho, thereby assisting the government of Lesotho and its development stakeholders in mapping out strategies for enhanced execution of transformative sustainable development (Mashinini 2019). The adoption of SDGs, therefore, forms part and parcel for the acceleration of decent and productive job creation and inclusive growth, energized and dynamic private sector, modernized public service that supports the private sector and is built on the principle of "people-centred development" and good governance and accountability system (Lesotho VRN 2019).

Although Lesotho is struggling to introduce policies and strategies that would empower it to successfully implement the SDGs as its development programme towards 2030, as envisioned in Agenda 2030 (Mashinini 2019), the country's commitment to SDGs generates both a challenge and an opportunity for the government to act differently, vigorously and with a clear purpose on investment in people-focused development (Lesotho Country Analysis 2017). Lesotho does not have a comprehensive set of labour laws compared to South Africa. However, the Constitution of Lesotho as the supreme law of the country protects the right to work by providing that "Lesotho shall endeavour to ensure that every person has the opportunity to gain his living by work which he freely chooses or accepts" (Constitution of Lesotho). The Labour Code Order 24 of 1994 is the principal law on labour and employment in Lesotho (Mosito 2014). The Code promotes employment by prohibiting unfair dismissals of employees in sect. 66, by allowing orders of reinstatement in sect. 73(1) and providing compensation in cases of unfair dismissal in sect. 73(2) (Labour Code 1992). The Labour Code provides a wide range of measures and policies that are aimed at ensuring the continued employment of employed persons, such as procedures to be followed prior to a dismissal (Mosito 2014).

5.1 Lesotho and SDG8

Lesotho continues to grapple with high rates of unemployment despite its efforts to alleviate poverty and create employment through the country's long-term policies (Lesotho DCWP 2018/19). The continuing retrenchments of mine workers from South African mines have worsened the unemployment situation in Lesotho (Lesotho DWCP 2018/19). The employment of Basotho mineworkers in South Africa is the result of the migrant labour system which was introduced by the South African government as a system that would source mine workers from surrounding countries in the Southern African hemisphere as a means to optimize production during the apartheid regime (Bezuidenhout and Buhlungu (2015). This system has remained in place despite significant labour reforms in South Africa. The creation of additional decent jobs and faster economic growth are primary determiners of poverty alleviation in Lesotho (Lesotho VRN 2019). The Government of Lesotho recognises that creating decent employment to address the increasing number of people in need of jobs is one of the major challenges the country is facing (Lesotho VRN 2019).

6 Measures Put in Place to Achieve SDG Goal Number 8 in Lesotho

Lesotho is highly devoted to the implementation of the SDGs. The country has put in place programmes and initiatives that are intended to stimulate economic growth and decent job creation particularly for the youth (Lesotho VRN 2019). Lesotho has endorsed Decent Work Agenda which was launched by the ILO in 1999 as its contribution to UN country programmes and as one core instrument to better integrate regular budget and extra-budgetary technical cooperation (Rantanen et al. 2020). The Decent Work programme calls for quality jobs, dignity, equality, a fair income, and safe and healthy working conditions and environments as well as putting people at the centre of development and creating a future that is inclusive and sustainable (Rantanen et al. 2020). Lesotho as a member of the United Nations (UN), International Labour Organisation (ILO), African Union (AU) and Southern African Development Community (SADC), is committed to complying with International Labour Standards for Decent Work and inclusive growth (Lesotho DWCP 2018/19). The Lesotho DWCP encompasses a planning framework for achieving decent work for all by the government of Lesotho, employers' and workers' organisations as well as key stakeholders (Lesotho DWCP 2018/19). The DWCP's key priorities include employment creation, social protection and good governance.

Apart from that, Lesotho has developed a National Strategic Development Plan (NSDP) as a medium-term implementation strategy (Makoa 2014). The NSDP implements the SDGs by focusing on the need to pursue sustainable inclusive growth as the most effective route for poverty alleviation (Lesotho VRN 2019). In particular, the NSDP outlines the growth strategy built on four productive sectors that have the highest potential to create decent jobs and attain sustainable inclusive growth (Lesotho VRN 2019).

7 Challenges to Achieving Decent Work and Economic Growth in Lesotho

As indicated, Lesotho continues to battle the intensifying unemployment rate as a result of limited employment opportunities and low absorption rate in labour markets (Lesotho DWCP 2018/19). Although the country has pursued poverty alleviation through job-creation growth as the primary object of the government's economic policies over the past two decades, unemployment rate is estimated to be 32.8 percent, with a higher prevalence for females and youth at 39.7% and 32.3% respectively, compared to males at 26.2% (Lesotho NSDP 2018/19). The government of Lesotho acknowledges that to attain economic growth and generate employment, a strong focus should be put on tapping investments in development of the private sector, particularly in the four National Strategic Development Plan identified growth generating sectors - commercial agriculture, tourism, mining and manufacturing (Makoa 2014).

Currently, the fight against the novel coronavirus disease (Covid-19) has left most countries' economies crippled. Lesotho is no exception. The country already has structural, macroeconomic and social challenges which have been intensified over time by environment-related shocks, political instability and decelerated economic growth, as a result, covd-19 is likely to have devastating socio-economic impacts on the country, including sharp surges in unemployment, vulnerability and poverty (UNDP 2020). With the onset of covid-19, South Africa announced a national lockdown and closed its borders with Lesotho, thereby bringing the travel and tourism industry, which provides formal employment to over 20,000 people in Lesotho, to a complete halt. According to the Lesotho Tourism Development Corporation, no revenues were generated during the months of March and April due to the covid-19 response measures (UNDP, 2020). A recent review suggests that, due to COVID-19, a number of vulnerable households has increased by almost 50 percent to 899,000 (179,000 in urban and 720,000 in rural areas) as a result of factors such as loss of employment and income sources including remittances and the loss of productive assets (UNDP, 2020). Given the current status quo, it is expected that multidimensional poverty and inequality will significantly increase, thus reversing the hard-won progress towards the attainment of decent work in Lesotho (UNDP 2020).

Other challenges facing the successful implementation of SDG8 include climate change and environmental degradation which represent a significant threat to poverty reduction and to achieving SDGs in Lesotho as well as corruption (Lesotho Country Analysis 2017). The country is extremely vulnerable to volatile weather conditions including floods, drought and heavy rainfalls which contribute to soil erosion and deteriorating conditions of range and arable land (Lesotho NSDP I 2012/13). This is particularly bad because most households in Lesotho practise "low input, low output traditional rain-fed crop farming and extensive livestock husbandry under a communal land tenure system," (Lesotho NSDP 2018/19).

These progressively irregular climate conditions, coupled with continuing environmental degradation, negatively impact production and productivity, leading to insufficient produce to meet these households' food requirements even in what could be regarded as good agricultural years (Lesotho NSDP 2018/19). To address the unreliable climate conditions that the country faces, Lesotho needs to promote sustainable agriculture strategies and climate-smart technologies to assist a high number of Basotho who depend on rain-fed subsistence farming (MDGs Status Report 2015).

In addition to that, Lesotho has recently experienced political instability as a result of the coalition government arrangement which culminated in a disruptive effect on the economy and developmental trajectory (Lesotho NSDP 2019/19). One more perturbing trend in the political landscape is the escalation of factionalism within Lesotho politics which seems to have shaped the conditions for political opportunism and rent-seeking amongst Members of Parliament, thereby threatening the continued existence of coalition governments and fuelling political instability and uncertainty (Lesotho Country Analysis 2017). Like in South Africa, corruption poses a massive threat to the economic development and democratic governance in Lesotho (Malephane and Isbell 2019). Although, the overall picture painted by public perception assessments about corruption in Lesotho carried out by external organisations have revealed that Lesotho is faring better than its African counterparts in combatting corruption, the challenges posed by corruption have always been eminent in Lesotho (OSISA 2017). One of the most notoriously documented corruption case which put Lesotho on the map involved bribery in the Lesotho Highlands Water Project (LHWP), the largest water transfer project in the world, between Lesotho and South Africa (OSISA 2017). Under this project, Lesotho signed a treaty with the Republic of South Africa in 1986 in which Lesotho agreed to export water to South Africa. Several multinational corporations were accused of having paid bribes to the chief executive of the LHWP, Mr. Masupha Sole, for the award of the project-related contracts (OSISA 2017). After a long-drawn-out court cases, Sole was found guilty of accepting bribes and sentenced to 15 years imprisonment. However, despite having a plethora of anti-corruption laws and having achieved some highly publicised victories over corruption, Lesotho still has significant room for improvement in the fight against corruption (Ardigo 2014).

The last challenge worth mentioning is the skills mismatch in the labour market. Lesotho has a persistent disparity between the skills required by labour market and those that are produced due to the country's education and training system that is not in line with the skills needs of the labour market (Lesotho DWCP 2018/19). Lesotho labour market is filled with a high percentage of low skilled literate labour (educated labour without employable skills) as well as graduates from post-secondary institutions who are job seekers rather than job creators (Lesotho DWCP 2018/19).

8 Opportunities to Achieving SDGs in Lesotho

The prospects of inclusive growth and decent job creation in Lesotho remain largely depended on opportunities the country possesses (Lesotho NSDP 2018/19). Lesotho has economy wide opportunities as well as sector specific opportunities that, if exploited, can yield the anticipated objectives enshrined in the NSDP II, Vision 2020, SDGs and SADC Regional Indicative Strategic Development Plan (RISDP) 2005–2020 (Lesotho NSDP 2018/19). Lesotho possesses approximately 80 percent of literate citizens, making it abundantly full of literate labour force and thereby providing it with a distinct advantage in both domestic and global markets in terms of educated workforce (Lesotho NSDP 2018/19). It is this literate labour force that gives Lesotho the greatest comparative advantage in that, if properly trained and employed, the labour force could be the driver for Lesotho's development (Lesotho NSDP 2012/13). The country also prides itself in well-established tripartite wages negotiation structures that encompasses representatives of workers, employees and government across sectors (Lesotho NSDP 2018/19). To increase productivity and compete in domestic and international markets, Lesotho needs to invest in effective skills development strategies (Lesotho NSDP 2012/13).

Moreover, as a least developed country, Lesotho enjoys a duty-free access for its goods to major world markets such as the United States and the European Union (Lesotho NSDP 2018/19). Because Lesotho's national output is derived from trade in manufactured products, the country's manufacturing sector constitutes the prime contributor to exports driven mainly by textiles and garments manufacturing, making it the leading textiles exporter to the US under African Growth and Opportunity Act (AGOA) (Makoa 2014). Lesotho has, therefore, taken advantage of the AGOA to become the largest exporter of garments to the US from sub-Saharan Africa (ILO EESEL 2014). However, the greatest source of potential for diversification and growth comes from South Africa and other regional markets and Lesotho's membership of bodies such as SACU and SADC provide access to the most advanced and prosperous markets in Africa (Lesotho NSDP 2012/13). In addition to that, Lesotho has a great potential for eco-tourism, which remains largely unexploited (Lesotho Country Analysis 2017). This potential lies in Lesotho's natural beauty, rich flora and fauna, and captivating prehistoric and cultural heritage, with mountains, valleys, and rivers providing a memorable scenery for tourists (Lesotho Country Analysis 2017). If exploited, the expansion of community-based tourism products can create more jobs without heavy investment costs (Lesotho Country Analysiss 2017). Lastly, Lesotho has very large and unexploited hydropower potential (ILO EESEL, 2014). Lesotho's electricity supply is already the greenest in the world and according to the pre-feasibility studies, the country has potential to produce about 6000 MW from wind, 4000 MW from pump storage and 400 MW from conventional hydropower (Lesotho NSDP 2012/13). Given that Lesotho is centrally located within South Africa, and that South Africa has a net power shortage, Lesotho has substantial potential to export electricity to South Africa.

9 Perspectives from Zimbabwe: Zimbabwe and SDG8

Zimbabwe has a high literacy rate of about 94 percent and an economically active labour population of over 60 percent of the total population (Zimbabwe ICDS 2017). In the 2019 budget statement, the Minister of Finance noted the high levels of unemployment particularly for the youths, informalisation of work, high import dependency, low industry capacity utilisation as some of the challenges facing the Zimbabwean economy (Ncube 2019). With a high literate rate, diverse resources and a huge population of economically active citizens, the expectation is that Zimbabwe would implement the SDGs, particularly SDG8 with less challenges than other African countries. However, as it will be shown below, this is not the case.

As will be shown below, SDG8 is interconnected to other SDGs and broadly to other labour standards. Thus, the success of SDG8 is interlinked to other rights, particularly participation rights including freedom of association, the right to collective bargaining and the right to strike.

10 Zimbabwe and SDG 8

Like Lesotho, Zimbabwe was a British colony until it gained its independence in 1980 (Decent Work Country Programme for Zimbabwe 2012–2015). After its independence, Zimbabwe inherited a dual economy encompassing a modern sector employing a fifth of the labour force, existing alongside a subsistence agricultural sector employing 80% of the labour force (Decent Work Country Programme for Zimbabwe 2012–2015). Like many countries in the world, Zimbabwe is also grappling with a daunting challenge of dealing with the scourge of unemployment and endemic poverty (ZiNEPF 2009). As a result, the Zimbabwe Congress of Trade Unions (ZTCU) and its affiliates have, for the recent past years, seized the opportunity to advocate and engage the government, employers and other national stakeholders for the achievement of the four pillars of the Decent Work Agenda namely: employment creation; workers' rights; social protection; and social dialogue (Chakanya 2017).

The Ministry of Macro-Economic Planning and Investment Promotion is responsible for coordinating the SDGs or the Agenda 2030, with supervision from the office of the President and Cabinet. In 2016, the Zimbabwean government aligned the SDGs to the Zimbabwe Agenda for Sustainable Socio-Economic Transformation (ZimAsset) and to the Interim Poverty Reduction Strategy (IPRSP) (Mpofu 2017). The government showed its commitment to decent work through several initiatives including the adoption of a strategic protectionist policy which was regulated by the Statutory Instrument 64 of 2016 with broader objectives of boosting local industry capacity (Mpofu 2017). The policy was designed to shield local industries from outside competitors, thus improving local job creation. In addition, the government removed several barriers for companies and persons seeking to do business in Zimbabwe prioritising key areas such as starting a business, protecting minority investors, enforcing contracts, trading across borders among others (Mpofu 2017). Despite the noble intentions underpinning this policy, it has not yielded positive results and its implementation was not complemented with genuine support for local industries. In addition, the policy also offended the SADC initiatives on regional integration as it restricted the movement of goods across borders.

Collaborations between the government and the United Nations Development Programme (UNDP) are also seen as paving way for decent work, with selected projects like the Young Farmers Innovation Lab being success stories. In addition, the ILO's Skills for Youth Employment and Rural Development includes the training of rural youths and economic empowerment on agriculture and rural development (ILO Zimbabwe 2013). The government initiatives are also informed by regional initiatives including the Africa Union Vision, Agenda 2063 and the SADC Regional Indicative Strategic Development Plan (RISDP) 2015–2020 and the SADC's Industrialization Strategy 2015–2063. Yet as it stands, most of the commitments expressed by the government remain on paper, with no serious attempt to implement the SDGs.

11 Barriers to Achieving of Goal Eight of SDGs in Zimbabwe

As indicated, Zimbabwe is grappling with high levels of unemployment, particularly among the youth. There are no accurate figures on the levels of unemployment, with most of the data available, including on the ILO website being outdated (Van Wyk 2014). Many companies have been shedding jobs citing the difficult economic environment and COVID-19, has increased unemployment. Many graduates leaving universities within the country and from foreign academic institutions are grappling to find work in the country (Mwenje 2016). About 30,000 students graduate annually from Zimbabwean institutions of higher learning and the unofficial unemployment rate stood at about 90% in 2016 (Mwenje 2016, p. 50). There is nothing to demonstrate that this unemployment rate has decreased, since no meaningful reforms or economic interventions have been developed. The following section demonstrates how achieving the decent work targets in Zimbabwe is interconnected to key rights, particularly participation rights and to the overall permissible environment conducive for corporations.

The Millennium Development Goals (predecessor to the SDGs) are regarded as having failed to achieve various targets for many reasons. Bad political environment, poor governance, unclear policies have all been cited as reasons for the failure, particularly in Africa (Ogujiuba and Jumare 2012). These factors are presented in starker terms when one assesses SDG8 with reference to Zimbabwe. The creation of meaningful jobs relies on the government to set up a clear business policy that safeguards the interests of parties involved, including property rights. In Zimbabwe, the economy and the political environment are often seen as the Siamese twins, with the political environment shaping the economy (Noyes 2020). As such, corporations can only play their role in creating meaningful jobs to the extent permitted by the politics of the day.

Moreover, the nexus between good jobs created by corporations and the quality of such jobs is enhanced in an environment where workers can exercise various rights, including participation rights. This requires an environment that allows workers to negotiate with their employer and to exercise the right to strike when necessary. Notably, Zimbabwe ratified the ILO Conventions, including all the core conventions. Despite having entrenched the right to freedom of assembly and association, the right to demonstrate and petition, freedom of conscience, political rights among other progressive rights in the Constitution, exercising these rights is severely restricted. Most organised strikes by trade unions representing various categories of workers, including public servants are hardly approved. The right to freedom of association and collective bargaining only remains on paper and workers cannot enforce or exercise these rights. Following these restrictions and the levels of violence unleashed on workers who exercise their rights, the Zimbabwe Congress of Trade Union (ZCTU) and the International Trade Union Confederation (ITUC) lodged complaints to the Committee of Experts on the Application of Conventions and Recommendations (CEACR). The 2019 CEACR Report highlighted issues of non-compliance by the Zimbabwean government on the implementation of the Convention on Freedom of Association and Protection of the Right to Organise, 87 of 1948 (ILO 2019b). The CEACR called upon the government to desist from arresting, detaining and harassing members of trade unions and to investigate allegations of violence against trade union members (ILO 2019b). There are further reports that the government is crushing professionals who are exercising their rights including journalists who have been exposing corruption (Amnesty International 2020). The undermining of these core rights further compromises not only the quality of jobs offered by corporations, but the conditions of employment. To achieve the decent work agenda in Zimbabwe, there is a therefore a need to create an environment where constitutionally guaranteed rights can be exercised without threats of violence or intimidation from the government. Employment creation, social dialogue, social protection and the rights at work must all be recognised as key elements of the decent work agenda in Zimbabwe.

There is therefore a clear linkage between political and economic policies and the performance of the overall economy. Economic growth in Zimbabwe has been hampered by the lack of political reforms. Some of the issues affecting decent work and corporate responsibility in Zimbabwe include the unclear monetary policy, high inflation, rampant corruption, lack of protection of property rights, lack of transparency in the exploitation of mineral resources and the shrinking of democratic space for workers to exercise their rights. Many companies continue to shut down, citing foreign currency shortages, the rising cost of doing business and the uncertain business environment among others. Thus, the success of corporations operating in Zimbabwe and by extension the success of SDG8 is hinged on the political climate prevailing at the time.

In addition to the above, corruption continues to undermine key transformative programmes.

12 Corruption and SDG8 in Zimbabwe

Zimbabwe is currently ranked 158/180 on the Corruption Perception Index – the most corrupt country outside a war zone in Sub-Saharan Africa (Transparency International 2019, p. 2). After ending colonial rule, Zimbabwe inherited the British public administration system characterised by hierarchy, departmentalism with mechanisms for checks and balances in handling unethical behaviour (Makumbe 1994, p. 45). With such a system of checks and balances and laws designed to curb corruption, the logical conclusion is that ending corruption and prioritising development is an easily achievable standard. However, since 1980, corruption has continued to grow with the ruling elite becoming the enablers and leaders of corruption in both the public and private space (Makumbe 1994, p. 46). There is close nexus between corruption and politics in Zimbabwe and institutions like the Anti-Corruption Commission have not been successful (Muzurura 2017, p. 106). In fact, those in public service have been accused of creating regulations and procedures designed to delay the issuing of permits trade licenses among other documents unless some form of incentive is offered Muzurura 2017, p. 106). In 2019, the Minister of Finance acknowledged the effect of corruption including creating an economic dissatisfaction within the general citizenry; wastage of public resources; undermining a healthy, investment environment; reinforcing political instability and breeding economic inequalities (Ncube 2019, p. 117). Despite this clear acknowledgment, the mechanisms that the government seeks to implement to tackle corruption are weak and do not address the lack of political will which is the main root cause (Ncube 2019, p. 117).

Several governmental initiatives and policies have all failed to create meaningful jobs and to raise the living standards of ordinary citizens, partly due to corruption (Chitongo et al. 2020, p. 13). Achieving SDG8 thus requires a genuine governmental commitment towards fighting corruption and to implement principles of transparency and accountability enshrined in the Constitution.

Recommendations for achieving decent work in South Africa, Lesotho and Zimbabwe.

13 CSR

Corporate social responsibility now forms part of the agenda within the sustainable development debates. Traditionally, the concept of corporate social responsibility denotes the voluntary obligations of companies and organisations to go beyond the requirements prescribed in law in promoting the advancement of socio-economic development by incorporating social responsibility measures in society (Amodu 2020). The process of implementing CSR within the SADC region has been a contentious subject since there is no uniformity on the applicable legal regimes on CSR among member states. In South Africa, several aspects of CSR that were traditionally voluntary are now mandated and prescribed by legislation (Howard 2014, pp. 13–15). The increase in legal prescriptions is particularly high in the mining industry, which notably still employs a substantial worker compared to other industries (Howard 2014), (Minerals Council South Africa 2017). Other CSR provisions are based soft law instruments, with voluntary codes of best practice such as the King Codes of good corporate governance being key towards ensuring that companies adhere to international best practice and standards (IODSA 2016). The Companies Act 71 of 2008 explicitly contains avenues which promote CSR through the creation Social and Ethics Committees (SECs) under sect. 72(4). The SECs are a paradigm shift in the governance of companies and in the recognition of companies' role in advancing CSR. One of the significant inclusions on the duties of the SECs is towards monitoring the company's standings under the ILO Protocol on decent work and promote the values of decent work. This is done through ensuring compliance of the company with ILO's protocol on decent work as well as with supporting instruments and laws that aim to promote social responsibility values within the company. Within the SADC region, Corporations can play a meaningful role in promoting decent work. Due to the interconnected nature of the SDGs, the failure to meet the targets for goal eight undermines these achievements of other goals.

14 Social Entrepreneurship

The concept of social entrepreneurship is hard to define, (Dees 1998) but it can be defined as marriage between the not for profit approaches combined with a value for profit making model in doing business and corporate sustainability through entrepreneurial skills and approaches (Dees 1998). It involves a process of using business skills and approaches in solving societal problems (Dees 1998). It mixes the elements of not for profit objectives with for profit strategies and processes. The value for social entrepreneurship in promoting the achievement of the decent work agenda is that it allows for the achievement of the rest of the 17 SDG goals to be pursued simultaneously as it addresses the challenges of society through entrepreneurship. Social entrepreneurship allows for social entrepreneurs to be change agents in the social sector by addressing societal problems with a mission to create and sustain social value (Dees 1998). This not only encourages job creation but allows for the engagement of society in issues that affect society whilst ensuring that innovative measures for entrepreneurship which encourage decent work and social justice are pursued.

15 Regional Integration and Partnerships

The above challenges and limitations faced by the three countries—South Africa, Lesotho and Zimbabwe in the implementation of SDG8 further undermine regional integration. Regional integration is based on the need to create a supra-national space between member states that facilitates the free movement of people, goods, services among other collaborative initiatives (Deacon et al. 2008). The SADC has aligned its developmental programmes to the African Union Agenda 2063 and to the UN 2030 Agenda for Sustainable Development (SADC 2019). As such, the SADC policies on economic growth have been centred on addressing employment creation, labour relations, labour market information and productivity, enhanced macro-economic environment, improved financial market systems and monetary cooperation. In addition, the SADC region is focusing on increasing the participation of the private sector in regional integration, thereby adopting a holistic approach to integration (SADC 2019). In the SADC region, several initiatives including the Southern African Power Pool, bring not only governments but independent power producers on board, thereby enhancing economic growth and job creation (Nagar and Mutasa 2017).

Despite having comprehensive policies on the implementation of SDGs, Africa is lagging in the implementation of the SDGs (Sitembo 2020). The implementation of the SDGs, particularly SDGs around economic growth (goals 8–11), has been slow, with levels of inequality increasing (Cerf 2018). The limitations in achieving the SDGs in the SADC region are not purely related to the economy, but are imbedded in poor governance, corruption and a general lack of prioritization of critical areas of economic growth (Ogujiuba and Jumare 2012).

16 Concluding Remarks

This short article introduced the challenges to decent work in the SADC region, with specific reference to three countries in the SADC community namely, South Africa, Lesotho and Zimbabwe. The article provided an overview of the opportunities and challenges for decent work in SADC. This of course is a distillation of a far more comprehensive body of work whose comprehension goes beyond the objective of the paper. Based on the brief discussions made on South Africa, Lesotho and Zimbabwe, it was argued that the challenges to decent work stem from the history of the countries and the socio-political setting in each country's jurisdictions. In South Africa, the core function of decent work is to redress the injustices of the past by addressing socio-economic inequality through laws and policy aimed at redressing the impact of colonization and apartheid. On the other hand, Lesotho as a least-developed country is facing challenges that are embedded with climate change, high levels of unemployment and its interdependency on South Africa. In other words, economic challenges faced by South Africa can have a drastic effect on the economy of Lesotho. This is despite of the nations' effort put in place under the National Strategic Development Plan (NSDP) as a medium-term implementation strategy. Lastly, Zimbabwe presents a unique opportunity for the realisation of the SDGs given the high levels of literacy, the number of graduates leaving universities yearly and the abundance of agricultural land and mineral resources. However, it is plagued with challenges of political instability and the shrinking of the democratic space. Due to the unstable political climate and the sluggish economic growth rate. Most skilled personnel with special and scarce skills have migrated to the diaspora and the countries in the neighbouring SADC region.

The Conclusion is that the objective towards achieving decent work in SADC is not a novel or new concept and the limitations faced by countries who are members of the SADC treaty persist despite the goals set out by the members of the SADC community under the SADC treaty. We therefore recommend that for the effective realisation of SDG 8 under the Agenda 2030, countries will require renewed efforts in promoting CSR as a real requirement in law and policy. The partnership between companies and government through CSR is pivotal in realising the decent work agenda. This is because private companies absorb most of the labour force in both developed and developing economies. Social entrepreneurship and improved efforts of regional integration is pivotal, and the collaboration of SADC member states is essential in re-engaging corporate social responsibility. However, for the SADC region to achieve the SDGs and to prevent another failure like the MDGs, amid the ameliorate the impact of the COVID-19 pandemic on decent work and economic growth, governments must prioritise facilitating employment creation and public-private partnerships. In addition, the SADC region must take decisive action against corruption and must make full commitment to regional integration and promote entrepreneurship across member states. We therefore suggest that CSR, social entrepreneurship and regional integration can be a good start towards achieving decent work within the remaining 9 years of the 2030 Agenda.

References

Alexandra, P., et al. (2013). *Marikana: Voices from South Africa's mining massacre*. Athens, OH: Ohio University Press.

Allen, V. (2003). *The history of black mineworkers in South Africa: Tracking strikes and protests from the 1920s to 1946 2*. London: Moor Press.

Amnesty International. (2020). Zimbabwe: Authorities continue their crackdown on dissent with arrest of investigative journalist and activist. Accessed November 20, 2020, from https://www.amnesty.org/en/latest/news/2020/07/zimbabwe-authorities-continue-their-crackdown-ondissent-with-arrest-of-investigative-journalist-and-activist/

Amodu, N. (2020). *Roadmap to embedding CSR in Africa* (1st ed., pp. 175–195). London: Routledge. https://doi.org/10.4324/9781003009825-9.

Ardigo, I. A. (2014). Overview of corruption and anti-corruption in Lesotho. Accessed November 24, 2014, from https://www.transparency.org.

Bendix, S. (2001). *Industrial relations in South Africa* (4th ed., rev. ed.). Cape Town: Juta.

Bezuidenhout, A., & Buhlungu, S. (2015). Enclave Rustenburg: Platinum mining and the post-apartheid social order. *Review of African Political Economy, 42*, 526–539.

Bhorat, H., Pauw, K., & Mncube, L. (2009). Understanding the efficiency and effectiveness of the dispute resolution system in South Africa: An analysis of CCMA Data. Development Policy Research Unit DPRU Working Paper, (09/137).

Budeli, M., (2007). Freedom of Association and Trade Unionism in South Africa: From Apartheid to the Democratic Constitutional Order (unpublished PhD thesis, University of Cape Town, 65.

Buhlungu, S. (2006). *Trade unions and democracy: Cosatu workers' political attitudes in South Africa*. Cape Town, Human Sciences Research Council.

Cameron, R. (1996). The reconstruction and development programme. *Journal of Theoretical Politics, 8*(2), 283–294.

Cerf, M. E. (2018). The sustainable development goals: Contextualizing Africa's economic and health landscape. *Global Challenges, 2*(8), 1800014.

Chakanya, N. (2017). The sustainable development goals (SDGs): A pathway towards inclusive economic development and promotion of decent work – Challenges, opportunities and future prospects. In *Background paper prepared for the ZCTU/LEDRIZ high level conference on SDGs, 1–2 November 2017*.

Chigwedere, A. (2001). *British betrayal of the Africans: Land, cattle, human rights: Case for Zimbabwe*. Marondera: Mutapa Publishing House.

Chinguno, C. (2013). Marikana massacre and strike violence post-apartheid. *Global Labour Journal, 4*(2), 160–166.

Chitongo, L., Chikunya, P., & Marango, T. (2020). Do economic blueprints work? Evaluating the prospects and challenges of Zimbabwe's transitional stabilization Programme. *AJGD, 9*, 7–20.

Chowdhury, G., & Koya, K. (2017). Information practices for sustainability: Role of iSchools in achieving the UN sustainable development goals (SDGs). *Journal of the Association for Information Science and Technology, 68*(9), 2128–2138. https://doi.org/10.1002/asi.23825.

Cluver, L., Pantelic, M., Orkin, M., Toska, E., Medley, S., & Sherr, L. (2018). Sustainable survival for adolescents living with HIV: Do SDG-aligned provisions reduce potential mortality risk? *Journal of the International AIDS Society, 21*(S1), e25056. https://doi.org/10.1002/jia2.25056.

Collier, D. (2015). Mind the gap: Widening income inequality in South Africa-an institutional failure or mission impossible? In *Labour Law Research Network 2nd Conference, Amsterdam*.

Connolly, S. (2001). *Apartheid in South Africa*. Oxford: Heinemann Library.

Consolidated Text of the Treaty of the Southern African Development Community. 2011. https://www.sadc.int/files/5314/4559/5701/Consolidated_Text_of_the_SADC_Treaty_-_scanned_21_October_2015.pdf.

Davenport, J. (2013). *Digging deep: A history of Mining in South Africa* (pp. 1852–2002). Jeppestown: Jonathan Ball.

Deacon, B. (2008). Global and regional social governance. In *Understanding global social policy* (pp. 25–48).

Dees, J. G. (1998). The meaning of social entrepreneurship.

Du Toit, D. (2003). *Labour relations law: A comprehensive guide* (4th ed.). Durban: LexisNexis Butterworths.

Esser, I. M., & Dekker, A. (2008). Dynamics of corporate governance in South Africa: Broad based black economic empowerment and the enhancement of good corporate governance principles. *J. Int'l Com. L. & Tech., 3*, 157.

Farlam, I. G., Hemdraj, P. D., & Tokota, B. R. (2015). Marikana commission of inquiry: Report on matters of public national and international concern arising out of the tragic incidents at the Lonmin Mine in Marikana, in the North West Province.

Fine, B. (2012). Assessing South Africa's new growth path: Framework for change? *Review of African Political Economy, 39*(134), 551–568.

Ghai, D. (2003). Decent work: Concept and indicators. *International Labour Review, 142*(2), 113–145. https://doi.org/10.1111/j.1564-913x.2003.tb00256.

Gocking, R. (1997). Colonial rule and the "legal factor" in Ghana and Lesotho. *Africa (London. 1928), 67*(1), 61–85. https://doi.org/10.2307/1161270.

Government of Lesotho. (2017) *Lesotho Country Analysis Working Document – Final Draft*. 79 pages.

Government of Lesotho. (2018). *National strategic development plan 2018/19–2022/23 – In pursued of economic and institutional transformation for private sector led job creation and inclusive economic growth*. Maseru: Lesotho Government, Ministry of Development Planning.

Government of Lesotho. *National Vision* 2020.

Government of Lesotho: Lesotho VRN. (2019). *The Kingdom of Lesotho Voluntary National Review on the Implementation of the Agenda 2030 Report 2019*.

Government of Lesotho: Ministry of Development Planning. (2012). *National Strategic Development Plan 2012/13–2016/17: Growth and development strategic framework – Towards an accelerated and sustainable economic and social transformation*. Washington, D.C.: International Monetary Fund.

Government of Lesotho: Ministry of Finance. 2018/19–*2020/21 Budget Strategy Paper* 34 pages.

Government of Lesotho: Ministry of Finance and Development Planning *Poverty Reduction Strategy* 2004/2005–*2006/2007*.

Government of Lesotho: Ministry of Labour and Employment *Decent Work Country Programme III* 2018/19–*2022/23 – Promoting Decent Work for All*.

Government of Zimbabwe, Zimbabwe Agenda for Sustainable Socio-Economic Transformation (Zim-Asset): Towards an Empowered Society and a Growing Economy October 2013 – December 2018); OM Mpofu 'Statement by the Minister of Macro-Economic Planning and Investment Promotion of the Republic of Zimbabwe' (2017) The 2017 High Level Political Forum: Voluntary National Review Process, New York.

Government of Zimbabwe: Minister of Labour and Social Services. (2012). *Decent work country Programme for Zimbabwe 2012–2015*. Harare: International Labour Organization.

Hans, P. (1935). Gold Mining in South Africa.

Horwitz, F. M., & Jain, H. (2011). An assessment of employment equity and broad based black economic empowerment developments in South Africa. *Equality, Diversity and Inclusion: An International Journal, 30*, 297–317.

Howard, J. (2014). Half-hearted regulation: Corporate social responsibility in the mining industry. *South African Law Journal, 131*(1), 11–27.

International Labour Organisation (ILO). (2013). *ILO Skills for Youth Employment and Rural Development Programme and the UNDAF (Zimbabwe) on a Joint mission.*

International Labour Organization (ILO). (2014). *Enabling environment for sustainable Enterprises in Lesotho* (p. 103). Geneva: International Labour Office, Enterprises Department.

International Labour Organisation (ILO). (2019a). Information supplied by governments on the application of ratified Conventions, Zimbabwe: Freedom of Association and Protection of the Right to Organise Convention, 1948 (No. 87).

International Labour Organisation (ILO). (2019b). Comments adopted by the CEAR: Zimbabwe (2019), Committee of Experts on the Application of Conventions and Recommendations: International Labour Organization (ILO). http://www.ilo.org/dyn/normlex/en/f?p=NORMLEXPUB:13202:0::NO::P13202_COUNTRY_ID:103183

IODSA. (2016). Report on Corporate Governance for South Africa 2016. Accessed November 24, 2020, from https://cdn.ymaws.com/www.iodsa.co.za/resource/collection/684B68A7-B768-465C-8214-E3A007F15A5A/IoDSA_King_IV_Report_-_WebVersion.pdf.

Luiz, J. (2002). South African state capacity and post-apartheid economic reconstruction. *International Journal of Social Economics, 29*(8), 594–614. https://doi.org/10.1108/03068290210434170.

Makoa, R. (2014). Lesotho National Policies, best practices and challenges in building productive capacities. In *Report to Workshop of National Focal Points of Least Developed Countries (LDCs) on July 2014, Cotonou, Benin.*

Makumbe, J. (1994). Bureaucratic corruption in Zimbabwe: Causes and magnitude of the problem. *Africa Development/Afrique et Développement, 19*, 45–60.

Malephane, L., & Isbell, T. (2019). Basotho see progress in fight against corruption but fear retaliation if they report incidents.

Mashinini, V. (2019). Sustainable development goals in Lesotho – Prospects and constraints. *Africa Insight*, 89–106.

Michie, J., & Padayachee, V. (1998). Three years after apartheid: Growth, employment and redistribution? *Cambridge Journal of Economics, 22*(5), 623–636.

Moodley, L., & Cohen, T. (2012). Achieving "decent work" in South Africa? *Potchefstroom Electronic Law Journal, 15*(2), 319–344. https://doi.org/10.17159/1727-3781/2012/v15i2a2490.

Mosito, K. E. (2014). A panoramic view of the social security and social protection provisioning in Lesotho. *PER/PELJ, 17*(4), 1572–1629.

Muzurura, J. (2017). Corruption and economic growth in Zimbabwe: Unravelling the linkages. *International Journal of Development Research, 7*(1), 1197–11204.

Mwenje, S. (2016). The challenge of graduate unemployment: A case of university graduates in the city of Mutare-Zimbabwe. *International Journal of Innovative Social Sciences and Humanities Research, 4*(4), 50–57.

Nagar, D., & Mutasa, C. (2017). The implementation gap of the regional integration agenda in SADC/Seminar report.

Ncube, M. (2019). Zimbabwe: The 2019 National Budget Speech. Accessed November 20, 2020, from https://www.parlzim.gov.zw/component/k2/2019-budgetspeech.

Noyes, A. (2020). *A new Zimbabwe: Assessing continuity and change after Mugabe.* Santa Monica, CA: Rand Corporation.

Ogujiuba, K., & Jumare, F. (2012). Challenges of economic growth, poverty and development: Why are the millennium development goals (MDGs) not fair to Sub-Saharan Africa?. *Journal of Sustainable Development, 5*(12), 52.

OM Mpofu. (2017). Statement by the minister of macro-economic planning and investment promotion of the republic of Zimbabwe (2017) The 2017 high level political forum: Voluntary national review process, New York.
Open Society Initiative for Southern Africa (OSISA). (2017). Effectiveness of anti-corruption agencies in Southern Africa Angola, Botswana, DRC, Lesotho, Malawi, Mozambique, Namibia, South Africa, Swaziland, Zambia and Zimbabwe. Rosebank, Johannesburg: OSISA, 26.
Pekmezovic, A. (2019). *'The new framework for financing the 2030 agenda for sustainable development and the SDGs', in [online]* (pp. 87–105). Chichester: Wiley.
Rantanen, J., Muchiri, F., & Lehtinen, S. (2020). Decent work, ILO's response to the globalization of working life: Basic concepts and global implementation with special reference to occupational health. *International Journal of Environmental Research and Public Health*, 1–27.
Rider, G., (2015). *Corporate Social Responsivity and decent work Social, C., & Social, D. C. SPECIAL ISSUE on corporate*.
Saul, J. (2012). The transition in South Africa: Choice, fate . . . or recolonization? *Critical Arts, 26*(4), 588–605. https://doi.org/10.1080/02560046.2012.723850.
Shuma, P. (2018). *Corruption costs SA's GDP R27 billion annually*. https://www.sabcnews.com/sabcnews/corruption-costs-sas-gdp-r27-billion-annually/
Simkins, C. (2011). Poverty, inequality, and democracy: South African disparities. *Journal of Democracy, 22*(3), 105–119.
Sitembo, H. (2020). Sustainable development goals (SDGs): Far from achievement for Sub-Saharan Africa.
Smit, N., & Mpedi, L. G. (2010). Social protection for developing countries: Can social insurance be more relevant for those working in the informal economy?. *Law, Democracy & Development, 14*.
South Africa. Marikana Commission of Inquiry. (2015). *Marikana commission of inquiry: Report on matters of public, national and international concern arising out of the tragic incidents at the Lonmin mine in Marikana, in the North West Province*. Government Printer, South Africa.
Southern African Development Community (SADC). Southern African Development Community, Decent work programme 2013–2019: Promoting decent work for all in the SADC region 9.
Stares, R. (1977). Black trade unions in South Africa: The responsibilities of British companies.
Statistics South Africa. (2020). Quarterly labour force survey (QLFS) – Q3:2020. http://www.statssa.gov.za/?p=13765
Tabata. (1973). *Industrial unrest in South Africa* (p. 1). Lusaka: All African Convention and Unity Movement of South Africa.
Theron, J. (2012). What is decent about decent work? http://www.ebe.uct.ac.za/usr/idll/resources/lep_opeds/LEP_Oped3_23July12.pdf
Transparency International. 2019. *The transparent international perception of corruption 2019 Report*.
Tshabalala, A. (2019). CCMA 2018/19 Report. Accessed November 24, 2020, from https://www.ccma.org.za/About-Us/Reports-Plans/Annual-Reports/Token/ViewInfo/ItemId/39.
United Nations Sustainable Development Goals. (n.d.).
Van Wyk, A., (2014). *Is Zimbabwe 's unemployment rate 4%, 60% or 95%? Why the data is unreliable*. Africa Check. Retrieved August 1, 2020, from https://africacheck.org/reports/is-zimbabwes-unemployment-rate-4-60-or-95-why-the-data-is-unreliable/
Webster, E. (1987). The two faces of the black trade union movement in South Africa. *Review of African Political Economy*, 33–41.
Wenckebach, A., Hanna, P., & Miller, G. (2019). Rethinking decent work: The value of dignity in tourism employment. *Journal of Sustainable Tourism, 27*(7), 1026–1043.

Mikovhe Maphiri is attorney of the High Court, South Africa. She is an academic (lecturer) at the Department of Commercial Law, University of Cape Town, South Africa. She holds LLB from the University of Limpopo and LLM from the University of Cape Town, where she is also currently a PhD candidate. She is an emerging researcher whose research interests are in corporate governance, CSR and sustainability in South Africa.

Matsietso Agnes Matasane holds LLB from the University of Venda, South Africa, LLM Mercantile Law from the University of Pretoria, South Africa, she is currently LLD Candidate at the University of Pretoria. She is an emerging researcher from the University of Pretoria. She is a PhD candidate in Banking Law at the University of Pretoria. Her area of research is in deposit guarantee system in the South African banking sector in which she interrogates the envisaged introduction of explicit deposit insurance system in South Africa. She is the recipient of the Absa Barclays Africa Chair in Banking Law in Africa scholarship which is awarded to PhD candidates in the field of Banking Law.

Godknows Mudimu is a Research Associate Mineral Law in Africa, a University of Cape Town PhD Graduand and currently works as a legal researcher, Office of the Chief Justice, Seychelles. He holds LLM from the UCT, LLB and BSocSc from Rhodes University, South Africa. His PhD thesis assesses the effectiveness of self-regulation on labour standards in selected South African mining sectors. His areas of interest include regulation, labour law, mineral law, occupational safety and health at mines, and transparency and accountability in the extractive industry.

Part II
Global South

CSR in the Global South: The Continuing Impact of Postcolonial Power and Knowledge

Stephen Vertigans

Abstract Recent years have witnessed a growth in international Corporate Social Responsibility (CSR) related activities. Analysis of these programmes has tended to focus upon intent and scale of ambition. Despite growing emphasis on stakeholder engagement corporate approaches are continuing to originate from northern hemisphere derived ethics, governance and standards that are being uncritically applied in the global south. By applying insights from postcolonial studies, this paper argues that CSR approaches are both reproducing colonial approaches and northern hemisphere sponsored development programmes that followed independence. Consequently, CSR programmes are based upon Northern hemisphere knowledge and are reinforcing power differentials within and between regions. The chapter concludes by tentatively proposing how to re-position CSR out of the post-colonial trajectory and towards the delivery of more sustainable improvements. To do so will require the more philanthropic approach of companies operating in the southern hemisphere (Amaeshi et al., *Corporate social responsibility in Nigeria: Western mimicry or indigenous influences?* 2006; Frynas, *International Affairs*, 81:581–598; Kuhn et al., *Business and Society* 54:1–44, 2015) to become more engaging, strategic and sustainable.

1 Introduction

The globalisation of products, markets and underpinning processes alongside national neo-liberalism programmes have had a huge impact upon regulation, jurisdictions and forms of business practices. Following the 1980s structural adjustment programmes, national and local social services contracted and there has been a shift in expectations concerning where responsibility lies for providing provision. Within this debate, the roles and responsibilities of largescale businesses have

S. Vertigans (✉)
Robert Gordon University, Aberdeen, UK
e-mail: s.vertigans@rgu.ac.uk

grown. Such organisations have developed and support social and environmental programmes, engage with NGOs and community networks under the rubric of corporate social responsibility (CSR). Blowfield and Frynas (2005, p. 499) explain that 'Government, civil society and business all to some extent see CSR as a bridge connecting the arenas of business and development, and increasingly discuss CSR programmes in terms of their contribution to development'. Trans-national Corporations (TNCs) are increasingly aware of CSR related expectations to invest in education, health, skills development, housing and community development. Moreover, in parts of the southern hemisphere corporate agendas are merging with national issues surrounding security, education, infrastructure in ways which can diminish corporate risk. Jamali et al. (2017) point out that this increase has been accompanied by a concomitant rise in related books, chapters and articles. They point out a number of gaps within the literature, including inattention to SMEs which they seek to fill. In this paper it is argued that greater attention also needs to be placed upon the historical trajectories that CSR connect into in order to better understand the wider impacts of policies that extend beyond short term outputs.

These concerns are compounded by limited knowledge about the precise effectiveness of TNCs' CSR approaches in the global south (Idemudia 2014; Vertigans et al. 2016). Certainly, the lack of established methodologies to capture effects is a considerable drawback. Despite considerable expenditure on related projects and programmes knowledge of long-term and broader impacts is patchy. These gaps can be attributed partly to uncertainty over what CSR should achieve allied to weak levels of monitoring, analysis and reporting (Idemudia 2016). And in the southern hemisphere, Jamali et al. (2015, p. 3) refer to the recent debate concerning whether 'CSR has indeed been living up to its stated promises in the developing world'. Leaving aside the judgmental reference to the 'developing world', in this chapter I argue that much closer attention also needs to be placed upon ways in which organisations have assumed power to influence local economies and habitus in ways that extend beyond production and consumption. In some respects, these influences appear to have emerged unintentionally not least because of a lack of appreciation both for the contexts in which CSR programmes are being introduced and the history of the regions and preceding forms of northern hemisphere interventions. These observations share similarities with multi-disciplinary post-colonial critiques of development processes and institutions and the continuing legacy of colonialism's inequitable relations. Helping to provide the base for this discussion, Young (2003, p. 7) argues that 'above all, postcolonialism seeks to intervene, to force its alternative knowledges into the power structures of the West as well as the non-west'. The primary task for this paper is to identify how contemporary CSR approaches have been implementing historical Western power structures and knowledge into other parts of the world before considering how these processes can be overcome. Some examples are drawn from extractive industries to supplement the analysis. Although this sector is not being put forward as representative of global corporate interests, in terms of product and activities, these organisations provide a continuum with colonial interests.

At this point it is important to clarify what is meant by CSR. Finding universal agreement on the definition will not be forthcoming anytime soon with the concept much contested (Idowu 2012). Consequently I am applying one of the most widely used definitions namely the World Bank definition that,

> A company's obligations [are] to be accountable to all of its stakeholders in all its operations and activities. Socially responsible companies consider the full scope of their impact on communities and the environment when making decisions, balancing the needs of stakeholders with their need to make a profit (cited in Doane 2005, p. 217).

The definition is underpinned by the triple bottom line of economics (profit), environment (planet) and social (people).

2 Critical Analysis of CSR's International Approaches

Opposing views about TNC approaches to CSR are spread across the positive to negative spectrum. Spencer (2018) outlines how sustainable development discourse has been co-opted by TNCs under the banner of CSR. In so doing, the concept is considered to be integral to the private sector adopting the role of agent of development agent both in terms of economic and human development (Idemudia 2009). By comparison, Kaplan and Kinderman (2017) explain how much of CSR has been reactive to isomorphic pressures from different stakeholders and regulatory bodies. Other companies have adopted pro-active responses that aim to alter the location in support of their business interests. Hence at the business level within the global south, there is discussion about the case for CSR enhancing economic prospects while also benefitting communities in 'win-win' situations. For instance, within the extractive industries, community investment is increasingly connected to the award of licences with reputation from preceding contracts often a factor (Frynas 2005, Idemudia 2014, Vertigans 2017). From the late 1990s Visser (2006) suggested that the private sector was considered well placed to improve economic, social and environmental conditions in Africa. Tan and Wang (2010), p. 373) refer to,

> literature in the past [that] generally posits an optimistic projection that the global triumph of MNCs ... will introduce a new mode of business practices to the developing countries by spreading a set of universal organizational patterns and business ethical standards.

The authors (2011: 382) go on to hope that 'MNCs will spread the common values and beliefs fine-tuned in the industrialized world to underdeveloped world'. Tan and Wang are critical about the intentions and commitments of many MNCs. However, typical of many CSR advocates, they do not question the appropriateness at the core of the optimism, namely that through CSR mechanisms Western values should spread to the 'developing world'. Nor is there a recognition as Blowfield and Frynas (2005) outline, that the adaptation of the concept of CSR within the West has been accompanied by the tendency to assume 'social' corporate values to be unique to the northern hemisphere, thereby neglecting the extremely long history in which localised social obligations have permeated other parts of the world.

Within what Bondy and Starke (2014) refer to as the 'universal' CSR discourse, TNCs have adopted worldwide CSR strategies into related global initiatives such as

Global Compact. The discourse is underpinned with the perception of values that transcend cultures and places the onus on TNCs to integrate global values with local pressures that often arise because of weak government engagement in the region. Moreover local histories, legalities and inequities help to shape the conditions in which CSR approaches are implemented and influence how TNC actions are interpreted. Hence TNCs can be seen to respond to different, often competing, stakeholder demands which could be incorporated within business and CSR strategies (Idemudia 2014). However Bondy and Starke (2014) and Khan and Lund-Thomsen (2011) explain how local needs and priorities often fail to be located within CSR. Cultural traditions are ignored or swept aside in the rush to implement TNC derived behaviours and policies that resemble northern hemisphere ideals. For Amaeshi et al. (2006), in Africa the CSR agenda has been more about Western mimicry rather than meaningful representation of indigenous interests. Instead of emerging from local requirements, business ethics are heavily influenced by the environments in which they are shaped and by the people who contribute. And because people who develop CSR approaches are frequently from, and work for companies based in northern hemisphere geographically based experiences and understandings shape the ways in which social and environmental programmes are devised and implemented in other parts of the world.

Alongside concerns about unintentional, inappropriate application of CSR programmes are questions over more intentional application of CSR for business means (Frynas 2010; Lompo and Trani 2013; Newell and Frynas 2007). For instance, Amaeshi et al. (2006) believe that CSR in Nigeria is practised under the guise of philanthropy as a way of addressing economic and social development. This point fits within Kaplan and Kinderman's (2017) examination of pro-active strategies that seek to change political and social environments. In their research they explore case studies in an earlier phase of CSR, from 1960s, onwards and observe how CSR has been applied to shape government and elite policies and the opinions of affected populations.

Picking up a different strand, Guzman and Becker-Olsen (2010: 203) have argued,

> Strategic CSR programmes can serve a marketing purpose. In these instances companies may choose to engage in CSR programmes as public relations opportunities, reputational insurance that will help in times of crisis, or brand building.

This quote should not infer that I am arguing that CSR as a PR exercise is necessarily problematic especially when connected within well considered, sustainable frameworks. And as Lompo and Trani (2013) observe, the fear of bad publicity is motivating corporations to adopt more socially responsible codes of practice for development. However too often projects have been ill conceived forms of philanthropy or short-term photo opportunities that fail to benefit the designated benefactors. In some cases, the photograph can be the last connection with the communities who do not possess the resources to deliver or resource the intended services (Idemudia 2009, 2016). This emphasis upon quick, visible outputs is part of wider approaches which are much easier to observe. Hence companies can measure progress by the number of local residents they employ and the health facilities and schools that are built. For example, glancing through mining companies corporate brochures and reports, such as BHP Billiton, Freeport-McMoRan, Glencoe, Konkola Copper Mines (KCM), Petra Diamonds and Rio Tinto place considerable emphasis on health, school and welfare provisions. Gulbrandsen and Moe (2007) have also reported on the corporate concentration on reportable benefits and communities in areas they operate and the 'micro-level' directing actions.

These criticisms of Western orientated assumptions that underpin CSR share similarities with failings within modernisation theories and development studies. Portrayals of the 'third world' or 'developing world' are seen to be without history, exotic locations on which to apply western derived assumptions with little committed stakeholder engagement. The remainder of the chapter explores ways in which CSR policies and knowledge have replicated problems and power relations associated with developmental approaches. In so doing postcolonial studies will be drawn upon to highlight commonalities between CSR today and the 'civilising mission' of the past.

3 Locating CSR within Development and Post-Colonial Studies

Building upon the preceding points this chapter will position CSR critically through the application of concepts from postcolonial studies, drawing out salient points in historical developments that continue to have contemporary resonance.

Often positioned in the inter-disciplinary philosophical space between Marxism and Foucauldian post structuralism, postcolonialism is a concept referring to the period after the end of 'direct rule domination' and the persistence of what is argued to be hegemonic economic power. Gandhi (1998) and Malreddy et al. (2015) outline how Marxism has influenced commitment to the analysis of power and global inequalities while literary contributions, influenced by poststructuralism, celebrate difference, subalternity, identity and the right to belong. The different strands have often been applied in isolation from, or on occasion in opposition to, other types of postcolonial insights, thereby confounding theoretical and methodological uniformity.

Nevertheless collectively postcolonial contributions provide a range of principles and critical insights into patterns of power, knowledge, language and economics and hierarchies of social life and culture that followed independence (Gandhi 1998, Loomba 2005, Malreddy et al. 2015, Willis 2011). Gandhi (1998) explains how the postcolonial critic needs to synthesis, or negotiate between, the critique of Western epistemology and theorisation of cultural and materialist philosophies. Together postcolonial analysis provides powerful, if not necessarily coherent, insights into the inequities embedded within global processes. In Said's (1978) seminal text Orientalism he exposed the reciprocal relation that existed between knowledge and power within European imperialism. As Gandhi (1998, p. 25) explains postcolonialism provides 'a very specific understanding of Western domination as the symptom of an unwholesome alliance between power and knowledge'. Consequently power and knowledge are the focal point of this chapter.

In light of the heavy emphasis on critical analysis and semantics, key concepts within postcolonial studies have aroused considerable debate. Willis (2011, p. 30) outlines how '"post-colonialism usually used to indicate a time period after colonialism, while "postcolonialism" describes an approach to understanding social, economic, political and cultural processes'. With the former indicating a phase after colonialism, the unhyphenated application is considered to emphasise more emphatically the continuation with the colonial period and ongoing consequences. For this reason, while being conscious of multiple differences that have followed independence, postcolonialism will be applied in this paper. Material, economic and cultural arrangements alongside ideas and discourses are considered to produce and reproduce colonial components within the continuum of postcolonial power. Hence the formation of nation-states did not bring independence for the majority but continuing and revised forms of dependence. The new indigenous elite acquired the colonial system of control, army, police, law, bureaucracy and prisons and many continued to use these mechanisms for their own, or collective, ends. Moreover the newly independent country remained under the heavy influence of former colonial masters to whom they were increasingly in debt.

There has been considerable debate over the extent to which former colonies continue to be shaped by legacies of the past. Networks, bonds and interwoven flows of goods, ideas and people that were integral to European colonialism became a base for what is today described as globalisation. The pervasive, expansive movements within globalisation have led to a focus upon interconnections and commonalities of economies, identities and cultures. Divisions and the roots of their formation and continuation have disappeared within the narrative of global flows (Loomba 2005). Nevertheless, inequalities between northern and southern hemispheres remain and were initially shaped through the application of colonial economic, political and social controls. The apparent disappearance of divisions from history fits within what Gandhi (1998, p. 4) describes as 'postcolonial amnesia' and 'historical reinvention' which 'erase painful memories of colonial subordination'. The deep rooted inequalities have contributed to ongoing northern hemisphere dominance while restraining developmental opportunities in the south. Moreover the dominance influenced perceptions both of the colonisers and the colonised.

Economic relationships that existed during colonialism continued to place the former colonies in subservient or dependent positions following political independence. Although newly formed states often sought to disown burdens of colonial inheritance and humiliations of race and racism some colonial practices were reproduced. Memmi (1968) refers to expectations surrounding independence of magical transformation which underestimate the psychological hold of the colonial past on the postcolonial present. Residue traces and memories of subordination remain. The colonial tradition 'chained the coloniser and the colonised into an implacable dependence, moulded their respective characters and dictated their conduct' (Memmi 1968, p. 45). Said (1978) has also argued that the longevity of the colonised is perpetuated through persisting colonial hierarchies of knowledge and power. To help further explain the control of individual bodies Gilroy (1994) applies the Foucauldian concept of biopolitics through work, language, regulation of education, gender and sexuality. These controls are interwoven with political, administrative, educational, religious and legal arrangements which continue both to reflect earlier European 'civilised' influences and shape contemporary behaviour and attitudes in the global south.

Hence while power and interdependencies have shifted across states, regions, generations and other demographics the extent to which power differentials have narrowed has been restricted. Wouters (2007: 187) explains how,

> shifts in power balances at times offered collective power chances to groups of relative outsiders, and these chances came to be realized and expressed in social ranking but usually not without some delay. The process of realizing collective power chances has a psychic (mental) and a social side. Individuals need to become aware of these chances and to shake off the submissiveness that the old balance of power demanded. This is not as easy as process as it may sound because this submissiveness is usually ingrained rather firmly in the personality.

Today post-colonial contributions have argued that power and influence continues to flow through different processes across inequitable international trade and relations that enable control of other countries. Seeking to address this imbalance through highlighting inequalities and injustice, postcolonial contributors also confront disparities in power relations and the transfer of knowledge that underpin the basis for self-projected superiority and imposed inferiority (Willis 2011). Therefore although lacking both a coherent theoretical stance and methodology, postcolonial critics explore empowerment of the poor and disadvantaged.

Against the backdrop of the perceived legacies of colonialism and the continuation of northern hemisphere post-independence development programmes, whether economic or judged philanthropic, have been subjected to critical analysis . For instance, international practitioners often address sensitive issues such as HIV/AIDS, poverty, inequality and injustice. However while perhaps possessing laudable intentions, such actions can replicate impressions of Western values, paternalism and southern hemisphere dependency. Moreover criticism that had previously been directed at earlier modernisation programmes applying Western processes to non-Western settings continues to resonate (Vertigans 2012). Escobar's (1995, p. 45) description of development as 'a historical construct that provides space in which poor countries are known, specified and intervened upon' reflects many international CSR approaches.

Criticism directed towards development studies is also applicable to the activities of TNCs and the implementation of CSR programmes. Comparisons can be drawn between colonial commercial activities and how TNCs heavily influence ways of organising market institutions, appropriate land and undermine local autonomy with colonial arrangements, especially indirect rule that was more decentralised, and the maintenance of core-periphery relations (Boussebaa and Morgan 2014). Moreover, corporate power and influence connects and disproportionately helps to shape localised administrative, political and economic arrangements. Within TNCs, Boussebaa and Morgan (2014) highlight, the division of activities between core producers of knowledge and skills and their peripheral recipients who follow instructions. As Blowfield and Frynas (2005, p. 499) explain,

> one of the reasons why CSR and international development practice have moved closer together is that they share and reinforce assumptions that poverty and marginalization are fundamentally matters of geography, identity or difference rather than structural phenomena.

Consequently, what is different about the global south becomes the basis for explaining the problems. Solutions to these fundamental issues are to be developed and delivered by discrete, identifiable actors who originate from, or are instructed by, the global north. And Lauwo and Otusanya (2014) have spotted that CSR approaches tend to neglect wider socio-political, economic and historical structures and global processes which are instrumental in shaping direction and scope.

All of these critical comments should not overlook that CSR can help alleviate poverty and sustainable development (Idemudia 2009). However, Adanhounme (2011) and Khan et al. (2010) both argue that approaches are partial and underdeveloped. Alternatively corporate ambition is exaggerated to the point of disbelief, exemplified in the energy sector by Lukoil's (2018) aim for social investment to be 'doing great things and changing the world for the better is an integral part of our lives' while Chevron are 'committed to helping the Angolan people improve their health, education and livelihoods'. Typically, the corporate ambition fails to acknowledge or realise that the scale of their CSR operations is directed at achieving only limited localised impacts with no programmes implemented to deliver widespread national and global ideals. Moreover, the connection with post-colonial conditions is not considered. Hence, the fundamental problems that result in environmental degradation, poverty and declining economy are neglected in the focus upon smaller scale 'solutions'.

Applying this critical edge to CSR suggests that there is a danger that TNCs cultivate particular values and practices that are underpinned by a sense of moral and cultural leadership within a postcolonial discourse. Idemudia (2011) points out CSR agendas have tended to reflect the priorities and concerns of the Northern hemisphere with Southern representation inadequate. The division of core and periphery or developed and developing remains albeit located within a more overtly financial framework. Understanding the basis for the division requires the history of colonial power relations and how these relations have become incorporated within international business practices to be analysed.

4 Power to the Corporate People

Within the former colonies, postcolonialist critics have argued that power structures have continued with policies shaped by the legacies from the past. Following independence the newly installed leaders reappropriated these channels across institutional and discursive spheres to maintain power for their, and the former colonials, ends (Mbembe 2001). Today the roles of large businesses are considered within postcolonial studies to be shaping power, material and discursive processes in many parts of the world (Malreddy et al. 2015). Levels of influence vary and the same is true for CSR connections.

The power balance between TNCs and communities are often stacked towards the former as the levels of interdependence are inequitable. Hilson (2012) explains there are variations according to location. In the northern hemisphere levels of regulation and extent of welfare provision provides the parameters for TNCs to operate. By comparison, other parts of less monitored regions with weak enforcement of law means that TNCs can be in position of self-regulation through introduction of voluntary mechanisms. Utting (2007) explores CSR as a mode of domination that enables TNCs to intervene to incorporate CSR measures that weaken opposition and regulatory demands while reinforcing their legitimacy and power.

TNC influences also stem in part from the imbalance between company and local stakeholders such as workers, communities and even governments. For instance, the structural adjustment programmes stemming from international financial transactions and constraints and imposed upon many southern hemisphere countries from the 1980s have, Adanhounme (2011) argues, provided grounds for TNCs to negotiate a post-colonial return to their considerable advantage. Moreover corporate reliance upon localised others is weak. This imbalance stems from workers lacking collective institutions to represent their interests, in part because their job related skills and knowledge can be easily replaced. Idemudia (2014) refers to the capacity of TNCs to reshape development agendas to fit with corporate interests and values. The power gap weakens regional and national governments bargaining positions. As Lauwo and Otusanya (2014) observe, 'developing' countries governments' desperation to attract foreign investment can create tensions between promoting socio-economic order and protecting citizens welfare from unethical corporate behaviour.

When regulations are introduced and reforms planned, TNCs can lobby to prevent or sponsor influential opinion shapers which weaken abilities of governments to implement changes. Kaplan and Kinderman (2017, p. 40) explain how proactive CSR strategies 'are designed to act as buffer against anti-corporate political threats and to exploit political opportunities to advance liberalization.' In so doing the authors highlight the connection between CSR and high level power relations. Conversely closer relations between TNCs and governments can result in communities' requirements being excluded from corporate approaches if community views are at odds with the government (Idemudia 2014). As Idemudia (2011, p. 3) observes, 'critics point out that in developing countries CSR is a domain of political contestation as opposed to the ideationally neutral terrain, which it is often made out to be by mainstream CSR proponents'. TNCs working with local leaders can contribute to inequitable, customary power relations and splits between civil and customary laws. Consequently, the involvement of TNCs can connect into the distribution of resources, inclusion and exclusion along existing regional, ethnic and religious channels and conflicts (Spencer 2018; Vertigans 2017). Moreover, just as colonialism reinforced and supported some local authorities and laws, community based CSR can contribute to strengthening the power of particular leaders and groups while weakening opportunities for civil unity that can be fragmented by TNC inclusionary and exclusionary criteria for the distribution of resources. Hence, CSR policies can be directed at particular communities and certain groups of people will benefit within those communities. For example, Lompo and Trani's (2013) research in Nigeria discovered that the educated highest wealth quartile was most likely to benefit from CSR activities while the poorest were not reported to be benefitting either financially or in terms of empowerment.

The Niger Delta, in Nigeria, is a high profile region where concerns about corporate approaches to communities have been particularly vocal. Oil and gas companies were involved in allocating resources to strategically positioned communities which resulted in significant changes in dynamics with neighbouring groups. Similarly money has been paid to youth groups who have the capability to damage pipelines and wells in a manner which has been described as protection money. These approaches to community relations were accompanied, as Ite (2007) outlines, by Shell's Community Assistance approach (1960–1997). Communities were portrayed as helpless victims, contributing to a dependency culture. This positioning of communities as what Banerjee (2002, p. 22) describes as 'passive recipients of the beneficiaries of development' has contributed to an extension of TNC roles. Consequently, like the colonialist programme before them, CSR is providing salvation from famine, deadly diseases, today often HIV/AIDS, corruption and 'backwardness'. Underpinning these programmes is the TNCs ability to devise the parameters for involvement and to define the problems to be resolved. Yet as critics of development approaches such as Escobar (1995) have argued, problems such as low life expectancy, poor housing and sanitation have recently emerged as problems. In other words, representations within colonialism and the postcolonial period, allied to campaigns against global poverty, have established the basis on which people live in poverty and those requiring assistance. Subsistence living and the informal economy are caught within these imposed poverty parameters. Northern hemisphere philanthropists, economists, academics, politicians, musicians, NGO and CSR practitioners are all involved, to various degrees in overcoming the problems they have determined exist.

TNCs tend to follow within this narrative, protecting local people from local conditions and improving human rights of other peoples which in essence can mean protecting people from their indigenous culture. Spivak's (1999, p. 192) pointed observation that 'white men are saving brown women from brown men' can be extended to also include white women as saviours but the essential point remains. Clearly, there are dangers within postcolonial criticism that the excesses of relativism can be overly applied to benevolent examples. Nevertheless, as Spivak (ibid.) explains, the logics of salvation that are being applied are often complacent, neglecting the rights and abilities of local peoples to make their own decisions or find indigenous solutions. Moreover, as McCarthy (2017) argues, CSR related gender empowerment programmes tend to lift women out of history and politics with the focus on individual female entrepreneurs separated from the deep rooted structural and institutional factors that continue to explain wider inequalities. Razack

(1998, p. 170) proposes key questions to be asked by human rights activists, postcolonialists, and to which CSR proponents could be added, such as 'Am I positioning myself as the saviour of less fortunate people? As the progressive one?...'. In essence, there are what Malreddy et al. (2015, p. 193) describe as 'human rights discourse ... strongly lodged within humanitarian discourses of Western benevolence'.

5 Knowledge of the Few for the Many

The construction, application and communication of 'knowledge' from colonial times onwards have been integral to postcolonial analysis. Following the acceleration of globalisation, ways in which knowledge is constructed across larger distances have been transformed. In this section, I will argue that Global North knowledge development and application are instrumental within CSR programmes.

Colonial knowledge was integral in the processes of domination and policy formation with academics instrumental in the Eurocentric self-regard for the superiority of their insights. Ways of understanding have been interwoven with benevolent approaches and power maintained and reinforced through knowledge applied within institutions and across everyday life. 'Colonization was undertaken by a workforce that from the start included knowledge workers: priests, clerks, engineers, map-makers, and soon enough lawyers, accountants, architects, teachers and researchers' (Connell et al. 2017, p. 24).

For Said (1978) the superiority of Western knowledge was accompanied by a degradation of knowledge on behalf of systems of thoughts such as colonialism. Early forms of 'knowledge' arose with anthropology and economics. Studies of indigenous peoples and development focussed upon difference explained by ill-founded primitive, backward caricatures and racial classifications. The colonised were largely attributed with childlike characteristics requiring paternal rulers who were best placed to help the helpless natives. These 'insights' into the colonised 'other' provided both the explanation and justification for colonial domination and the interwoven civilising intent outwardly portrayed as the benevolent face of European domination. Influence was to be long standing because 'the perverse longevity of the colonised is nourished, in part, by persisting colonial hierarchies of knowledge and value' (Gandhi 1998, p. 7).

Moreover, the lack of historical awareness and specificity within CSR is indicative of ways in which portrayals of Africa have become split between abstract universalism and intimate paternalism (Mamdani 1996). For example, the mining company Rio Tinto (2016, p. 32) in their corporate report under the heading 'Contributing to strong and prosperous communities' describe how 'through our investments in, for example, health and education services our business makes significant positive contributions to the growth of local economies and the improvement of living standards'. Similarly, BHP Billiton (2016, p. 36) discuss, playing an 'important role in developing economies and improving standards of living'. Less modestly, de Beers (2016) suggest that their partnership with the government 'has helped take Botswana from being one of the poorest countries in Africa to being one of its biggest modern economic success stories'. And the adoption of benevolence underpinned by northern hemisphere perceptions and salvational intentions have been delivered by people who Fanon (2001, p. 29) describes as 'moral teachers' who 'separate the exploited from those in power'.

The early colonial emphasis on the construction of 'universal' knowledge in Europe and North America, and exclusion of non-Western thought, has been replicated in subsequent generations of social science (Onwuzuruigbo (2017) and Connell et al. (2017). Chakrabarty (1992, p. 3) argues that,

> for generations now, philosophers and thinkers shaping the nature of social science have produced theories embracing the entirety of humanity; as we well know, these statements have been produced in relative, and sometimes absolute, ignorance of the majority of humankind i.e., those living in non-Western cultures.

Today although there is a more extensive epistemological critique of Western knowledge systems, significant gaps in the making of knowledge in Southern contexts remain. Moreover, with indigenous modes of knowledge undermined during colonialism and only weakly reintroduced following independence, European and North American insights remain dominant. Their universal application of knowledge and understanding continue to reverberate in settings and experiences that differ both in where they formed and are being applied (Drebes 2016). When, allied to colonial memories, TNCs create and represent places, practices and distances between north and south as discursive knowledge. And over recent decades 'professional services' organisations have appeared such as accountancies, law and management that sell Westernised knowledge as essential for business around the world (Boussebaa and Morgan 2014). The work of Foucault (1978) continues to influence perceptions of the interplay between postcolonial knowledge and power, despite his relative neglect of agency and ironically ethno-centric applications (Loomba 2005). By controlling the mode and distribution of knowledge the powerful can speak on behalf of sections of the population who are denied a voice by those who claim to represent and understand their interests and characteristics. Today, Onwuzuruigbo (2017) explains how indigenous knowledge continues to be marginalised despite attempts to produce localised insights. Consequently, global knowledge is shaped by northern hemisphere academic, business and political perceptions that stem from relatively narrow social milieu and nationalities. Moreover, 'knowledge' within many former colonies tends to be disseminated in English. For instance, CSR programmes, such as Chevron (2013) and Petra Diamonds (2016) provide English school education in countries where indigenous languages dominate. And as aside Chevron has also been involved in a Loaves and Fishes program with Christian religious connotations (Chevron 2013; Loaves and Fishes 2017).

The power of corporate discourse based on narrowly drawn wells of knowledge can appear morally justifiable, 'systematic and coherent, specifying the only reasonable and possible way in which people can act in order to realise their intentions' (Sharp 2006, p. 217). In light of the self-centred knowledge base within northern hemisphere activities there is an element of irony. Carroll and Shabana (2010) highlight critical concerns about corporate involvement including the lack of appropriate expertise within TNCs to address social activities. The global nature of organisations further complicates organisational approaches to extending knowledge parameters as they look to assimilate international vicissitudes in forms of behaviour, law, financial contributions, ethics, human rights and expectations. Yet TNCs appraisal of their strategies and knowledge of the regions is limited. And regions that have the most acute social and environmental crises are, as Visser (2006) points out, also the locations with the most rapidly expanding economics. Hence, these are the settings where development has the most dramatic social and environmental impacts yet where there is limited international knowledge on which to devise and implement corporate approaches.

With limited appraisal of the environment or community, TNCs are establishing new sites and businesses and have been applying their underlying expectations. Roberts (2006) compares the depth of investigations into new markets and competitors with limited investigation into understanding new locations or community impacts of activities. Furthermore, enhanced familiarity of local conditions could acknowledge that other forms of moral principles and voluntary regulation are already embedded within economic enterprises in different contexts. And as Newell and Frynas (2007, p. 669) have argued, CSR's misplaced focus can mean 'we fail to tackle, or worse, deepen, the multiple forms of inequality and social exclusion that characterise contemporary forms of poverty'. For the authors this is typical of the lack of knowledge about CSR's developmental potential and claims made about CSR's role in social and economic development are often weakly substantiated' (Newell and Frynas 2007, p. 671).

In essence, the adaptation of terms such as 'developing' and 'underdeveloped' within CSR approaches are indicative of postcolonial modernisation programmes that have been initiated by the former colonialists and international agencies. For Edward Said (1978) these terms are typical of 'imagined geographies' that homogenize people within underpinning connotations such as backward, barbarian etc. and moral, cultural, economic or socio-political attributes semantically developed judgements. These images are often simplified and distorted and not based upon actual experiences and lived relationships. Knowledge, as Said points out, becomes a form of power in the implementation process.

6 Conclusion

Exposing CSR to critical application of postcolonial adaptations of knowledge and power within a longer term trajectory draws comparisons with colonial civilising missions. Both colonial and corporate missions have been underpinned by unequal power differentials and include aims to raise health, educational and living standards. At first glance these aims can appear laudable. However the reality is that programmes are contributing to dependence and reinforcement of northern hemisphere solutions to southern hemisphere problems. In so doing, TNCs continue to reinforce perceptions of global north dominance and values which become integral to individual progress. And by selecting programmes to be funded, and often the individuals to be sponsored and community leaders to engage, TNC strategies create the parameters for community and individual 'advancement' and the barriers for the excluded majority. That the TNCs are able to do so is indicative of the dominance of their power and knowledge levels that are reinforced by the advancement of the beneficiaries of their support who are better positioned to progress into corporate, media and political positions of responsibility. Moreover as Sharp (2006) argues, the limited focus highlights the implications and consequences, often unintentional (Vertigans 2017), of businesses becoming drivers of development. The lack of relevant expertise in development allied to limited regional knowledge has contributed to TNCs investing in ways that fail to consider how their actions impact not only upon the immediate communities but unintentionally help shape regional political and economic dynamics.

The imbalance in power relations partly stems from the partiality of Northern hemisphere knowledge. Arguably, more detached, inter-disciplinary approaches would provide more rounded insights that are better able to position activities within historical and contemporaneous contexts. However, Kuhn et al. (2015) argues that this potential is significantly underdeveloped. Despite considerable common interests, methodological differences, different behavioural assumptions and modes of inquiry allied to tensions between academic disciplines have resulted in limited exchange between disciplines such as economics and development studies.

Nevertheless, despite this critique, it is important to acknowledge signs of improvement in better informed, broader approaches. For instance, a tendency towards generic global applications is changing. In Africa GIZ (2014) identified large business CSR programmes in Cameroon, Ghana Mozambique, Nigeria, Senegal, Uganda and Zambia that focus primarily on employee matters or philanthropic projects in health, education and poverty and sport (in Ghana), 'beyond compliance' and social emergency situations (Senegal). In part, these situations may be influenced by demand for greater CSR involvement against a backdrop of weak governments and governance. By comparison, in South America, more robust regulatory pressures can be found with programmes that connect into history of protection of workers' rights that can be to the disadvantage of those not employed in protectionist industries.

Greater emphasis on difference and flexibility is in part a consequence of enhanced commitment to meaningful stakeholder engagement. By comparison with colonial times, Adanhounme (2011) observes how contemporary local actors can negotiate and influence corporate narratives and practices that was not previously possible. More inclusive networking has spread power. Another crucial difference is that intentions behind CSR are usually to influence the lives of a fraction of nations' populations rather than rule over the majority. However as explained above, TNCs can inadvertently have a much wider impact than intended.

Rolling out CSR into communities contributes to a social and business environment that is conducive to, and supportive of approaches becoming embedded within surrounding relationships. CSR 'can only have real substance if it embraces all the stakeholders of a company' with social and environmental sustainable goals 'embedded across the organization horizontally and vertically' (Frankental 2001, p. 23). By opening up channels for meaningful communication, affected peoples are able to contribute to policy direction. Inevitably decisions will be made that do not suit all stakeholders. In these circumstances, open, transparent decision-making and communication processes will enhance the likelihood of keeping different stakeholders on board. Moreover, TNCs are better informed about peoples who are in disagreement and can take steps to further engage in order to manage any subsequent difficulties.

Despite extensive procedures, considerable cynicism remains about the level of, and intent behind, stakeholder engagement. However, even if undertaken comprehensively and in good faith CSR strategies struggle, as Amaeshi (2011) suggests, when isolated from other complementary institutional configurations. When encountering considerable localised difficulties, corporate solutions can be severely restricted. Therefore, rather than look to provide short term, isolated solutions programmes need integration within broader activities including material resources and external factors. A greater concentration of targeted resources will better enable deeper underlying causes to be addressed. Moreover, the introduction of more extensive collaborations will overcome some of the concerns regarding corporate legitimacy when engaged in political, social and wider economic matters. By opening up the partnership framework to consolidate different stakeholders and accommodate different expectations and demands, power differentials are diminished in part through the greater resonance of localised knowledge. When TNCs realise that indigenous knowledge is invaluable to the effectiveness of their operations, these insights both improve corporate working and contribute to stronger mutual interdependence and empowerment of local knowledge providers. In turn this reliance on indigenous insights diminishes the power differential with the identification of both problems and solutions no longer the sole responsibility for TNCs. Through TNCs working in partnership together and with the range of stakeholders more meaningful, representative aims can be created within a sustainable direction that is owned by the necessary partners in the delivery of respective and mutual interests. And as Buckler's following chapter explains, the implementation of local content within many Global South countries is resulting in TNCs having to comply with local regulations which help to reshape some power relations and levels of interdependency.

Finally, greater familiarity of local conditions can lead to an awareness of historical failings and sensitivities. Such insights should also acknowledge that other forms of moral principles and voluntary regulation are already embedded within different contexts. For this to occur, Amaeshi et al. (2006, p. 13) explain that CSR 'can be neutrally positioned within the non-market environmental strategy without any moral taints, and without losing its objectives' although I would add that this depends on the objectives. Hence, rather than seek to replace indigenous forms with Western ethical principles and activities, TNCs can incorporate different strands of social responsibility within an expanding, inclusive framework.

References

Adanhounme, A. (2011). Corporate social responsibility in postcolonial Africa: Another civilizing mission? *Journal of Change Management, 11*(1), 91–110.

Ahmad, A. (1992). *Theory, classes, nations, literatures*. London: Verso.

Alam, S. M. S., Hoque, S. S., & Hozen, Z. (2010). Corporate social responsibility (CSR) of MNCs in Bangladesh: A case study on GrameenPhone ltd. *Journal of Potuakhali University of Science and Technology*. Retrieved from http://ssrn.com/abstract=1639570.

Amaeshi, K. M. , Adi, A. B. C., Ogbechie, C., &, Amao, O. O.. (2006). Corporate Social Responsibility in Nigeria: Western mimicry or indigenous influences? Retrieved June 11, 2019, from http://ssrn.com/abstract=896500

Amaeshi, K. M. (2011). International financial institutions and discursive institutional change: Implications for corporate social responsibility in developing economies. *Journal of Organizational Change Management, 11*, 111–128.

Asaolu, T. O., & Ayoola, T. J. (2014). Multinational corporations and CSR in the Nigerian oil and gas sector. In S. O. Idowu, A. S. Kasum, & A. Y. Mermod (Eds.), *People, planet and profit*. Farnham: Gower Publishing.

Banerjee, S. B. (2002). Reinventing Colonialism: Biotechnology, intellectual property rights and the new economics of sustainable development. *Presented at 9th Biennial Conference for the Study of Common Property, June 17–21, 2002*. Retrieved August 28, 2019, from https://dlc.dlib.indiana.edu/dlc/handle/10535/1813

BHP Billiton. (2016). *Integrity Resilience Growth: Sustainability Report 2016*. Retrieved December 19, 2018, from http://www.bhp.com/~/media/bhp/documents/investors/annual-reports/2016/bhpbillitonsustainabilityreport2016.pdf

Blowfield, M., & Frynas, J. G. (2005). Setting new agendas: Critical perspectives of corporate social responsibility in the developing world. *International Affairs, 81*(3), 499–513.

Bondy, K., & Starke, K. (2014). The dilemmas of internationalization: Corporate social responsibility in the multinational corporation. *British Journal of Management, 25*(1), 4–22.

Boussebaa, M., & Morgan, G. (2014). Pushing the frontiers of critical international business studies: The multinational as a neo-imperial space. *Critical Perspectives on International Business, 10*(1/2), 96–106.

Burchell, J. (Ed.). (2008). *The corporate social responsibility reader*. Routledge: Abingdon.

Carroll, A., & Shabana, K. (2010). The business case for corporate social responsibility. *International Journal of Management Reviews, 10*, 85–105.

Chakrabarty, D. (1992). Postcoloniality and the artifice of history: Who speaks for "Indian" Imperial fantasies and postcolonial histories. *Representations, 37*(1), 1–26.

Chevron, (2013). *Corporate Responsibility Report*. Retrieved April 7, 2018, from https://www.chevron.com/-/media/shared-media/documents/Chevron_CR_Report_2013.pdf

Connell, R., Collyer, F. & Maia, J. (2017). Toward a global sociology of knowledge: Post-colonial realities and intellectual practices. *International Sociology, 32*(1), 21-37

Crane, A., Matten, D., & Spence, L. (2014). *Corporate social responsibility: Readings and cases in a global context*. Abingdon: Routledge.

De Beers (2016). *Turning finite resources into enduring opportunity*. Retrieved from https://www.debeersgroup.com/content/dam/de-beers/corporate/images/impact-report/DEB081_01_Foreword.pdf

Doane, D. (2005). Beyond corporate social responsibility: Minnows, mammoths and markets. *Futures, 37*, 215–229.

Drebes, M. (2016). Including the 'other': Power and postcolonialism as underrepresented perspectives in the discourse on corporate social responsibility. *Critical Sociology, 42*(1), 105–121.

Escobar, A. (1995). *Encountering development: The making and unmaking of the third world*. Princeton: Princeton University Press.

Fanon, F. (2001). *The wretched of the earth*. London: Penguin.

Foucault, M. (1978). Politics and the study of discourse. *Ideology and Consciousness, 3*, 7–26.

Frankental, P. (2001). Corporate social responsibility: A PR invention? *Corporate Communications, 6*(1), 18–23.
Frynas, J. G. (2010). *Beyond corporate social responsibility: Oil multinationals and social challenges.* Cambridge: Cambridge University Press.
Frynas, G. (2005). The false developmental promise of corporate social responsibility. *International Affairs, 81*(3), 581–598.
Gandhi, L. (1998). *Postcolonial theory: A critical introduction.* Edinburgh: Edinburgh University Press.
Gilroy, P. (1994). "After the love has gone": Bio-politics and Etho-poetics in the black public sphere. *Public Culture Fall 1994, 7*(1), 49–76.
GIZ (2014). Shaping corporate social responsibility in sub-Saharan Africa: Guidance notes from a mapping survey. Bonn: Deutsche Gesellschaft fur Internationale Zusammanenarbeit. Retrieved June 14, 2019, from http://www.giz.de/fachexpertise/downloads/giz2013-en-africa-csr-mapping.pdf
Gulbrandsen, L., & Moe, A. (2007). BP in Azerbaijan: A test case of the potential and limits of the CSR agenda. *Third World Quarterly, 28*(4), 813–830.
Guzman, F., & Becker-Olsen, K. (2010). *Strategic corporate social responsibility: A brand building tool.* In Louche et al. (Eds.), *Innovative CSR: From risk management to value creation.* Sheffield: Greenleaf Publishing.
Hilson, G (2012). Corporate Social Responsibility in the extractive industries: Experiences from developing countries. *Resources Policy, 37*(2), 131-37
Idemudia, U. (2009). Oil extraction and poverty reduction in the Niger Delta: A critical examination of partnership initiatives. *Journal of Business Ethics, 90*, 91–116.
Idemudia, U. (2011). Corporate Social Responsibility and developing countries: moving the critical CSR research agenda in Africa forward. *Progress in Development Studies, 11*(1), 1–18
Idemudia, U. (2014). Corporate social responsibility and development in Africa: Issues and possibilities. *Geography Compass., 8*(7), 421–435.
Idemudia, U. (2016). Environmental business-NGO partnerships in Nigeria: Issues and prospects. *Business Strategy and the Environment., 26*(2), 265–276.
Ite, U. (2007). Changing times and strategies: Shell's contribution to sustainable community development in the Niger Delta, Nigeria. *Sustainable Development, 15*(1), 1–14.
Jamali, D., Lund-Thomsen, P., & Khara, N. (2015). CSR institutionalized myths in developing countries: An imminent threat of selective decoupling. *Business and Society, 54*, 1–33.
Jamali, D., Lund-Thomsen, P., & Jeppesen, S. (2017). SMEs and CSR in developing countries. *Business and Society, 56*(1), 11–22.
Kaplan, R., & Kinderman, D. (2017). The business-led globalization of CSR: Channels of diffusion from the United States into Venezuela and Britain, 1962-1981. *Business and Society, 59*(3), 1–50.
Khan, F. R., Westwood, R., & Boje, D. M. (2010). 'I feel like a foreign agent': NGOs and corporate social responsibility interventions into third world child labour. *Human Relations, 63*(9), 1–22.
Khan, F. R., & Lund-Thomsen, P. (2011). CSR as imperialism: Towards a phenomenological approach to CSR in the developing world. *Journal of Change Management, 11*(1), 73–90.
Kuhn, A.-L., Stiglbauer, M., & Fifka, M. (2015). Contents and determinants of corporate social responsibility website reporting in sub-Saharan Africa: A seven country study. *Business and Society, 54*(1), 1–44.
Lauwo, S., & Otusanya, O. J. (2014). Towards a political economy perspective on CSR in a developing country. In S. O. Idowu, A. S. Kasum, & A. Y. Mermod (Eds.), *People, planet and profit.* Farnham: Gower Publishing.
Loaves and Fishes@ccloavesfishes (2017). Retrieved October 27, 2017, from https://twitter.com/ccloavesfishes?lang=en-gb (posted 3 October).
Lompo, K., & Trani, J.-F. (2013). Does corporate social responsibility contribute to human development in developing countries? Evidence from Nigeria. *Journal of Human Development and Capabilities, 14*(2), 241–265.

Loomba, A. (2005). *Colonialism/Postcolonialism*. London: Routledge.
Lukoil. (2018). *Social investment*. Retrieved August 30, 2018, from http://lukoil.com/Responsibility/SocialInvestment
McCarthy, L. (2017). Empowering women through corporate social responsibility: A feminist Foucauldian critique. *Business Ethics Quarterly, 27*(4), 603–631.
McEwen, C. (2009). *Postcolonialism and development*. Abingdon. Routledge.
Malreddy, P. K., Heidemann, B., Laursen, O. B., & Wilson, J. (2015). *Reworking Postcolonialism: Globalization, labour and rights*. Basingstoke: Palgrave Macmillan.
Mamdani, M. (1996). *Citizen and subject: Contemporary Africa and the legacy of late colonialism*. Princeton: Princeton University Press.
Maon, F., Lindgreen, A., & Swaen, V. (2010). Organizational stages and cultural phases: A critical review and a consolidation model of corporate social responsibility development. *International Journal of Management Reviews, 12*(1), 20–38.
Mbembe, A. (2001). *On the Postcolony*. Berkeley: University of California Press.
Memmi, A. (1965). *The colonizer and the colonized*. Boston: Beacon Press.
Memmi, A. (1968). *Dominated man: Notes towards a portrait*. Orion Press.
Munshi, D., & Kurian, P. (2007). The case of the subaltern public: A postcolonial look at public relations, greenwashing and the separation of public. *Public Relations Review, 31*(4), 513–520.
Newell, P., & Frynas, J.-G. (2007). Beyond CSR? Business, poverty and social justice: An introduction. *Third World Quarterly, 28*(4), 669–681.
Onwuzuruigbo, I. (2017). Indigenising Eurocentric sociology: The 'captive mind' and five decades of sociology in Nigeria. *Current Sociology., 66*, 1–18.
Petra Diamonds (2016). *Strong under pressure*. Retrieved July 11, 2018, from https://www.petradiamonds.com/wp-content/uploads/Petra-Diamonds-Limited-Sustainability-Report-2016.pdf
Prieto- Carron, M., Lund-Thomsen, P., Chan, A., Muro, A., & Bhushan, C. (2006). Critical perspectives on CSR and development: What we know, what we don't know and what we need to know. *International Affairs, 82*(5), 977–987.
Razack, S. (1998). *Looking white people in the eye: Gender, race, and culture in courtrooms and classrooms*. Toronto: University of Toronto.
Rio Tinto (2016). *Partnering for Progress: 2016 Sustainable development report*. Retrieved July 18, 2018, from http://www.riotinto.com/documents/RT_SD2016.pdf
Roberts, J. (2006). Beyond rhetoric: Making a reality of corporate social responsibility. In J. Allouche (Ed.), *Corporate social responsibility: Concepts, accountability and reporting*. Basingstoke: Palgrave.
Robinson, A., & Tormey, S. (2009). Resisting 'global justice': Disrupting the colonial 'emancipatory' logic of the west. *Third World Quarterly, 30*(8), 1395–1409.
Sachs, W. (Ed.). (1992). *The development dictionary*. London: Zed Books.
Said, E. (1994). *Culture and Imperialism*. New York: Vintage Books.
Said, E. (1995). *Orientalism*. London: Penguin.
Sharp, J. (2006). Corporate social responsibility and development: An anthropological perspective. *Development Southern Africa, 23*(2), 213–222.
Shiva, V. (1988). *Staying alive: Women, ecology and development*. London: Zed Books.
Simon, D. (1998). Rethinking (post)modernism, postcolonialism and post-traditionalism: South-north perspectives. *Environment and Planning: Society and Space, 16*, 219–245.
Spencer, R. (2018). CSR for sustainable development and poverty reduction? Critical perspectives from the anthropology of development. In S. Idowu & R. Schmidpeter (Eds.), *CSR, sustainability, ethics and governance*. Heidelberg: Springer International.
Spivak, G. C. (1999). *A critique of postcolonial reason: Toward a history of the vanishing present*. London: Harvard University Press.
Sumner, A., & Tribe, M. (2008). *International development studies*. London: SAGE.
Tan, J., & Wang, L. (2010). MNC strategic responses to ethical pressure: An institutional logic perspective. *Journal of Business Ethics, 98*, 373–390.

Utting, P. (2007). CSR and equality. *Third World Quarterly, 28*(4), 697–712.
Utting, P., & Marques, J. C. (2009). Introduction: The intellectual crisis of CSR. In P. Utting & J. C. Marques (Eds.), *Corporate social responsibility and regulatory governance*. Basingstoke: Palgrave.
Vertigans, S. (2017). Unintentional social consequences of disorganised marketing of corporate social responsibility: Figurational insights from the oil and gas sector in Africa. In J. Connolly & P. Dolan (Eds.), *The social organisation of marketing: A figurational approach to people, organisations and markets*. London: Palgrave.
Vertigans, S., Idowu, S., & Schmidpeter, R. (2016). Corporate social responsibility in sub-Saharan Africa: From dependency to socially responsible African development. In S. Vertigans, S. Idowu, & R. Schmidpeter (Eds.), *CSR in Africa*. Heidelberg: Springer.
Vertigans, S. (2012). Paying the Price for corporate social responsibility: Social costs and dividends of oil and gas company approaches in Nigeria. *Social Responsibility Review., 1*, 35–48.
Visser, W. (2006). Corporate social responsibility in developing countries. Mendeley: London. Retrieved July 20, 2017, from http://www.mendeley.com/research/corporate-social-responsability-developing-countries/
Willis, K. (2011). *Theories and Practices of Development*. Abingdon. Routledge.
Wouters, C. (2007). *Informalization: Manners and emotions since 1890*. London: Sage.
Young, R. (2003). *Postcolonialism: A very short introduction*. Oxford: Oxford University Press.

Stephen Vertigans is a sociologist who is currently Head of School of Applied Social Studies at Robert Gordon University, Aberdeen, UK. His research interests include corporate social responsibility, political violence and communitydevelopment. At present, he is researching into processes of resilience in East African informal settlements during the COVID-19 pandemic. In his wide-ranging publications, he applies a sociological approach to try help levels of knowledge and understanding into social, political, economic and cultural processes that shape individual and community behaviours and activities. Positioning CSR within these wider processes can provide better insights into the successes and failures of CSRpolicies.

CSR, Local Content and Taking Control: Do Shifts in Rhetoric Echo Shifts in Power from the Centre to the Periphery?

Sarah Buckler

Abstract In the current climate of increasing rhetoric around protectionism, nationalism and border security versus free movement, transnational corporations are having to negotiate some particularly tricky issues. One of these is the increasing prevalence of local content regulations which are impinging more and more upon the ways those corporations operate, including having an impact upon the scope and nature of corporate social responsibility and sustainable development activities.

In this chapter I examine the historical, political and economic context of local content policies, exploring their roots in conflict and the contemporary, contested discourses that lie behind the development of different local content requirements. As local content requirements have become increasingly adopted by countries in the developing world they have displaced activities more generally associated with corporate social responsibility, a move which is synchronous with claims that CSR is a neo-colonial means by which the developed world attempts to continue to exert power over its erstwhile colonies. I explore how this has worked in different contexts, highlighting the rhetorical nature of policy setting, reflecting power struggles on the international stage rather than meaningful or sustainable developments in terms of national or local economies.

In 2014, returning to Ghana for a short visit after having lived there for a while, I managed to speak to a senior manager from one of the oil companies operating in the country, looking to establish a firm foothold in the fairly new landscape of Ghanaian commercial oil production. We had an interesting conversation over tea and coconut water in which he confessed to the major difficulties his company was having trying to anticipate and plan for the Local Content Regulations that were about to be made law. It turned out that the large companies who were operating in Ghana at that time had grave concerns over the Local Content Regulations that were being proposed

---

S. Buckler (✉)
Robert Gordon University, Aberdeen, UK
e-mail: e.s.buckler@rgu.ac.uk

© Springer Nature Switzerland AG 2021
S. Vertigans, S. O. Idowu (eds.), *Global Challenges to CSR and Sustainable Development*, CSR, Sustainability, Ethics & Governance,
https://doi.org/10.1007/978-3-030-62501-6_5

and this sparked my interest and persuaded me to dig further in an attempt to understand what the significant tensions were and how they had come about.

In this chapter I explore the issue of Local Content, what it aims to achieve and why it is often problematic, both for companies that have to ensure compliance and also for countries that expect to benefit. I do this through an examination of the tension between Corporate Social Responsibility (CSR) and Local Content showing how the shift from one set of principles to another marks a shift in power from the developed centre towards the periphery and how that shift is now a site of global tussles regarding where and to whom the benefits of the global economy do, and should, accrue. More specifically I will trace the ways in which CSR, Local Content regulations and political rhetoric are closely intertwined and how these shift according to the ebbs and flows of the global economy. To begin the chapter I will trace the origin and development of contemporary local content policies, their link to rhetoric around resource nationalism in the developing world and the shifts now taking place in the populist rhetoric of developed nations. I will then go on to show how this shift in rhetoric is leading to a shift in understandings of the relationship between business and society and how that, in turn, is resulting in changes of both approach and language when it comes to CSR, with the CSR agenda being completely dropped at times. I will argue that local content became a means of developing nations asserting their authority over the development agenda and control over the resources that fall within their geographic boundaries. I will also show an echoing rhetoric is now being adopted by politicians in the developed world wanting to establish a populist foothold amongst the electorate and resisting the apparent shifts in power that are emerging. I will finish with some reflections on whether shifts in language really do reflect shifts in power or whether it is merely rhetoric.

Corporate Social Responsibility has long been understood to be a voluntary action that corporates take in response to various social and environmental pressures and orientations. It has also long been noted that there is no clear definition of what CSR is. These two factors combine to make CSR and related activities a fluid set of responses that corporates can draw upon to manage their economic and political relationships in the global economy. In this chapter I note that neither CSR nor local content requirements can achieve sustainable development without integration into a meaningful economic strategy. Conversely, whilst governments point towards the benefits they claim can be gained from the operation of the global economy through implementation of local content policies, the increasing nationalist rhetoric across the world (i.e. emanating from both developed and developing nations) is attempting to reframe the relationship between business, government and the global economy. As part of this process expectations and delivery of CSR policies and projects have shifted as the ways in which corporations contribute to the societies within which they operate become more defined by the expectations of the governments of those countries. At such time the resources that may have been directed towards CSR initiatives are now being ploughed into programmes intended to meet Local Content requirements (both formal and informal). Ultimately I argue that CSR appears not as a clear cut practice but as one kind of response to political and economic contexts without any meaningful substance or philosophy underpinning it.

1 Local Content: A Brief Introduction

Speaking in 2016 Silvana Tordo, Lead Energy Economist of the World Bank said of Local Content:

> Local content in the extractive industries is being given ever higher priority by host governments through a wide array of policy instruments. And oil, gas and mining companies now rate local content among the most significant expectations in the communities in which they operate.—World Bank (2016)

Local Content Regulations relate to a variety of measures that national governments put in place with the intention of ensuring benefits to the national economy. Such regulations can relate to numbers of employees from the national workforce, goods sourced from within the country and services sourced from national companies and they can range from very prescriptive percentage quotas that companies are expected to achieve to more fluid processes that companies are expected to comply with, such as procurement processes.

Contemporary local content regulations and the debates around both their legitimacy and effectiveness have their origins in the world recession prior to World War II. This period of history saw an increase in nationalist, protectionist rhetoric which blamed the recession on 'others' and attempted to shore up national economies by measures intended to bolster national companies and the jobs they created (Ikenson 2009; Heffernan and Thorpe 2018). Importantly, the perceived link between the vagaries of the global market and the fortunes of nation states inevitably informed the policies of struggling governments and attempts to protect local (i.e. national) economies and associated businesses and jobs led to demonization of 'others' and eventually leading to the conflicts of the Second World War.[1]

As the world started to emerge from the war the international community began to look at the factors that had led to that war and consider what might be done to try to prevent a recurrence. One element that was noted was how the rise in nationalist rhetoric was tied to increasingly protectionist economic policies. Partly in response to this, and in an attempt to prevent such tensions resurfacing, the General Agreement on Trade and Tariffs (GATT) was created, stating:

> Recognizing that their relations in the field of trade and economic endeavour should be conducted with a view to raising standards of living, ensuring full employment and a large and steadily growing volume of real income and effective demand, developing the full use of resources of the world and expanding the production and exchange of goods.

---

[1]The literature on the economic policies that underpinned the tensions leading up to WW2 is huge and it is not my intention to review it here but see, for instance Ahamad (2009) and Kitchen (1988).

> Being desirous of contributing to these objectives by entering into reciprocal and mutually advantageous agreements directed to the substantial reduction of tariffs and other barriers to trade and to the elimination of discriminatory treatment in international commerce.—World Trade Organisation, GATT papers (2020a)

In effect the stated intention behind the GATT was for nation states to come to a cooperative agreement around how to manage international trade and persuade nations not to enact policies which would (deliberately or otherwise) impact negatively on other states, making them more vulnerable to the kind of protectionist reaction that had been seen prior to the war. At the outset, with only 23 signatories,[2] it was generally those economies connected to the allied powers and which were also considered to be more powerful or influential that were involved. Clearly there still needed to be some bridge building and development of relationships before nations that had been engaged in mortal warfare could come together to sign a trade agreement. Furthermore, those countries which were still colonies[3] were not involved in signing the agreement, it was the colonial powers who were negotiating with one another. It is perhaps worth noting that it was the nations of the developed world which had created the conflict that led to the war, conflicts in the underdeveloped world tended to be extensions of conflicts between colonial powers or wars asserting the power and dominance of the coloniser over the colonised. As such they were geographically focused and minimised any sense of the significance of the colonised peoples or nations.[4] This was an attitude that continued for some time after the war and into the time these 'peripheral' countries gained independence and asserted their significance. Nonetheless, countries that had been colonised by the more powerful were gradually gaining independence, including trying to establish independent economies and influence over the raw materials that they were so rich in, and for which they were so valued by the west. Gradually more and more countries became signatories to GATT and as the years went by a variety of agreements and amendments were made which exercised significant controls over the ways in which the newly independent former colonies could establish trading relations (Copelovitch and Ohls 2012).

Perhaps inevitably, given the global situation at the end of the war and the impact of the cost of the war on national economies, the GATT can also be considered as a gateway to a new kind of colonialism. It has enabled erstwhile powerful, colonial nations to continue to have some kind of foothold in countries becoming newly independent and to exercise ongoing control over a variety of natural resources found in those former colonies (Davis and Wilf 2017; Weissman 1991). At the same time the nascent economies of newly independent nations were attempting to flex

---

[2] Australia, Belgium, Brazil, Burma (Myanmar), Canada, Ceylon (Sri Lanka), Chile, China, Cuba, Czechoslovakia, France, India, Lebanon, Luxembourg, Netherlands, New Zealand, Norway, Pakistan, Southern Rhodesia, Syria, South Africa, UK, USA.

[3] With the exception of Burma and Ceylon which became an independent nation as part of the Commonwealth shortly after signing the GATT, in February 1948 and Southern Rhodesia which was a self-governing territory of the UK.

[4] Again, literature on colonial conflicts is large and it is not my intention to review it here.

their power in terms of being able to sit at the negotiating table and have some say over the ways their wealth was extracted and value added by the powers of the developed world.

In fact the GATT acknowledges that there were power discrepancies between developed and developing states which might prevent developing states from participating fully in the GATT process. To counter these differentials there are various exemptions made as regards provisions which might favour local companies when companies are national to the developing world. These exemptions are made given an understanding that in order to participate fully in the global market then developing countries may have some need to invest in their own growing economies before opening up fully to competition. It was recognised that some countries in the developing world needed to have some level of differential treatment in order to be able to develop to a position whereby they could be on more or less equal terms as the developed nations who instigated the GATT. For example, it states in the TRIM agreement[5]:

> Developing countries are permitted to retain TRIMs that constitute a violation of GATT Article III or XI, provided the measures meet the conditions of GATT Article XVIII which allows specified derogation from the GATT provisions, by virtue of the economic development needs of developing countries.

Nevertheless, in the eyes of some, as the GATT developed, these principles ended up benefitting multinational corporations and the developed nations in which they were registered rather than developing states from whom they extracted resources (Weissman 1991).

In 1995 the GATT became the World Trade Organisation, an organisation perhaps better equipped to deal with the new face of global economic trade in a 'post-colonial' world in that it is a permanent institution with more powers to negotiate and reach agreements in trade disputes (WTO 2020b). The WTO also reaches further than the GATT in that it includes agreement over trade in services and intellectual property, not just trade in goods (ibid). Since its inception the WTO has set certain standards and requirements that all members are required to sign up to—its basic operating principle being that access to markets and opportunities should be both liberalised and equally available to all. However it does, as does the GATT, also recognise that the global economy is not a level playing field and there are some states that should be enabled to have some preferential, or at least differential, treatment in order to enable them to come closer to being able to compete with already established national economies (WTO 2020c).

---

[5]TRIM agreement refers to Trade Related Investment Measures negotiated under the umbrella of GATT (see https://www.wto.org/english/tratop_e/invest_e/trims_e.htm).

1.1 Link to Growth of Resource Nationalism

There is clearly a tension between discouraging protectionism and allowing some level of encouragement for developing countries to do just that—develop—and that tension is further reflected in the growth of resource nationalism and the identity politics that have accompanied it (Ndlovu-Gatsheni 2009; Wilson 2015). Commonly understood as a move by a nation state to bring under its control the raw materials and produce that come from within its borders, resource nationalism is in effect a statement that the world's resources do not belong equally to humanity and the planet as a whole, they primarily belong to the nation state within whose geographical boundaries they are found. By implication this means also that control over these resources rests with the government of whichever nation state has geographical sovereignty over locations where those resources are found.

In fact, even this rather complex understanding over-simplifies the issue, not least because, as is often pointed out, national borders are permeable and unfixed (Barth 1998; Elden 2006), especially in the parts of the developing world where many valuable natural resources are found. For instance the war-ravaged, mineral-rich edges of the Democratic Republic of Congo or the oil-rich waters off the coast of Ghana and Cote d'Ivoire. In fact it would seem that rhetoric around resource nationalism is often rooted in the desire of government to establish a strong identity and, with that, reliable electoral support (Bremmer and Johnston 2009; Andreasson 2015). On the other hand, resource nationalism can be concerned with maintaining a negotiating position at the table of multinational deliberations around trade and so on with nations using their resources as bargaining tools in order to get the best deal for itself (Childs 2016). Whichever way it manifests, resource nationalism is about identity politics writ large and concomitant power relations far more than it is about identifying who owns what resource and this is significant when looking at the shift towards local content legislation in resource rich nations.

Further, the desire to establish a firm, national, political identity is also linked to a sense that one's nation, or one's rule, is under threat in some way (Mann 1997; Abbink 2014; Schmidt 2019). So, a rise in resource nationalism can also be linked to increasing national uncertainties in a world where much of the global economy seems to operate in spaces that don't belong to nations—virtual spaces and offshore spaces, for instance. This sense of threat can be exacerbated in the developing world where boundaries are contested and nebulous. Increasing insecurity develops around the ways in which governments can exercise some kind of control over such a globalised and dislocated economy, where multinational and transnational corporations seem to have the upper hand. In such an economic set up the benefits from the operations of corporations can appear to fade away as company registrations move from one state to another or to tax havens and out of the purview of countries who may need to benefit from the taxes and fees they might expect to receive had those companies been registered in the countries in which they operated.[6] In fact tracing the rise of massive multinational corporations is crucial to understanding the emergence of first CSR and more lately local content requirements which is where I will turn my attention to next.

2 The Rise of the Multinational Corporation

Links between the emergence of the post-colonial world and the emergence of a new form of multinational corporation have been noted (Ohmae 1990; Rajak 2011, 2014) and it certainly is no accident that as more countries became independent, companies which were based at the centre of the colonial powers developed strategies to maintain a presence in the newly independent states by establishing bases there. Hence a new kind of colonialism was borne, one not rooted in political power and rule but one rooted in economic power. Furthermore these companies were largely 'extractive' corporations—not only those involved in mining and increasingly those involved in oil but also those concerned with trading agricultural commodities which would be 'extracted' from the countries where they were grown and transported to the developed world for processing and turning into goods that could make huge profits—see, for instance cocoa, coffee and cotton.

This era of expanding neoliberal policies was a time that saw the rise of huge multinational and transnational corporations that could use their supply chains to ensure that maximum profits were directed to their investors and away from producers and nation states. It was during this time that discussions around Corporate Social Responsibility increasingly began to be the subject of debate and when 'classic' CSR theorists such as Milton Friedman defended the interests of shareholders above all others (Friedman 1970). A responding increase in thinking around stakeholder theory (Freeman 1984) led to mounting pressure on multinationals from the public and a growing awareness in the developing world that their governments, their producers and their own, national businesses were all potential stakeholders in the operations of global corporations and all had the potential to impact upon business practices for their own benefit.

Whilst the global economy developed more and more of a tension between the sovereign claims of nations and the machinations of multinationals, so too emerged debates around the nature of the relationship between business and society (Donaldson and Preston 1995; Mitchell et al. 1997; Phillips 2003). As these discussions continued an accompanying growth in activities which could be labelled CSR also grew. Understanding that in an increasingly networked and 'virtual' world it was important to manage how people understood and perceived of your business

---

[6]That this is an issue that reaches beyond the developing world can be seen in the fact that a non-profit organisation, the Fair Tax Mark, has been established in the UK and is working on standards applicable on a global level for 2021. See https://fairtaxmark.net/.

activities became of increasing concern to multinationals. A variety of events which perhaps in the past would not have gained so much attention and influence, came to put pressure on corporates' activities (see, e.g. Brown 2008). In some part this was also a means by which nations could exert their authority and their own power in the situation. In this environment the concept of CSR quickly gained a foothold and became a way that huge corporations could demonstrate their values as regards their various stakeholders (Porter and Kramer 2006; Pang et al. 2018; van den Heijkant and Vliegenthart 2018). We can note that as publics began to exert their influence over perceptions of multinational corporations those same corporations began to put in place programmes intended to improve the relationship between the company and its various stakeholders. At first these initiatives were largely philanthropic and tended to replace projects which had been the province of the colonial governments and their associated development NGO spin-off, the aid agencies. However, the manifest failures of many of these schemes (Porter and Kramer 2006; Utting 2008; Jamali and Keshishian 2009) led to an increasing awareness of the difficulty of delivering CSR programmes for companies whose business was in a different sphere and it also led to accusations from the host governments of neo-colonial attitudes and the suspicion that this was just another way of the developed world attempting to control and suppress the developing world.

So, intially CSR activities tended towards the voluntarily philanthropic (Carroll 2016), drawing on a colonialist narrative where the poor and disenfranchised needed the helping hands of others to lift them out of their misery.[7] Projects focused on building health care centres, hospitals, schools and so on, with little thought given to how the ongoing running and maintenance costs would be met. There was very much a sense that businesses were doing something that they didn't need to, out of the 'goodness of their heart' and because they felt some kind of responsibility towards society (even if they weren't paying taxes to that society). This kind of language and the activities which accompany it perpetuate the disenfranchisement of people in the developing world and there have been numerous accusations that the CSR activities of multinationals are often more damaging than is acknowledged and instead of acknowledging the history and traditions of the people intended to benefit it actually perpetuates an agenda in the interests of those carrying out the initiatives (Gilberthorpe and Banks 2012; Katamba and Nkiko 2015).

Many of these early CSR projects not only did not work, in some instances they backfired and corporates left a trail of white elephants and a bad taste in the communities where they had operated and then pulled out with little or no regard for the unintended consequences of their largesse. Corporations were criticised by a variety of stakeholders for wasting resources (time, expertise and money) on projects that were poorly thought through and were not linked to any meaningful, strategic business aims. In response to this, gradually the face of CSR changed becoming more strategic and more tied in to corporates' stated values. CSR projects began to be seen as ways in which corporates could benefit their business activities—gaining

---

[7]For a useful illustration of this process see Pearson et al. (2019).

social license to operate, ensuring suitably trained employees, a healthy workforce and so on.

In the context of poorly managed business initiatives with suspect rationale behind their implementation and arguments about neo-colonialist attitudes towards the countries in which these projects were being delivered, a move towards taking control through development of local content requirements became ever more inevitable and apparent. This move became even more predictable when considered alongside the potential to exercise some identity politics through linking local content to resource nationalism. I will now go on to explore how this move has played out in specific cases with a focus on business response to changing times and a shift from CSR to local content related initiatives.

In an age when massive corporations can have a turnover greater than some national economies, and when those employed by such corporations can exceed the populations of some countries, an insecurity around the future of nations and the exercise of political and economic power grows. Alongside that burgeoning sense of insecurity and threats to national identity comes a rise in nationalist rhetoric which often centres upon a form of resource nationalism with nation states trying to maintain control over resources in the face of pressure from hugely powerful multinationals. It is in this context that CSR and Local Content Requirements begin to impact upon one another.

2.1 Local Content and Competition

In terms of Local Content the GATT and later the WTO have established some basic principles preventing protectionist policies or at least trying to limit their impact. Signatory nations are expected to treat one another equally and not to favour one over another by imposing differential tariffs. However, power imbalances between the developed and the developing world are very apparent when considering the emergence of local content regulations. For instance, with the discovery of commercial quantities of oil in the North Sea, Norway successfully established Local Content requirements which meant that it could ensure benefits from the discovery would go to Norway rather than be distributed across the world and into the pockets of the multinationals and the countries in which they were registered. Norway's Local Content requirements also underwrote investment in the national economy and infrastructure, requiring as they did a certain amount of uptake of national firms, workforce and services (Columbia Centre on Sustainable Investment 2016; Asiago 2017).

Norway's particular manifestation of local content requirements was, in general, not considered to be protectionist as such, instead it was considered to be in line with the socialist/democratic values common in Scandinavia and generally acceptable to the developed world. Furthermore, the language which framed it did not appeal to a populist resource nationalism which was fast emerging in the developing world and becoming a significant force in terms of the shift from CSR to local content requirements as I will discuss below. In Norway, in contrast to the ex-colonial states, there was less identification between raw materials and the identity of the nation and more focus on skills and human attributes (Heum 2008). Nevertheless, what Norway's legislation and requirements did demonstrate was that governments could work successfully with private, multinational corporations in order to ensure that national economies benefitted without damaging international trading agreements. It also showed that if this was to happen, the relationship between government and corporation needed to be quite sophisticated—to simply expect business to fill the gaps that government couldn't fill themselves was unrealistic and unachievable (Santos and Milanez 2015; Eikeland and Nilsen 2016). Norway also demonstrated that GATT need not stand in the way of national economies benefitting from the resources that lay within their own geographical boundaries.

In brief, it could be pointed out that whilst CSR was contending with a perception of neo-colonial interference and public image management, local content was establishing itself as a way of developing business responsibly, enabling a level of sustainable development and political control that CSR couldn't.

3 Rhetoric Around Local Content Today

Having described some of the historical and policy context behind the development of local content regulations I will now go on to explore the ways in which they are manifesting today (2000 to present). This is particularly pertinent because as we move away from the years of Cold War post colonialism and the enmities of those times we have moved into a new war of words (and increasingly, actions) whereby some influential politicians scrabble around trying to find any visible target to 'other' and serve as a scapegoat for the insecurities and fears of our lives today. Such scapegoats at the present time tend to be migrants, or Muslims, they have tended not to be multinational companies that pay little to no corporate tax, despite social movements trying to force this. Nor does the media often find itself scapegoated, instead it is often the media that is doing the 'othering' and profiting from it through sales to an already convinced readership. In this context protectionist, populist rhetoric is gaining a foothold in the developed world whilst attempts to force local content regulations are gaining popularity in the developing nations. Whilst they are not diametrically opposed to one another (and indeed look remarkably similar to one another from some perspectives) these two forces do demonstrate different currents that are working in different ways towards different ends, and this creates some tension between them.

In a relatively benign way I was witness to one aspect of this tension in Ghana where the various oil and gas companies had all been practicing their own versions of CSR (as had other multinationals from telecoms companies to agricultural commodity exporters). Such CSR programmes ranged from modest grants given to community organisations to entire departments concerned with managing the interface between business activity and communities. Each company had its own approach to CSR and rarely, if ever, were they coordinated with similar or complimentary programmes operating in the same location or community. In many communities this could lead to immense confusion amongst the local populace who weren't sure who was doing what, or why. It was with this as the background that local content requirements were being drawn up that were to be applied to the extractives sector. These local content requirements were detailed and numerous and would impact upon all aspects of oil and gas operations. One key complaint from a number of companies at the time was that they had not been involved in consultations about what these Local Content requirements should be, the timescale over which they should be achieved, or even what was practicable.

Whilst the local content requirements were being drawn up aid agencies and government departments were involved in discussions around how CSR could be made both more effective and more efficient. They had come up with an idea of trying to get corporations to work cooperatively together and combine CSR approaches and funds. After some discussions the Western Coastal Foundation (WCF) was established with the intent that it would serve as a foundation into which companies could place their CSR funds. It was also the stated intention of the WCF that it would consult with communities to identify what kinds of initiatives were most needed. In the context of diverse, uncoordinated and frequently unsuccessful CSR programmes this seemed like an eminently sensible approach. In the context of imminent local content requirements it enabled CSR funds to resource training facilities and programmes that should allow corporations to meet those requirements in the near future. Hence in a very reasonable, responsive and orderly way CSR funding was diverted from a voluntary engagement between the corporation and the community to a business investment to ensure the sustainability of corporations in the light of new legal requirements. It was a move well supported by the government[8] and one which meshed well with the government's moves at the time to become more vocal in asserting their rights over offshore waters that were being contested by Cote d'Ivoire—they had a plan, they had a strategy and the government was in control.

Adding to the narrative, these events took place shortly after the 2013 trial regarding accusations of corruption in the previous general election which was followed by a great deal of national pride that the law had served its purpose and the country had not descended into civil conflicts as had happened in numbers of neighbouring countries in similar situations.[9] There was a definite sense of Ghana being 'on the up'—handouts from aid agencies or multinational corporations did not fit well with this mood whereas Local Content regulations felt a whole lot more appropriate and in keeping with a strengthening sense of national self-confidence.

---

[8]During my visit I was able to arrange interviews with ministers and other officials involved in drawing up the local content regulations and also with the representative of DfID who was developing the work on the WCF.

[9]I was resident in Accra at the time of the trial and witnessed first-hand both the nervousness and the sense of pride that followed the trial.

This brief illustration demonstrates the ways in which resources for CSR can subtly shift and become more regulated and prescriptive as a result of shifts in the wider national, or international, mood and the rhetoric accompanying that shift. In this case the move was generally well received by all parties, and whilst there is still some debate as to the effectiveness or otherwise of local content (Ablo 2015, 2017, 2019) it was felt that these developments were likely to be more beneficial than the rather ad hoc approach that there had been previously. In short, business and government were partnering up to deliver development, rather than relying on the interventions of agencies from erstwhile colonial powers. Elsewhere in the world similar shifts that re-formulate the expected relationship between business and society can also be observed and it is to these that I now turn my attention.

4 Current Moves Towards Populism and Protectionism

Whilst the world has moved on from the cold war and the national tensions that developed around that, and whilst it has also moved on from the tensions which erupted following the collapse of the Iron Curtain, it has moved into a period whereby the rhetoric from the developed world, perhaps especially the US, has developed into a clear 'us vs them' narrative which has its roots in the attacks on the Twin Towers in 2001. At that time President Bush made a speech outlining the new world order—an order in which the US was embattled by the dark forces of Islam and the 'axis of evil' (Bush 2001; Carrithers 2005).

Insecurities created at this time were exacerbated by the global economic crash caused by the implosion of the banking and finance sector in 2008—what had previously been the certainties and assurances of the neoliberal world started to look rather flimsy in the light of the restructuring, job losses, housing collapse and so on that occurred at this time.

The above should also be seen in the context of a global economy in which national borders are becoming eroded and less relevant (Enderwick 2011). Business operates more and more offshore, meanwhile global conflicts have led to movements of population across the world and the pressures of refugees and migrants looking for a better, more stable life. In other words, the global economy has become seen as a threat to states who are insecure and who are struggling to establish a sense of self-confident certainty in this new world order (Nakano 2004; Kalyuzhnova et al. 2016). Once it was the lot of 'peripheral' states to suffer major economic crises and lack of confidence whilst the privileged states at the 'centre' imposed on them numerous interventions and 'aid'. In this new world order resource rich states are not only exercising more control over how their resources are extracted they are also starting to compete in terms of wealth-generating industries in both manufacturing and services (Sirkin et al. 2008; Cooper 2019). Now the expensive countries of the developed world struggle to generate the value they had taken for granted and alongside that struggle, comes a change in the political rhetoric and accompanying policies that intend to reshape the relationship between business and society.

Recent years have seen a well-documented rise in populist rhetoric and protectionist tendencies in the global economy, which from the writer's perspective are best seen in the rise of Trump in the USA and the debates around Brexit in the UK (Van Reenen 2017). These, the USA and UK, are of course countries that have been used to wielding significant economic and political power and influence on the world stage—the issues are not the same as those facing the countries of the developing world such as Ghana and yet some of the arguments they express seem similar, at least on the surface. Much of the rhetoric is around identity politics; being a great country again, taking back control and so on. However much of the rhetoric emanating from the UK and USA ignores the crucial significance of the global economy to the industries they are so dependent upon—Trump's spats with China seem to disregard the vital nature of Chinese supply to Walmart and Apple whilst his focus on 'homegrown' steel disregards the economics of the automotive industry that needs to maintain trade relations across borders, both to ensure a cost-effective supply of materials and also to provide access to the emerging markets of South and Central America. Meanwhile the rhetoric in the UK about removing the 'bureaucratic red tape' that emanates from the EU ignores the importance of the European market for much of its agricultural and fisheries output and also disregards the vital nature of European supply to the pharmaceutical and health industries. This rhetoric is not quite the same as the 'resource nationalism' of the developing world, but it echoes it as a kind of 'produce nationalism' and the sense that there is a political need to establish a strong identity that resonates with the electorate is very much present.

There is another side to this rhetoric too, and one which echoes further the development of local content regulations in the developing world but in a kind of inverse, topsy-turvy way. There is a suggestion that national identities are being eroded due to porous borders and that this foreign threat is denying nationals their livelihoods. These observations are well rehearsed by some politicians and some branches of the media; less well examined is the impact upon expectations around the relationship between multinationals, CSR and local content. We have seen how expectations for national benefits play out in terms of local content in the developing world, and how that can come to impact upon delivery of CSR. However, we don't often think about these debates around national identity and self-confidence in the context of the developed centre—these are the province of the periphery, yet the impact is there, it just looks rather different, let us see how, with a focus on the UK.

5 Business and Government: Who Owes What to Society?

At the time of writing major businesses with locations in the UK are trying to understand how they will be impacted by Brexit as the UK tries to establish new trading relations and agreements outwith those of the EU. It is (May 2020) a time of great uncertainty exacerbated by the Covid-19 pandemic. How this will play out in economic terms is as yet unknown. News coming from the Sunderland Nissan plant is not promising and there are suggestions that a combination of uncertainty related to Brexit and nervous markets responding to the pandemic mean that it cannot sustain its operations in the North East of England. When it first arrived in Sunderland, Nissan was the object of great hope both from the City Council and from the local populace. Having suffered through numerous recessions, the manufacturing and extractive industries of Sunderland (glass, ship building and coal) had died a death, there was very little industry remaining. Nissan was a beacon of hope, allowing Sunderland to imagine itself as once again part of a powerful global economic power, rather than a forgotten back water (Evening Chronicle 2016).

As the plant became established it provided both jobs and a sense of pride and focus for the city's aspirations. Sunderland also became a recipient of the Nissan corporate citizenship scheme—it's way of addressing CSR issues, incorporating environmental and social issues. In terms of the communities of Sunderland, however, the most important thing was that the plant would continue to exist and that it would continue to provide jobs (and associated training and apprenticeships) for locals. It could be said that 'local content' was more important to Sunderland than CSR.[10] Recently, however, things have shifted. First, with the events leading up to (and emerging out of) Brexit, the future of the plant has by no means been certain. In this context the expectation of both locals and their political representatives is that government will step in to persuade Nissan to continue investing in the area. Far from corporations being expected to invest in communities, here local communities (or their governments) are expected to invest in corporations.

More recently, with the developments of the COVID-19 pandemic, Nissan has been manufacturing and supplying personal protective equipment (PPE) using a scaled back facility. In what could be seen as a traditional manifestation of CSR a multinational corporation is now voluntarily providing a benefit to society that the government cannot provide. This is an ironic echo of how corporations were once perceived to operate in the developing world with a corporate gaining purchase and power in a political sense by carrying out actions that go beyond its core business but which are designed to develop and maintain favourable relationships.

6 Conclusions

We now have a situation whereby nations in the developing world are trying to assert their identity, at least partly through control over the resources which lie within their geographical borders. Meanwhile an echoing rhetoric is emerging from the developed world whereby politicians trying to capitalise on the difficulties faced by ordinary people attempt to exert authority by control over their manufacturing and importing power and influence through protectionist and populist rhetoric. In effect

---

[10]The author was a senior manager at Sunderland City Council from 2003–2011 with responsibility for community engagement and development—Nissan was notably absent from any activities around community issues being far more engaged with business development.

these two rhetorics are pushing the global economy in the same direction—towards a more fractured and fractious state of affairs whereby corporations struggle to maintain a smooth flow of various capitals in the face of increasing demands from nations that they should benefit most.

CSR is getting caught in the middle of this—as corporations struggle to meet the various requirements of governments they spend more resources on meeting and managing those requirements. If there is a demand for employees to come from a nation's workforce then monies that might have been spent on a variety of CSR initiatives now go towards specific training programmes. Monies that might have been spent on community programmes now go towards entrepreneur development in order to meet requirements that services and goods are sourced from national companies. Meanwhile in the developed world, when it seems that governments are no longer able to provide the skilled workforce that corporations require, those corporations increasingly invest in educational and technical training initiatives (see e.g. Shell and BP's 'Techfest' in Aberdeen and Nissan's apprenticeship schemes in Sunderland) to ensure an appropriately skilled workforce into the future.

Perhaps it is no accident that whilst the political language is becoming increasingly populist and nationalist the corporate language around CSR is also changing. Corporations now have budgets and operations related to 'social investment', 'sustainability', 'stakeholder relations', 'corporate citizenship' and local content. Fewer and fewer are investing any resources in something called CSR. Activities have also shifted focus with programmes becoming more geared towards seeing quantifiable and verifiable results that enable local content requirements or expectations to be met.

At the time of writing the political and economic relationships between corporations and nation states are defined by attitudes which veer towards the populist and nationalist and corporates are operating in an environment hemmed in by increasing local content expectations. In truth this does not mean an enormous change in direction for corporates as many of the activities engaged in to address these expectations are similar to activities previously engaged in and labelled CSR. However it does mean that the language around CSR has changed and is now less redolent of the colonial past and starts to reflect the increasing flexing of power that comes with the ownership of precious resources. Local content talk represents a shift away from the colonialist past and places more power and initiative in the hands of developing countries. In response to that the developed world is kicking back and demanding preferential trade deals, placing embargoes and becomes increasingly insular in terms of rhetoric.

CSR, then, could be seen as a reflection of the privileged position of the developed world and the orientation of multinational corporations towards that privilege. Now things are shifting and the language of local content is about the power of the resource rich developing world. This is not happening unopposed, as could be predicted and the response from the privileged, developed world, as reflected in political rhetoric, looks very much like a 'tooth for a tooth' reaction. Ultimately CSR appears not as a clear cut practice but as one kind of response to political and economic contexts without any meaningful substance or philosophy underpinning it.

References

Abbink, J. (2014). Religious freedom and the political order: The Ethiopian 'secular state' and the containment of Muslim identity politics. *Journal of Eastern African Studies, 8*(3), 346–365.

Ablo, A. D. (2015). Local content and participation in Ghana's oil and gas industry: Can enterprise development make a difference? *The Extractive Industries and Society, 2*, 320–327.

Ablo, A. D. (2017). The micromechanisms of power in local content requirements and their constraints on Ghanaian SMEs in the oil and gas sector. *Norsk Geografisk Tidsskrift-Norwegian Journal of Geography, 71*(2), 67–78.

Ablo, A. D. (2019). Actors, networks and assemblages: Local content, corruption and the politics of SME's participation in Ghana's oil and gas industry. *International Development Review, 41*(2), 193–214.

Ahamad, L. (2009). *Lords of finance: The bankers who broke the world*. London: Penguin.

Andreasson, S. (2015). Varieties of resource nationalism in sub-Saharan Africa's energy and minerals markets. *The Extractive Industries and Society, 2*, 310–319.

Asiago, B. C. (2017). Rules of engagement: A review of regulatory instruments designed to promote and secure local content requirements in the oil and gas sector. *Resources, 6*, 46.

Barth, F. (1998). *Ethnic groups and boundaries; The social organization of culture difference*. Long Grove, IL: Prospect Heights, Ill: Waveland.

Bremmer, I., & Johnston, R. (2009). The rise and fall of resource nationalism. *Survival, 51*(2), 149–158.

Brown, R. (2008). Sea change: Santa Barbara and the eruption of corporate social responsibility. *Public Relations Review, 34*(1), 1–8.

Bush, G. (2001). *The text of President Bush's address Tuesday night, after terrorist attacks on New York and Washington: Transcript from CNN*. Retrieved June 14, 2020, from http://edition.cnn.com/2001/US/09/11/bush.speech.text/

Carrithers, M. (2005). Anthropology as a moral science of possibilities. *Current Anthropology, 46*(3), 433–456.

Carroll, A. B. (2016). Carroll's pyramid of CSR: Taking another look. *International Journal of Corporate Social Responsibility, 1*(1), 3.

Childs, J. (2016). Geography and resource nationalism: A critical review and reframing. *The Extractive Industries and Society, 3*, 539–546.

Columbia Centre on Sustainable Investment. (2016). *Local content: Norway – petroleum*. Retrieved June 15, 2020, from http://ccsi.columbia.edu/files/2014/03/Local-Content-Norway-Petroleum-CCSI-May-2016.pdf

Cooper, A. F. (2019). 'Rising' states and global reach: Measuring 'Globality' among BRICS/MIKTA countries. *Global Summitry, 4*(1), 64–80.

Copelovitch, M. S., & Ohls, D. (2012). Trade, institutions and the timing of GATT/WTO accession in post-colonial states. *Review of International Organisations, 7*(1), 81–107.

Davis, C. L., & Wilf, M. (2017). Joining the club: Accession to the GATT/WTO. *Journal of Politics, 79*(3), 964–978.

Donaldson, T., & Preston, L. (1995). The stakeholder theory of the modern corporation: Concepts, evidence and implications. *Academy of Management Review, 20*, 65–91.

Eikeland, S., & Nilsen, T. (2016). Local content in emerging growth poles: Loal effects of multinational corporations' use of contract strategies. *Norsk-Geografisk Tidsskrift – Norwegian Journal of Geography, 70*(1), 13–23.

Elden, S. (2006). Contingent sovereignty, territorial integrity and the sanctity of borders. *SAIS Review of International Affairs, 26*(1), 11–24.

Enderwick, P. (2011). Understanding the rise of global protectionism. *Thunderbird International Business Review, 53*(3), 325–336.

Evening Chronicle. (2016). *30 years of Nissan: How the Sunderland plant transformed the North East*. Retrieved June 25, 2020, from https://www.chroniclelive.co.uk/business/business-news/30-years-nissan-how-sunderland-11854830

Freeman, R. E. (1984). *Strategic management: A stakeholder approach*. Boston: Pitman.
Friedman, M. (1970). *The responsibility of business is to increase its profits New York Times Magazine 13/09/1970*.
Gilberthorpe, E., & Banks, G. (2012). Development on whose terms?: CSR discourse and social realities in Papua New Guinea's extractive industries sector. *Resources Policy, 37*(2), 185–193.
Heffernan, M., & Thorpe, B. J. (2018). 'The map that would save Europe': Clive Morrison-bell, the tariff walls map and the politics of cartographic display. *Journal of Historical Geography, 60,* 24–40.
Heum, P. (2008). *Local content development: Experiences from oil and gas activities in Norway*. Bergen: The Institute of Research and Economics and Business Administration.
Ikenson, D. (2009). Protectionist pandemics? The durability of free trade. *Georgetown Journal of International Affairs, 10*(2), 15–22.
Jamali, D., & Keshishian, T. (2009). Uneasy alliances: Lessons learned from partnerships between businesses and NGOs in the context of CSR. *Journal of Business Ethics, 84*(2), 277–295.
Kalyuzhnova, Y., Nygaard, A., Omarov, E. S., & Saparbayev, A. (2016). *Local content policies in resource rich countries*. London: Palgrave Macmillan.
Katamba, D., & Nkiko, C. M. (2015). The landscape of corporate social responsibility in Uganda: Its past, present and future. In S. O. Idowu, S. Vertigans, & R. Schmidpeter (Eds.), *Corporate social responsibility in Sub-Saharan Africa*. London: Springer.
Kitchen, M. (1988). *Europe between the wars: A political history*. London: Routledge.
Mann, M. (1997). Has globalization ended the rise and rise of the nation state? *Review of International Political Economy, 4*(3), 472–496.
Mitchell, R. K., Agle, B. R., & Wood, D. J. (1997). Toward a theory of stakeholder identification and salience: Defining the principle of who and what really counts. *The Academy of Management Review, 22*(4), 853–886.
Nakano, T. (2004). Theorising economic nationalism. *Nations and Nationalism, 10*(3), 211–229.
Ndlovu-Gatsheni. (2009). Making sense of Mugabeism in local and global politics: 'So Blair, keep your England and let me keep my Zimbabwe'. *Third World Quarterly, 30*(6), 1139–1158.
Ohmae, K. (1990). *The borderless world: Power and strategy in the interlinked economy*. London: Collins.
Pang, A., Lwin, M. O., Ng, C. S. M., Ong, Y. K., Chau, S. R. W. C., & Yeow, K. P. S. (2018). Utilization of CSR to build organizations' corporate image in Asia: Need for an integrative approach. *Asian Journal of Communication, 28*(4), 335–359.
Pearson, Z., Ellingrod, S., Billo, E., & McSweeney, K. (2019). Corporate social responsibility and the reproduction of (neo)colonialism in the Ecuadorian Amazon. *The Extractives Industries and Society, 6*(3), 881–888.
Phillips, R. A. (2003). *Stakeholder theory and organizational ethics*. San Francisco: Berrett-Koehler Publishers.
Porter, M. E., & Kramer, M. R. (2006). Strategy and society: The link between competitive advantage and corporate social responsibility. *Harvard Business Review, 84*(12), 78–93.
Rajak, D. (2011). *In good company: An anatomy of corporate social responsibility*. Palo Alto: Stanford University Press.
Rajak, D. (2014). Corporate memory: Historical revisionism, legitimation and the invention of tradition in a multinational mining company. *Political and Legal Anthropology Review, 37*(2), 259–280.
Santos, R. S. P. d., & Milanez, B. (2015). The global production network for iron ore: Materiality, corporate strategies, and social contestation in Brazil. *The Extractive Industries and Society, 2,* 756–765.
Schmidt, I. (2019). The populist race: Neoliberalism falling behind, new right forging ahead, the left stumbling along. *Perspectives on Global Development and Technology, 18*(1–2), 61–78.
Sirkin, H. L., Hemerling, J. W., & Battacharya, A. K. (2008). *Globality: Competing with everyone from everywhere for everything*. Business Plus.

Utting, P. (2008). The struggle for corporate accountability. *Development and Change, 39*(6), 959–975.

van den Heijkant, L., & Vliegenthart, R. (2018). Implicit frames of CSR: The interplay between the news media, organizational PR, and the public. *Public Relations Review, 44*(5), 645–655.

Van Reenen, J. (2017). *"Brexit and the future of globalization?" centre for economic performance, special paper 35*. London School of Economics.

Weissman, R. (1991). The real purpose of GATT: Prelude to a new colonialism. *The Nation, 252*(10), 336–338.

Wilson, J. D. (2015). Understanding resource nationalism: economic dynamics and political institutions. *Contemporary Politics, 21*(4), 399–416.

World Bank. (2016). *Local content in oil, gas and mining*. Retrieved June 15, 2020, from https://www.worldbank.org/en/topic/extractiveindustries/brief/local-content-in-oil-gas-and-mining

World Trade Organisation. (2020a). *GATT papers*. Retrieved May 29, 2020, from https://www.wto.org/english/docs_e/legal_e/gatt47_01_e.htm

World Trade Organisation. (2020b). *Understanding the WTO*. Retrieved May 29, 2020, from https://www.wto.org/english/thewto_e/whatis_e/tif_e/fact3_e.htm

World Trade Organisation. (2020c). *Developing countries*. Retrieved May 29, 2020, from https://www.wto.org/english/thewto_e/whatis_e/tif_e/dev1_e.htm

Sarah Buckler is a social anthropologist with an interest in the relationships between people and organisations, organisational cultures, human creativity and development. A particular interest in narrative and rhetoric and the ways these are used by individuals and organisations in order to achieve particular ends began with her PhD research with Gypsy Travellers and continued into her subsequent employment carrying out research for local and national government departments. More recently this interest has expanded to focus on the ways different CSR and other development projects are delivered, especially in West Africa where she spent some time working on evaluation of a variety of CSR projects. In the light of Brexit and increasing populist rhetoric across the world, this interest has become a fascination with the ways power shifts are reflected in language and how this in turn influences national and international economic policies.

Corporate Social Responsibility Practices and Its Implementation after the Legal Mandate: A Study of Selected Companies in India with Special Emphasis on the Mining Sector

Sumona Ghosh

Abstract Mining operations frequently involve a high degree of environmental and social impacts and hence categorized under Red category according to Indian Ministry of Forest and Environment. It is a very sensitive sector thus requiring the companies of this sector to be very responsive towards its social responsibilities. The paper investigated the corporate social responsibility (CSR) practices adopted by the selected companies in India belonging to the mining sector, post-2013 Companies Act mandate from 2014–2015 to 2018–2019 and their implementation during the said time period using Longitudinal Qualitative Document Analysis and constructing a Corporate Social Responsibility Index (CSRI). The study was based on secondary sources, i.e., by analyzing the company's annual reports. We observed that education was the most preferred CSR activity undertaken by the companies belonging to the mining sector and the least preferred activity was disaster relief measures. The study revealed an average CSRI of 29.66% which happened to be extremely low. A detailed assessment of the overall CSR expenditure made by the selected Indian companies belonging to the mining sector for the time period 2014–2015 to 2018–2019 showed that the companies have not spent the mandated amounts excepting in the year 2018–2019. Detailed assessment of CSR expenditure activity wise amongst selected Indian companies belonging to the mining sector for the time period 2014–2015 to 2018–2019 showed that CSR expenditure was the highest in the field of education, followed by health and rural upliftment. Despite governmental regulations'and pressure from other stakeholders majority of the companies failed to emphasis concerning areas like energy, climate change, marine and terrestrial ecosystem. Thus strict monitoring and accountability of the projects undertaken in the name of CSR has to be initiated by the government along with proper encouragement so that the companies become socially responsive.

S. Ghosh (✉)
Department of Commerce, St. Xavier's College, Kolkata, India
e-mail: sumonaghosh@sxccal.edu

1 Introduction

1.1 Facets of CSR

The first time period of 1920–1950s saw some pioneering works like Chester Barnard's *"The Functions of the Executive"* (1938), J. M. Clark's *"Social Control of Business"* (1939) and Theodore Kreps's *"Measurement of the Social Performance of Business"* (1940), to point out just a few and these works mostly concentrated on the social responsibilities of businessmen. In the second time period of 1950s Corporate Social Responsibility started evolving as a theory, as a concept. Howard Bowen's *"Social Responsibilities of the Businessmen"* (1953) was the first attempt to theorize the relationship between companies and society. According to Bowen (1953) CSR refers to the obligations of businessmen" to pursue those policies, to make those decisions, or to follow those lines of action which are desirable in terms of the objectives and values of our society". The meaning of CSR became more accurate and prominent in the decade of 1960s. This decade also marked the contributions made by some of the most prominent writers during that time like Keith Davis, Joseph W McGuire, William C Frederick and Clarence C Walton. The notion of Business Ethics to CSR was introduced during this era. The time period of 1970s saw a breakthrough in conceptual development of CSR when a new study on CSR was commissioned by the Committee for Economic Development. The Committee described CSR as being 'related to products, jobs and economic growth; related to societal expectations; and related to activities aimed at improving the social environment of the firm' (US Committee for Economic Development in Wheeler et al. 2003). The era of 1980s saw the development of the concept of sustainable development (World Commission on Environment and Development 1987).The era of 1990s was that of imbibing the concept of CSR in the field of strategic management by strategic management scholars such as Philip Kotler, Nancy Lee, Michael Porter, Rosabeth Moss Kanter and Stuart Hart. During this era Corporate Social Performance (CSP), Stakeholder-theory, Business Ethics and Corporate Citizenship were the major themes that took center stage. In the twenty first century theoretical contributions to the concept of CSR paved the way for more empirical research on this area thus leading to related topics such as Stakeholder theory, Business Ethics, Sustainability, and Corporate Citizenship. The most optimistic perspective during this era was depicted well by Lydenberg (2005) in his book *"Corporations and the Public Interest: Guiding the Invisible Hand"*, where he saw CSR as 'a major secular development, driven by a long-term reevaluation of the role of corporations in society'. Thus CSR as a concept has been evolving through decades.

1.2 CSR in India

If we look at the development of CSR in India, we find that it can be divided into four main phases. The first phase was an era where there was charity and philanthropy which motivated CSR. The pioneers of industrialization such as the Tata, Birla, Bajaj, Lalbhai, Sarabhai, Godrej, Shriram, Singhania, Modi, Naidu, Mahindra and Annamali, who were strongly devoted to philanthropically motivated CSR. These families were leaders in the economic, as also in the social fields. The second phase of Indian CSR (1914–1960) was influenced fundamentally by Gandhi's theory of Trusteeship. Schools and colleges, training and scientific institutes were setup by well-established family businesses through their trusts. They took part in social reforms, poverty alleviation, women empowerment and eradication of caste systems. The overall socio-political goal focused on building a solid industrial base while nurturing the Indian cultural traditions. The third phase (1960–1980) was characterized by strict legal and public regulation of business activities. During this era stakeholder dialogues, social accountability and transparency was emphasized. In the fourth phase (1980 until the present) companies started integrated CSR into a sustainable business strategy, partly adopting the multi-stakeholder approach mainly due to economic liberalization in 1992 and greater levels of privatization. At present we find that CSR and it's reporting in India which was voluntary till 2013, has been made mandatory as per section 135 of the Companies Act of 2013, perhaps the only country in the world to do so (MCA 2013).

The CSR journey in India took a turn on February 27, 2014 when the Ministry of Corporate Affairs amended section 135 of the Companies Act of 2013. This provision mandates eligible companies having net worth of rupees five hundred crores or more or turnover of one thousand crore or more or net profit of rupees five crores or more during any financial year shall contribute 2% of its average net profits towards CSR activities. The activities under which companies can allot its CSR funds have been specified under Schedule 7of the Act. The Act came into effect from April 1, 2014.

The Ministry of Environment, Forest and Climate Change (MoEFCC) had brought out notifications in 1989, with the purpose of prohibition/restriction of operations of certain industries for ecological protection. The notification introduced the concept of categorization of industries as "Red", "Orange" and "Green" with the purpose of facilitating decisions related to location of these industries. Subsequently, the application of this concept was extended in other parts of the country not only for the purpose of location of industries, but also for the purpose of Consent management and formulation of norms related to surveillance/inspection of industries. The concept of categorization of industries continued to evolve and the Working Group developed the criteria of categorization of industrial sectors based on the Pollution Index which is a function of the emissions (air pollutants), effluents (water pollutants), hazardous wastes generated and consumption of resources. For the purpose of categorization of industrial sectors the following criteria on "Range of Pollution Index" was finalized-

- Industrial Sectors having Pollution Index score of 60 and above—Red category
- Industrial Sectors having Pollution Index score of 41 to 59—Orange category
- Industrial Sectors having Pollution Index score of 21 to 40—Green category
- Industrial Sectors having Pollution Index score including and up to 20—White category

Mining operations frequently involve a high degree of environmental and social impacts and hence categorized under Red category according to The Ministry of Environment, Forest and Climate Change (MoEFCC). It is a very sensitive sector thus requiring the companies of this sector to be very responsive towards its social responsibilities. The paper thus aims to identify the corporate social responsibility (CSR) practices adopted by the selected companies in India belonging to the mining sector, post-2013 Companies Act mandate from 2014–2015 to 2018–2019 and their implementation during the said time period. With the inclusion of the introduction the remaining paper is further structured as follows. Section 2 provides a brief literature review on the issue. The third section discusses the objectives that we would try to examine through our study. Section 4 describes the methodology. The results and discussions of empirical research are provided in the &&fifth section. The article ends with observations and concluding remarks provided in Sect. 6.

2 Corporate Social Responsibility and Mining Industry

Kapelus (2002), Imbun (2007), Arko (2013), Bice (2013), Diale (2014), Franks et al. (2014), Sharma and Bhatnagar (2015) studied the geographical dimension of mining company engagement with communities and observed that since local communities are the most affected and therefore have the most credible claims mining companies use CSR as a tool to claim that local communities were benefiting from their operations in order to legitimize mining operations, protect them from other external claims, and thus obtain license to operate. The authors also focused on the corporate social responsibility activities of the mining companies with respect to the governance structure put in place and the funds allocated for such activities. They proposed that mining companies could have decentralized CSR management approaches, which created gaps between headquarters and management at the community level. The notion of CSR as a discourse and its understanding within the mining industry was also explored. The authors presented a framework for responsible businesses within the mining industry. Lastly an analysis of the progress achieved was also made. Their research brought into focus the fact that the development prospects of the mining industry depended on the extent to which the private sector internalized the environmental and social aspects of development into business decision making.

Hamann (2003), Conway (2003), Slack (2012) had observed that Company-community relations were at the heart of mining company CSR discourse, particularly around the issue of sustainable development however in the same year through his research he presented his observation that perceived benefits of mining activities at the national level sometimes was in contradiction with the negative impacts of mining projects in the most immediate communities. They had observed that many large mining companies had their own initiatives towards environmental and social development. However, a structured CSR policy and planning was missing especially among the small and medium players in the industry. The gap between the rhetoric and reality of CSR existed because companies had failed to fully integrate CSR into their business models. Banerjee (2004), Hamann and Kapelus (2004), Jamali and Mirshak's (2007) through their research observed that for carrying on mining activity adequate attention would have to be given to the social dimensions of mining. Their research engaged with wider debates and ambiguities surrounding mining, CSR, and development at national and local levels. The authors found that, despite these companies positive CSR discourse, the actual approach to CSR was very 'amateurish and sketchy'.

Yakovleva (2005), Muller (2006), Lahiri-Dutt (2008), Viviers and Boudler (2008), Welker (2014) explored recent trends in the reporting of social and environmental impacts and issues in the global mining industry. He studied world's ten largest mining companies. The results showed that there was evidence of increasing sophistication in the development of social and environmental disclosure but there was considerable variation in the maturity of reporting content and styles of these companies. They observed the lack of coordination between corporate headquarters and a mine site. Poor attention to community development and engagement with the landowners were causing the closing down of mines. The authors also observed that, those things which have economic impact like HIV/AIDS were given priority in the CSR of mining companies, whereas CSR issues related to empowerment received less attention.

Kemp (2009) through their research brought into focus the fact that the development prospects of the mining industry depended on the extent to which the private sector incorporated the environmental and social aspects into business decision making. Boon (2009) devoted his thesis to understanding how the 'home' and 'host' government can help to make mining company CSR initiatives more effective. His study of the Peruvian mining industry highlights the local context and power relationships. There has been a substantial body of literature concerning studies related to mining communities in the United Kingdom, considering issues like the economic and social impacts of mine closure; the impact of closure on health and wellbeing; and efforts to regenerate former mining communities (Beatty and Fothergill 1996; Waddington 2004; Waddington et al. 2001). Similar work has been undertaken examining these sorts of issues in mining communities in North America (Keyes 1992; Randall and Ironside 1996), mainland Europe (Critcher et al. 1995) and Australia (Ingamells et al. 2011; Storey 2001). Anguelovski (2011), Buenar and Hevina (2011) examined dialogue processes between mining companies and communities. She explained how communities may resist spaces created to address issues or concerns about the mine operation. For Mining firms, *the authors* proposed practical internal engagement mechanisms that would include synergies between leadership action and other mechanisms such as learning and culture in order to be responsive with their CSR.

Hilson (2012), Yakovleva and Vazquez-Brust (2012), Dobele et al. (2014) through research showed that CSR could reduce conflicts and maintain security around mine sites. Through their work the authors provided a rare insight into multi-stakeholder perceptions of CSR in the industry that heavily impacted the natural environment and local economic and social structures. Through his work they provided evidence of the importance of corporate commitment at different organizational levels for a successful stakeholder relationship. Dong et al. (2014) provided evidence that mining companies give priority to different stakeholders in different countries. They studied the influence of different stakeholder groups on the Chinese mining industry's CSR practice, where the highest importance was given to the Government as stakeholder. They provided evidence of important differences with Western-based mining companies, in which the focus was on customers and affected communities. This raised questions about the importance of location, culture and political views on how mining companies prioritized their stakeholders. The study undertaken by Kepore et al. (2013) examined how one indigenous community in the Western Province of Papua New Guinea (PNG) viewed the social responsibility initiatives of OK Tedi Mining Ltd. (OTML).The study by Mzembe and Meaton (2014) examined the drivers of the CSR agenda pursued by Paladin (Africa), a subsidiary of an Australian multinational mining company (MNC) operating the first uranium mine in Malawi. The findings suggested that the CSR agenda in the mining industry in Malawi was strongly influenced by externally generated pressures such as civil society organization activism and community expectations.

Bodruzic (2015) studied the role governments should play in CSR projects of multinational mining companies. She argued that the Canadian Government was involved in those projects based on Canadian commercial interests, rather than a real interest in promoting international development. ICMM (2015) presented the fact that the wider trend towards increased CSR spending was a response to growing conflict surrounding mine sites globally. Kotilainen et al. (2015) had studied the Peruvian mining industry and they highlights on the local context and power relationships prevailing there. Mzembe (2016) studied processes and practices that Malawi mining companies utilized to engage with stakeholders. Abuya (2016, 2018) examined the place of CSR in Kenya's nascent titanium mining industry. Through an ethno-ecological lens, it examined the extent to which CSR has managed to assuage the disaffection of the local community with the mining operations of the company Tiomin (K). The author suggested a shift in thinking from "corporate responsibility" to "reciprocal responsibility" as a way to minimize mining conflicts. *The author* highlighted that CSR activities in developing countries, especially in Africa, have had a questionable reputation. The few programs rolled out under this program have done little in meeting the needs of the affected mining communities. CSR in Kenya's mining industry had received very little attention. This work also reviewed the mining conflicts in Africa and examined how CSR could assuage mining community disaffection over mining projects.

Thus reviewing the literature the observation is that, even though there are studies where researchers have dealt with various CSR related issues dealing with the mining sector but there are few studies which have made an in-depth analysis of CSR activities undertaken by the mining sector and corporate CSR fund utilization by the mining sector. Therefore, this study would attempt to fill the existing gaps and overcome the limitations of the literature by investigating the corporate social responsibility activities undertaken by the mining companies in developing countries with special reference to India.

3 Objectives

The Ministry of Environment, Forest and Climate Change (MoEFCC) has categorized the mining sector as the Red category sector since it involve a high degree of environmental and social impacts hence this sector has to be very responsive towards its social responsibilities. It becomes extremely crucial for the companies to be socially responsive. With this in the background, the present study would focus on the analysis of corporate social responsibility practices and its implementation for selected companies in India belonging to the mining sector for the time period from 2014–2015to 2018–2019. It would also be concerned with the following objectives:

- Objective 1: To estimate the percentage of companies involved in different CSR activities during 2014–2015 to 2018–2019.
- Objective 2: To determine the most preferred and the least preferred CSR activity during the time period under consideration.
- Objective 3: To construct Corporate Social Responsibility Index (CSRI) for different time periods across all companies and to differentiate the companies on that basis.
- Objective 4: Detailed assessment of overall CSR expenditure amongst selected Indian companies belonging to the mining sector for the time period 2014–2015 to 2018–2019.
- Objective 5: Detailed assessment of CSR expenditure activity wise amongst selected Indian companies belonging to the mining sector for the time period 2014–2015 to 2018–2019

4 Methodology

4.1 Data Source and Study Design

The paper is based on an empirical and analytical study undertaken for the financial years 2014–2015 till 2018–2019 to give us an overview about the corporate social responsibility practices and its implementation amongst selected companies in India belonging to the mining sector. The study was based on secondary sources, i.e., by analyzing the company's annual reports since Rule 8 of the Companies (Corporate Social Responsibility Policy) Rules, 2014 provides the following:

- The Board's Report (which is a mandatory section to be included in the annual report) of a company covered under these rules pertaining to a financial year commencing on or after the first day of April, 2014 shall include an annual report on CSR containing particulars specified in Annexure (Annexure 1). Section 134 (3)(o) of the Companies Act of 2013 requires that the report by the Board of Directors shall include the details of the policy developed and implemented by the company on corporate social responsibility initiatives taken during the year. If the company fails to spend 2% of average net profits during a financial year, the Board shall, in its report specify the reasons for not spending the amount.

Data was then generated from such an analysis using Longitudinal Qualitative Document Analysis for the 5 year period. Three categories have been identified for the study. Detailed exposition of these categories is provided below.

Content Category I
Through this category we have tried to analyze which are the corporate social responsibility activities that the companies have shown their responsiveness to and have implemented.

Content Category II
Through this category we have tried to explore which are the most preferred and the least preferred CSR activity/activities undertaken by the companies of this sector for the aforesaid time period. We have divided the CSR activities specified by the Companies Act of 2013, Schedule VII, into nine groups namely—education, environment, disaster relief, health, rural upliftment, empowerment and skill development, livelihood, drinking water and sanitation and others (sports, art and culture). (Refer to Annexure 2 for details).

Content Category III
Through this category we have tried to make a detailed assessment of not only the overall CSR expenditure undertaken by the selected Indian companies belonging to the mining sector for the time period 2014–2015 to 2018–2019 but also of the CSR expenditure activity wise by such companies for the aforesaid time period.

4.2 Extent of Responsiveness to CSR Implementation Category

We have tried to capture the extent of responsiveness towards CSR implementation for the financial years 2014–2015 till 2018–2019. We have identified four main stages of responsiveness to CSR implementation—high level of responsiveness towards implementation of CSR activities, medium level of responsiveness towards implementation of CSR activities, progressive responsiveness and low level of responsiveness towards implementation of CSR activities. Companies whose CSRI (Corporate Social Responsibility Index) score was higher than the grand average score for each of the 5 years; we regarded such companies to be highly responsiveness to CSR implementation. Similarly companies whose score was lower than the grand average score for each of the 5 years, we regarded such companies to have a low responsiveness to CSR implementation, companies with scores gradually improving over the time period were regarded as progressive responsiveness to CSR implementation and the remaining companies were regarded to have moderate responsiveness to CSR implementation.

4.3 Selection of Companies

India has two major stock exchanges: Bombay stock exchange of India (BSE) and National stock exchange of India (NSE). The BSE was established in 1875 and is Asia's first stock exchange. It is the world's 11th largest stock exchange. More than 5500 companies are publicly listed on the BSE. The NSE was founded in 1992 and started trading in 1994, as the first demutualised electronic exchange in India. The NSE is the world's 12th largest stock exchange. Our study is based on firms listed on the BSE's and NSE's indexes as on 6th August 2019. We have studied all the firms belonging to the sector "mining and minerals" as classified by the BSE. This gave us a set of 29 firms. The annual reports of these companies were analyzed for the time period 2014–2015 to 2018–2019 i.e. 5 years.

4.4 Method

With respect to "*Content Category I*" we calculated the proportion of companies who had undertaken the CSR activities specified by the Companies Act of 2013. With respect to "*Content Category II*" we calculated the most preferred and the least preferred CSR activity/activities undertaken by the companies from the annual reports. With respect to "*Content Category III*" the budgeted CSR expenditure and the actual amount spent were compared with the minimum prescribed CSR expenditure of each company of this sector. The data about budgeted and actual amount spent were available for particular projects of the company specified in the CSR report annexed to the Directors Report disclosed in the annual reports, which, when clubbed together, became comparable with the minimum prescribed amount to be spent. Longitudinal Qualitative Document Analysis was employed for the period 2014–2015 to 2018–2019. A Corporate Social Responsibility Index (CSRI) was constructed by examining the annual reports of the 29 companies belonging to the "mining and minerals" sector. The CSRI was calculated in the following manner:

$$\text{CSRI} = \left(\sum di/n\right) \times 100 = (TS/n) \times 100$$

Where:
CSRI = Corporate Social Responsibility Index (CSRI).
di: 1 if CSR activity i is implemented; 0 if CSR activity i is not implemented.
n = no of activities = Maximum Score.
TS = Total Score.

Since there are nine CSR activities to be implemented (based on Companies Act of 2013), the maximum score for each sampled company would be nine. Therefore the maximum CSRI score is "100" and the minimum score is "0". Hence a score of 100 or closer to it suggests high level of responsiveness to CSR implementation by the company. A score of "0" or closer to it suggests low level responsiveness to CSR implementation. We also calculated the grand mean score for each of the years. From here we identified the four levels of integration which are as follows:

- Companies with score above the grand average score for each year reflected high level of responsiveness to CSR implementation
- Companies with score below grand average score for each year reflected low level of responsiveness to CSR implementation
- Companies with scores gradually improving over the time period considered for the analysis reflected progressive companies with respect to responsiveness to CSR implementation
- The remaining companies reflected moderate level of responsiveness to CSR implementation.

5 Results and Discussions

Table 1 gives us an overview of the companies forming a part of the data set. Three are state government companies, four are union government companies and 22 are public limited or public incorporated companies. The table highlights the year of incorporation, the type of company their location and the areas where they are carrying on their CSR activities for the period under consideration.

Table 2 focused on the percentage of companies involved in different CSR activities during 2014–2015 to 2018–2019. A diagrammatic representation has been produced in Fig. 1. For the years 2014–2015 till 2016–2017 we observe that a high percentage of the companies had taken up projects involving education and health. For the year 2017–2018 we observe that a high percentage of the companies had taken up projects involving education and rural upliftment and for the year 2018–2019 it was education and "others" involving art, culture, preservation of heritage, sports. For all the years we observe that a very minimum percentage of the companies had taken up projects involving disaster relief measures. Along with

Corporate Social Responsibility Practices and Its Implementation after the... 115

Table 1 Overview of the companies forming a part of the data set

Company	Year of incorporation	Registered as	Location/ headquarters	States/city/district where CSR projects have been undertaken
COAL INDIA	1975	State Government company	West Bengal	Bihar, Assam, West Bengal, Orissa, Mumbai, Rajasthan, New Delhi, Jharkhand, Madhya Pradesh, Delhi, Haryana, Uttarakhand, Jammu and Kashmir, Uttar Pradesh, Kerala, Andhra Pradesh, Karnataka
VEDANTA	1979/1965	Public incorporated	Maharashtra	Tamil Nadu, Goa, Karnataka, Jharkhand, Orissa Rajasthan, Gujarat, West Bengal, Andhra Pradesh, Delhi NCR, Haryana
MOIL	1962	Union government company	Maharashtra	Nagpur (Maharashtra), Chhindwada (Madhya Pradesh), Faziabad, Bhandara, Balaghat, Jaunpur
GUJ minerals	1963	State government company	Gujarat	Rajkot, Gujrat, Kutch, surat, Dwaraka, Bharuch, Ahmedabad, Tarkeswar, Bharnagar, Phandhro
MAITHAN ALLOYS	1985	Public incorporated	West Bengal	West Bengal, Tamilnadu, Bangaluru, Chennai, Haryana, Faridabad, Vrindavan, Uttar Pradesh
SANDUR MANGANES	1954	Public incorporated	Karnataka	Sandur, Kammathuru, Sandur (ballari)
INDIA METALS	1961	Public incorporated	Orissa	Orissa, Jajpur
OMDC	1918	Union government company	West Bengal	Orissa (keojhar, Katak, puri, sudargah), Jajpur
DECCAN GOLD	2003	Public incorporated	Maharashtra	NO CSR spent
ASHAPURA MINES	1960	Public incorporated	Maharashtra	NO CSR spent
20 MICRONS	1987.	Public incorporated	Gujarat	Vadodara, Bhuj, Nandesari, Mumbai, Udaipur
RSMML	1949	State government company	Rajasthan	Udaipur, Jaipur, Nagpur, Jodhpur, Barmer, Jaisalmer, Jaipur
INDSIL HYDRO	1990	Public incorporated	Tamil Nadu	Tamilnadu, Kerala, Coimbatore, Chattisgarh
RAW EDGE INDUSTRIAL SOLUTIONS LTD	2005	Public incorporated	Maharashtra	NO CSR spent

(continued)

Table 1 (continued)

Company	Year of incorporation	Registered as	Location/ headquarters	States/city/district where CSR projects have been undertaken
FERRO ALLOYS	1955	Public incorporated	Orissa	Orissa, Dhanekhali, Bhadrak, Jajpur, NewDelhi, Keonjhar
SHYAM CENTURIES	2011	Public limited company	Meghalaya	Haryana, Uttar Pradesh
SHIRPUR GOLD	2001	Public Ltd. company	Maharashtra	NO CSR spent
NAGPUR POWER	1996	Public incorporated	Maharashtra	NO CSR spent
FACOR ALLOYS	2004	Public incorporated	Andhra Pradesh	NO CSR spent
IMPEX FERRO TECH	1995	Public incorporated	West Bengal	NO CSR spent
NMDC	1958	Union government company	Telangana	Madhyapradesh, Telangana, Maharashtra, Andhrapradesh, Karnataka, Chhattisgarh, New Delhi, Uttar Pradesh
HINDUSTAN ZINC LTD	1966	Public incorporated	Rajasthan	Rajasthan, Uttarakhand, Udaipur, Haridwar, Ajmer, Chittorgarh
CEETA INDUSTRIES LTD	1984	Public incorporated	Karnataka	NO CSR spent
GLITTEK GRANITE LTD	1990	Public incorporated	Karnataka	NO CSR spent
KIOCL LTD	1990	Union government company	Karnataka	Karnataka, Orissa, Mangalore, Bangalore, New Delhi, Kerala
FOUNDRY FUEL PRODUCTS LTD	1964	Public incorporated	West Bengal	NO CSR spent
ASI INDUSTRIES LTD	1945	Public incorporated	Maharashtra	Rajasthan(kota), Maharashtra
PACIFIC INDUSTRIES LTD	1989	Public incorporated	Karnataka	Udaipur (Rajasthan)
ORIENTAL TRIMAX LTD	1996	Public incorporated	Delhi	NO CSR spent

Source: Authors own compilation

disaster relief for the year 2018–2019 very limited number of companies undertook projects involving creation of livelihood opportunities.

From Table 3 we get an idea about the most preferred and the least preferred CSR activity during the time period 2014–2015 to 2018–2019 and the observation is the

Table 2 Percentage of companies involved in different CSR activities during 2014–2015 to 2018–2019

	Year					
	2014–2015	2015–2016	2016–2017	2017–2018	2018–2019	Total
Education	**48.28%**	**55.17%**	**48.28%**	**51.72%**	**51.72%**	51.03%
Health	**48.28%**	51.72%	**48.28%**	44.83%	41.38%	46.90%
Environment	24.14%	31.03%	24.14%	27.59%	31.03%	27.59%
Drinking water and sanitation	31.03%	27.59%	17.24%	20.69%	24.14%	24.14%
Empowerment	34.48%	27.59%	24.14%	27.59%	24.14%	27.59%
Livelihood	10.34%	10.34%	10.34%	10.34%	**6.90%**	9.66%
Disaster relief	**0.00%**	**0.00%**	**0.00%**	**6.90%**	**6.90%**	2.76%
Rural upliftment	31.03%	41.38%	41.38%	**48.28%**	41.38%	40.69%
Others	34.48%	27.59%	31.03%	41.38%	**48.28%**	36.55%

Source: Authors own computation
Bold values indicate highest and lowest % of companies involved in different CSR activities

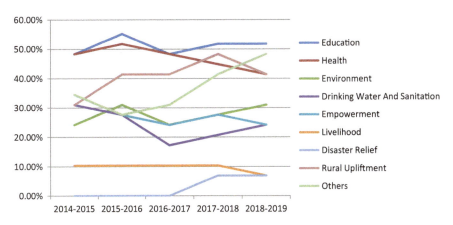

Fig. 1 Percentage of companies involved in different CSR activities during 2014–2015 to 2018–2019

education was the most preferred CSR activity undertaken by the companies belonging to the mining sector and the least preferred activity was disaster relief measures. A diagrammatic representation has been produced in Fig. 2.

Tables 4 and 5 discloses the individual Corporate Social Reporting Index (CSRI) of the companies for the time period 2014–2019 and the extent of responsiveness to CSR implementation of the companies for the time period 2014–2019. The study revealed an average CSRI of 29.66%. (Table 3). A diagrammatic representation has been produced in Fig. 3 Table 4 showed the extent of integration whereby the companies were differentiated on the basis of their CSRI into high level of responsiveness to CSR implementation (HLR), low level of responsiveness to CSR implementation (LLR), progressive level of responsiveness to CSR implementation (PLR)

Table 3 The most preferred and the least preferred CSR activity during the time period 2014–2015 to 2018–2019

	% companies
Education	51.03%
Health	46.90%
Rural upliftment	40.69%
Others	36.55%
Environment	27.59%
Empowerment	27.59%
Drinking water and sanitation	24.14%
Livelihood	9.66%
Disaster relief	2.76%

Source: Authors own computation

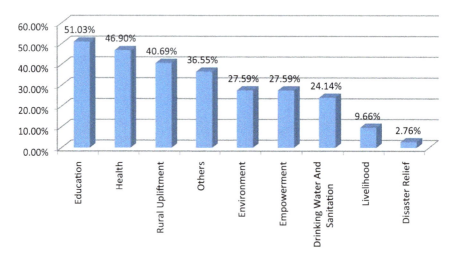

Fig. 2 The most preferred and the least preferred CSR activity during the time period 2014–2015 to 2018–2019

and moderate level of responsiveness to CSR implementation (MLR). From Table 4 we observe that ten companies fell into the category of high level of responsiveness to CSR implementation (HLR), five companies fell into the category of moderate level of responsiveness to CSR implementation (MLR)., two companies fell into the category of progressive level of responsiveness to CSR implementation (PLR) and 12 companies fell into the category of low level of responsiveness to CSR implementation (LLR).

Here Red: below Average CSRI; Green: above Average CSRI.

The above table (Table 6) gives us an assessment of overall CSR expenditure amongst the selected companies belonging to the mining sector for the time period 2014–2015 to 2018–2019. We observe that average only 87% of the 2% mandatory expenditure that is required to be spent on CSR have been spent over a span of 5 years, of which the highest was in the year 2018–2019 and only an average of 60%

Table 4 Individual CSRI for the companies (%)

Name of the companies	CSRI					
	2014-2015	2015-2016	2016-2017	2017-2018	2018-2019	Overall
COAL INDIA	66.67	77.78	55.56	66.67	77.78	68.89
VEDANTA	88.89	88.89	88.89	88.89	88.89	88.89
MOIL	66.67	66.67	66.67	55.56	55.56	62.23
GUJ MINERALS	77.78	55.56	77.78	88.89	55.56	71.11
MAITHAN ALLOYS	22.22	33.33	44.44	44.44	44.44	37.77
SANDUR MANGANES	11.11	33.33	22.22	11.11	33.33	22.22
INDIA METALS	66.67	66.67	55.56	66.67	66.67	64.45
OMDC	55.56	33.33	11.11	11.11	0.00	22.22
DECCAN GOLD	0.00	0.00	0.00	0.00	0.00	0.00
ASHAPURA MINES	0.00	0.00	0.00	0.00	0.00	0.00
20 MICRONS	0.00	0.00	0.00	55.56	33.33	17.78
RSMML	33.33	44.44	33.33	44.44	77.78	46.66
INDSIL HYDRO	33.33	33.33	33.33	33.33	22.22	31.11
RAW EDGE INDUSTRIAL SOLUTIONS LTD	0.00	0.00	0.00	0.00	0.00	0.00
FERRO ALLOYS	44.44	44.44	33.33	55.56	44.44	44.44
SHYAM CENTURIES	0.00	11.11	11.11	11.11	22.22	11.11
SHIRPUR GOLD	0.00	11.11	0.00	0.00	0.00	2.22
NAGPUR POWER	0.00	0.00	0.00	0.00	0.00	0.00
FACOR ALLOYS	0.00	0.00	0.00	0.00	0.00	0.00
IMPEX FERRO TECH	0.00	0.00	0.00	0.00	0.00	0.00
NMDC	88.89	66.67	66.67	66.67	55.56	68.89
HINDUSTAN ZINC LTD	77.78	77.78	77.78	77.78	77.78	77.78
CEETA INDUSTRIES LTD	0.00	0.00	0.00	0.00	0.00	0.00
GLITTEK GRANITE LTD	0.00	0.00	0.00	0.00	0.00	0.00
KIOCL LTD	33.33	55.56	33.33	44.44	66.67	46.67
FOUNDRY FUEL PRODUCTS LTD	0.00	0.00	0.00	0.00	0.00	0.00
ASI INDUSTRIES LTD	55.56	66.67	66.67	66.67	66.67	64.45
PACIFIC INDUSTRIES LTD	22.22	11.11	11.11	11.11	0.00	11.11
ORIENTAL TRIMAX LTD	0.00	0.00	0.00	0.00	0.00	0.00
Average CSRI	29.12	30.27	27.20	31.03	30.65	29.66

Source: Authors own computation
Note: Red: below Average CSRI; Green: above Average CSRI

of the total budgeted amount for the 5 years have been spent, of which the highest was in the year 2016–2017.

Table 7 gives a more detailed picture about the individual companies overall CSR expenditure for the time period 2014–2015 to 2018–2019.The red shade is more indicating that majority of the company's overall CSR expenditure was below the average CSR expenditure for that year.

The above table (Table 8) gives us an assessment of CSR expenditure (in lakhs) activity wise amongst selected Indian companies belonging to the mining sector for the time period 2014–2015 to 2018–2019. We observe that in the years 2014–2015, 2016–2017 and 2018–2019 CSR expenditure was the highest in the field of education. In 2015–2016 CSR expenditure was the highest in the field of health and in 2017–2018 it was rural upliftment . CSR expenditure was the least in the area of disaster relief. A more detailed assessment of CSR expenditure for each of the companies for each of the CSR activities can be observed in Tables 9, 10, 11, 12, 13, 14, 15, 16, 17 (see Annexure 3). Amongst the companies the highest amount spent in the field of education and health (Tables 9 and 10) was by National Mineral Development Corporation (NMDC) followed by Hindustan Zinc Ltd. for the year 2018–2019. The highest amount of CSR expenditure in the field of environment and

Table 5 Classification of companies by extent of responsiveness to CSR implementation for the time period 2014–2019.

Name of the companies	CSRI					Level of Responsiveness
	2014-2015	2015-2016	2016-2017	2017-2018	2018-2019	
COAL INDIA	66.67	77.78	55.56	66.67	77.78	HLR
VEDANTA	88.89	88.89	88.89	88.89	88.89	HLR
MOIL	66.67	66.67	66.67	55.56	55.56	HLR
GUJ MINERALS	77.78	55.56	77.78	88.89	55.56	HLR
MAITHAN ALLOYS	22.22	33.33	44.44	44.44	44.44	PLR
SANDUR MANGANES	11.11	33.33	22.22	11.11	33.33	MLR
INDIA METALS	66.67	66.67	55.56	66.67	66.67	HLR
OMDC	55.56	33.33	11.11	11.11	0.00	MLR
DECCAN GOLD	0.00	0.00	0.00	0.00	0.00	LLR
ASHAPURA MINES	0.00	0.00	0.00	0.00	0.00	LLR
20 MICRONS	0.00	0.00	0.00	55.56	33.33	MLR
RSMML	33.33	44.44	33.33	44.44	77.78	HLR
INDSIL HYDRO	33.33	33.33	33.33	33.33	22.22	MLR
RAW EDGE INDUSTRIAL SOLUTIONS LTD	0.00	0.00	0.00	0.00	0.00	LLR
FERRO ALLOYS	44.44	44.44	33.33	55.56	44.44	MLR
SHYAM CENTURIES	0.00	11.11	11.11	11.11	22.22	PLR
SHIRPUR GOLD	0.00	11.11	0.00	0.00	0.00	LLR
NAGPUR POWER	0.00	0.00	0.00	0.00	0.00	LLR
FACOR ALLOYS	0.00	0.00	0.00	0.00	0.00	LLR
IMPEX FERRO TECH	0.00	0.00	0.00	0.00	0.00	LLR
NMDC	88.89	66.67	66.67	66.67	55.56	HLR
HINDUSTAN ZINC LTD	77.78	77.78	77.78	77.78	77.78	HLR
CEETA INDUSTRIES LTD	0.00	0.00	0.00	0.00	0.00	LLR
GLITTEK GRANITE LTD	0.00	0.00	0.00	0.00	0.00	LLR
KIOCL LTD	33.33	55.56	33.33	44.44	66.67	HLR
FOUNDRY FUEL PRODUCTS LTD	0.00	0.00	0.00	0.00	0.00	LLR
ASI INDUSTRIES LTD	55.56	66.67	66.67	66.67	66.67	HLR
PACIFIC INDUSTRIES LTD	22.22	11.11	11.11	11.11	0.00	LLR
ORIENTAL TRIMAX LTD	0.00	0.00	0.00	0.00	0.00	LLR
Average CSRI	29.12	30.27	27.20	31.03	30.65	29.66

Source: Authors own computation

Note: HLR: high level of responsiveness to CSR implementation, PLR: progressive level of responsiveness to CSR implementation, MLR: moderate level of responsiveness to CSR implementation, LLR: low level of responsiveness to CSR implementation.

Here Red: below Average CSRI; Green: above Average CSRI

health (Table 11) was made by National Mineral Development Corporation (NMDC) followed by Coal India Ltd. for the year 2015–2016. National Mineral Development Corporation (NMDC) followed by Coal India Ltd. and Guj Minerals spent the highest amount in the field of drinking water and sanitation (Table 12). The highest amount of expenditure in the field of empowerment was done by National Mineral Development Corporation (NMDC), Coal India Ltd., Guj Minerals, MOIL Ltd. and 20 Microns (Table 13). National Mineral Development Corporation (NMDC) spent the highest amount in the field of livelihood (Table 14). Guj Minerals and Kudremukh Iron Ore Company (KIOCL) spent the highest amount in the field of disaster relief (Table 15). Amongst the companies the highest amount spent in the field of rural upliftment (Table 16) was by National Mineral Development Corporation (NMDC) followed by Orissa Minerals Development Company Ltd. (OMDC) for the year 2017–2018. National Mineral Development Corporation (NMDC)

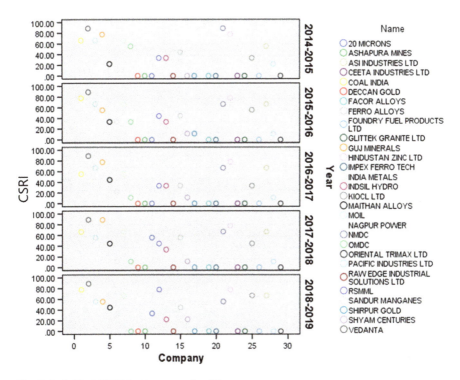

Fig. 3 Individual CSRI for the companies (%)

Table 6 Overall CSR expenditure (in lakh) amongst select companies for the time period 2014–2015 to 2018–2019

Year	CSR (mandatory 2% expenditure)	Budget amount	Actual amount spent	Spend % of 2% of CSR	Spend % of budget
2014–2015	1485.749310	2155.843352	1257.338016	85%	58%
2015–2016	1464.679655	2321.475095	1349.014750	92%	58%
2016–2017	1310.927586	1516.446810	1138.767845	87%	75%
2017–2018	1576.111379	2051.557535	1175.630431	75%	57%
2018–2019	1173.593793	2147.954310	1209.113586	103%	56%
Average	1402.212345	2038.655420	1225.972926	87%	60%

Source: Authors own computation

followed by Coal India Ltd. and Hindustan Zinc Ltd. spent the highest amount in the field which has been marked others" (Table 17).

Table 7 Individual companies overall CSR expenditure (lakhs) for the time period 2014–2015 to 2018–2019

	Actual amount spent(lakhs)					
	2014-2015	2015-2016	2016-2017	2017-2018	2018-2019	Overall
COAL INDIA	2462.32	7322.53	4568.36	1141.37	2733.12	3645.54
VEDANTA	2428.66	1675.21	48.48	43.86	48.62	848.97
MOIL	1357.57	1447.39	1143.10	961.63	480.25	1077.99
GUJ MINERALS	4761.87	3.36	1067.00	2121.16	943.87	1779.45
MAITHAN ALLOYS	50.00	100.59	127.13	281.91	489.22	209.77
SANDUR MANGANES	102.91	97.14	69.79	76.85	176.60	104.66
INDIA METALS	405.00	626.00	545.00	395.00	473.00	488.80
OMDC	32.95	72.81	2919.00	2937.89	0.00	1192.53
DECCAN GOLD	0.00	0.00	0.00	0.00	0.00	0.00
ASHAPURA MINES	0.00	0.00	0.00	0.00	0.00	0.00
20 MICRONS	0.00	0.00	0.00	20.20	41.38	12.32
RSMML	176.00	228.18	360.90	190.18	254.01	241.85
INDSIL HYDRO	8.96	4.80	4.09	8.41	2.00	5.65
RAW EDGE INDUSTRIAL SOLUTIONS LTD	0.00	0.00	0.00	0.00	0.00	0.00
FERRO ALLOYS	53.13	72.87	44.54	39.44	49.32	51.86
SHYAM CENTURIES	0.00	6.00	7.00	15.00	17.52	9.10
SHIRPUR GOLD	0.00	0.00	0.00	0.00	0.00	0.00
NAGPUR POWER	0.00	0.00	0.00	0.00	0.00	0.00
FACOR ALLOYS	0.00	0.00	0.00	0.00	0.00	0.00
IMPEX FERRO TECH	0.00	0.00	0.00	0.00	0.00	0.00
NMDC	18864.72	21009.04	17418.03	16937.03	16724.13	18190.59
HINDUSTAN ZINC LTD	5667.00	6325.00	4609.00	8843.00	12562.00	7601.20
CEETA INDUSTRIES LTD	0.00	0.00	0.00	0.00	0.00	0.00
GLITTEK GRANITE LTD	0.00	0.00	0.00	0.00	0.00	0.00
KIOCL LTD	53.02	64.00	38.19	31.96	32.51	43.94
FOUNDRY FUEL PRODUCTS LTD	0.00	0.00	0.00	0.00	0.00	0.00
ASI INDUSTRIES LTD	25.00	45.01	44.04	44.67	36.75	39.09
PACIFIC INDUSTRIES LTD	13.70	21.50	10.62	3.72	0.00	9.91
ORIENTAL TRIMAX LTD	0.00	0.00	0.00	0.00	0.00	0.00
Average CSR expenditure	**1257.34**	**1349.01**	**1138.77**	**1175.63**	**1209.11**	**1225.97**

Source: Authors own computation
Note: Red: below the average CSR expenditure, Green: above Average CSR expenditure

Table 8 Detailed assessment of CSR expenditure (in lakhs) activity wise amongst selected Indian companies belonging to the mining sector for the time period 2014–2015 to 2018–2019

	Year					
	2014-2015	2015-2016	2016-2017	2017-2018	2018-2019	Total
Education	441.15	295.87	333.99	393.67	244.03	341.74
Health	244.20	449.50	173.86	135.64	126.49	225.94
Environment	23.19	33.25	104.94	102.73	42.60	61.34
Drinking Water And Sanitation	34.72	5.58	25.89	92.42	16.40	35.00
Empowerment	160.39	69.95	16.81	203.24	42.44	98.57
Livelihood	32.79	10.41	10.97	42.82	56.39	30.68
Disaster Relief	0.00	0.00	0.00	34.86	0.14	7.00
Rural Upliftment	394.14	354.66	270.11	4480.59	235.27	1146.96
Others	75.94	120.64	83.79	158.72	117.17	111.25

Source: Authors own computation
Note: Greenest is Highest/Most Red is lowest

6 Observation and Conclusion

This is an exploratory study to understand the corporate social responsibility activities undertaken by the select companies and the trends in CSR investments (expenditures) with respect to such activities after the legal mandate. We assembled a detailed data set comprising of 29 listed companies belonging to the mining sector, considering data for 5 years, and with all firm year data available. The focus of this study was to analyze and compare the trends in CSR investments for companies in the mining sector in India for the time period from 2014–2015 to 2018–2019. The reason for choosing the mining sector is due to the fact that mining activities in emerging economies have tremendously affected the principal elements of the environment (land, water and air) culminating into serious consequences for the health of indigenes.

The CSR journey in India took a turn when the Ministry of Corporate Affairs introduced section 135 to the Companies Act of 2013. India became the first country to include provisions of CSR in Company Law and make CSR expenditure mandatory for corporates based on pre-specified criteria. The distinguishing feature of Section 135 is that it not only makes the reporting of CSR activities mandatory, but goes a step further to mandate CSR activities in the first place. The companies are required to spend at least 2% of the average net profits made during the three immediately preceding financial years on CSR activities specified in Schedule 7 of the Act. We have categorized the activities specified in Schedule 7 of the Act into 9 broad areas—education, environment, disaster relief, health, rural upliftment, empowerment and skill development, livelihood, drinking water and sanitation and others (sports, art and culture).

The study revealed that over a span of 5 years (2014–2015 to 2018–2019), 51.03% of the total number of companies in this sector had invested in the field of education. Only 27.59% of the total number of companies in this sector had invested in the field of environment which was disturbing despite the fact that this sector has a higher environmental footprint and impact than other sectors. Despite governmental regulations'and pressure from other stakeholders majority of the companies failed to emphasis concerning areas like energy, climate change, marine and terrestrial ecosystem. Focusing on areas of human education, health, and livelihood generation is important but with a population of over a billion people, corporations also need to focus on reducing the exploitation of natural resources more vigorously. For all the years we also observe that only 2.76% of the companies had taken up projects involving disaster relief measures.

The study also revealed an average CSRI of 29.66% which happened to be extremely low. 38.48% of the companies chosen for our analysis, proved to be highly responsive to CSR implementation, 17.24% of the companies chosen for our analysis showed moderate level of responsiveness to CSR implementation, 6.9% of the companies chosen for our analysis proved to be progressive in nature with respect to CSR implementation and 41.37% of the companies chosen for our analysis, proved to be having an extremely low level of responsiveness to CSR implementation. Average CSRI happened to be extremely low since out of 29 companies, 10 (34.48%) failed to have any CSR related projects during the time period under consideration. This was because of the fact that these companies due to continuous losses were unable to spend at least 2% of the average net profits of the company made during the three immediately preceding financial years on CSR activities as recommended by the Companies Act 2013, Section 135. This leads us to question ourselves whether the spirit behind the concept of CSR has just got reduced to mere legal compliances. These companies may not have been in a position to spend the required amount as per the legal mandate but they could have at least shown their responsiveness towards the society by investing in any of the areas specified by the Act to whatever extent possible for them. A good example on this aspect would be Orissa Minerals Development Company ltd (OMDC) who due to losses could not spend the required amount as per the legal mandate in the year 2017–2018 but on their own accord spend 2937. 89 lakhs on education, health, empowerment, skill development and sanitation projects.

A detailed assessment of the overall CSR expenditure made by the selected Indian companies belonging to the mining sector for the time period 2014–2015 to 2018–2019 showed that the companies have not spent the mandated amounts excepting in the year 2018–2019.This may be because of two reasons—some of the companies due to continuous losses were unable to spend at least 2% of the average net profits of the company made during the three immediately preceding financial years on CSR activities as recommended by the Companies Act 2013, Section 135 so they decided not to undertake any CSR projects where as some companies like 20 microns did have the required profits but they were undecided as to which CSR project they should invest in. This highlights that the legal system needs to be further tightened and regulated. This finding also indicates that the companies may be lacking focus or proper planning, as the amounts allocated to these activities was not being spent to the optimal capacity.

It was interesting to note that CSR expenditure was noticeably high only amongst the major companies of the mining sector—National Mineral Development Corporation (NMDC), Hindustan Zinc Ltd., Coal India Ltd., Guj Minerals, MOIL Ltd., Kudremukh Iron Ore Company (KIOCL), and Orissa Minerals Development Company Ltd. (OMDC). Barring Hindustan Zinc Ltd. all others are government companies either state owned or union owned. Barring Kudremukh Iron Ore Company (KIOCL) which was incorporated in 1990, others have a long legacy dating from 1918 (refer to Table 1). Lower performing CSR operators were public incorporated companies and apart from Ashapura Mines and Foundry Fuel Products Ltd. which were incorporated in 1960 and 1964 respectively others were fairly young with respect to its incorporation. Besides better performing CSR operators were those companies with consistent profit during the period under consideration where as lower performing CSR operators either had inconsistent profits over the years under consideration or negligible profits or no profits at all as a result of which they failed to meet that 2% mandatory spending on CSR .

Detailed assessment of CSR expenditure activity wise amongst selected Indian companies belonging to the mining sector for the time period 2014–2015 to 2018–2019 showed that CSR expenditure was the highest in the field of education, followed by health and rural upliftment . Since there is evidence of land disposition, displacement and resettlement of local communities whereby the rural population is denied their basic access to land resulting in negative impact on livelihood and consequent food insecurity, the companies should have focused on areas such as livelihood projects, empowerment projects, skill development projects which could have improved the standard of living of the people. The companies have also failed to address issues such as climate change, conserving flora and fauna, marine life, and sustainable consumption and production since they have not undertaken any such projects. The Swachh Bharat Abhiyan that came into being on October 2, 2014, took up the agenda to achieve Swachh Bharat (Clean India) by 2019, got a lot of corporate attention and funding in general. The primary agenda of this mission is to provide clean drinking water, build toilets, and promote proper sanitation in India. But study of the mining sector revealed that companies have spent negligible amount of 35 lakhs on an average during 2014–2015 to 2018–2019 on such an area. Thus strict monitoring and accountability of the projects undertaken in the name of CSR has to be initiated by the government along with proper encouragement so that the companies become socially responsive.

Our study was on corporate social responsibility practices and its implementation after the legal mandate in India with special emphasis on the mining sector. This study would help future researchers or practitioners to understand corporate social responsibility practices and its implementation with a special focus on the mining sector. This study would encourage future researchers to carry out empirical work on such areas since we are far behind in terms of research on such an area as discussed under literature review.

Annexure 1

Format for the Annual Report on CSR Activities to be Included in the Board's Report

1. A brief outline of the company's CSR policy, including overview of projects or programs proposed to be undertaken and a reference to the web-link to the CSR policy and projects or programs.
2. The Composition of the CSR Committee.
3. Average net profit of the company for last three financial years.
4. Prescribed CSR Expenditure.
5. Details of CSR spent during the financial year:

    (a) Total amount to be spent for the financial year;
    (b) Amount unspent, if any;
    (c) Manner in which the amount spent during the financial year is detailed below:

1	2	3	4	5	6	7	8
Serial No	CSR Project or activity identified	Sector in which the activity is covered	Projects or programmes local area or specify the state or district where project or programme was undertaken	Amount Outlay	Amount spent on the project or programme	Cumulative expenditure up to the reporting period	Amount spent direct or through implementing agency
1							
2							
3							
	TOTAL						

[a]Give details of implementing agency

6. In case the company has failed to spend the 2% of the average net profit of the last three financial years or any part thereof, the company shall provide the reasons for not spending the amount in its Board report
7. A responsibility statement of the CSR Committee that the implementation and monitoring of CSR Policy, is in compliance with CSR objectives and Policy of the company

Annexure 2

Schedule VII

Activities specified in Schedule VII to the 2013 Act specifies activities which companies may include in their CSR Policies. These are activities relating to:

- Eradicating hunger, poverty and malnutrition—Promoting health care including preventive healthcare and sanitation including contribution to the Swatch Bharat Kosh set up by the Government for promotion of sanitation—Making available safe drinking water
- Promoting education, including special education and employment enhancing vocation skills especially among children, women, elderly and the differently abled—Livelihood enhancement projects
- Promoting gender equality—Empowering women—Setting up homes and hostels for women and orphans—Setting up old age homes, day care centres and such other facilities for senior citizens—Measures for reducing inequalities faced by socially and economically backward groups
- Ensuring environmental sustainability—Ecological balance—Protection of flora and fauna—Animal welfare—Agro forestry—Conservation of natural resources—Maintaining quality of soil, air and water including contribution to the Clean Ganga Fund set up by the Central Government for rejuvenation of river Ganga.
- Protection of national heritage, art and culture including restoration of buildings and sites of historical importance and works of art—Setting up of public libraries—Promotion and development of traditional arts and handicrafts
- Measures for the benefit of armed forces veterans, war widows and their dependants

- Training to promote—Rural sports—Nationally recognized sports—Paralympic sports—Olympic sports
- Contribution to—The Prime Minister's National Relief Fund or—Any other fund set up by the Central Government for socioeconomic development and relief and welfare of SC/ST/OBC, minorities and women
- Contributions or funds provided to technology incubators located within academic institutions which are approved by Central Govt.
- Rural development projects.
- Slum area development -the term slum area shall mean any area declared as such by the Central Government or any State Government or any other competent authority under any law for the time being in force

Annexure 3

Table 9 Detailed assessment of CSR expenditure on education by the selected companies for the time period 2014–2015 to 2018–2019 (in lakhs)

Company	Education					
	2014-2015	2015-2016	2016-2017	2017-2018	2018-2019	Overall
COAL INDIA	0.27	593.80	21.89	256.35	51.08	923.39
VEDANTA	616.78	690.83	5.71	9.71	10.50	1333.53
MOIL	437.42	159.73	0.00	0.00	359.12	956.27
GUJ MINERALS	419.06	0.32	137.64	710.35	0.00	1267.37
MAITHAN ALLOYS	50.00	25.00	115.11	271.00	399.00	860.11
SANDUR MANGANES	102.91	48.95	61.51	76.85	115.00	405.22
INDIA METALS	276.00	503.00	446.00	232.00	270.00	1727.00
OMDC	0.00	0.00	0.00	0.00	0.00	0.00
DECCAN GOLD	0.00	0.00	0.00	0.00	0.00	0.00
ASHAPURA MINES	0.00	0.00	0.00	0.00	0.00	0.00
20 MICRONS	0.00	0.00	0.00	1.28	35.00	36.28
RSMML	36.00	71.75	90.52	24.12	89.28	311.67
INDSIL HYDRO	0.50	0.60	1.29	0.41	0.75	3.55
RAW EDGE INDUSTRIAL SOLUTIONS LTD	0.00	0.00	0.00	0.00	0.00	0.00
FERRO ALLOYS	6.00	12.19	0.75	5.34	10.71	34.99
SHYAM CENTURIES	0.00	6.00	7.00	15.00	7.50	35.50
SHIRPUR GOLD	0.00	5.00	0.00	0.00	0.00	5.00
NAGPUR POWER	0.00	0.00	0.00	0.00	0.00	0.00
FACOR ALLOYS	0.00	0.00	0.00	0.00	0.00	0.00
IMPEX FERRO TECH	0.00	0.00	0.00	0.00	0.00	0.00
NMDC	9040.99	5244.35	7044.60	6448.97	572.26	28351.17
HINDUSTAN ZINC LTD	1795.00	1180.00	1730.00	3350.00	5140.00	13195.00
CEETA INDUSTRIES LTD	0.00	0.00	0.00	0.00	0.00	0.00
GLITTEK GRANITE LTD	0.00	0.00	0.00	0.00	0.00	0.00
KIOCL LTD	1.99	11.50	14.04	1.00	15.00	43.53
FOUNDRY FUEL PRODUCTS LTD	0.00	0.00	0.00	0.00	0.00	0.00
ASI INDUSTRIES LTD	10.53	27.33	9.52	14.07	1.57	63.02
PACIFIC INDUSTRIES LTD	0.00	0.00	0.00	0.00	0.00	0.00
ORIENTAL TRIMAX LTD	0.00	0.00	0.00	0.00	0.00	0.00
TOTAL	12793.45	8580.35	9685.58	11416.45	7076.77	49552.60

Source: Authors own calculation
Note: Green – highest expenditure, Red – lowest expenditure

Table 10 Detailed assessment of CSR expenditure on Health by the selected companies for the time period 2014–2015 to 2018–2019 (in lakhs)

Company	Health					
	2014-2015	2015-2016	2016-2017	2017-2018	2018-2019	Overall
COAL INDIA	1958.84	2764.48	716.13	128.47	1320.67	6888.59
VEDANTA	635.68	338.14	7.86	4.36	12.52	998.56
MOIL	47.98	85.04	78.72	0.00	22.62	234.36
GUJ MINERALS	165.15	0.00	0.00	0.00	0.00	165.15
MAITHAN ALLOYS	0.00	75.24	11.26	5.22	70.00	161.72
SANDUR MANGANES	0.00	25.34	8.28	0.00	0.00	33.62
INDIA METALS	62.00	82.00	56.00	132.00	157.00	489.00
OMDC	5.34	61.27	0.00	0.00	0.00	66.61
DECCAN GOLD	0.00	0.00	0.00	0.00	0.00	0.00
ASHAPURA MINES	0.00	0.00	0.00	0.00	0.00	0.00
20 MICRONS	0.00	0.00	0.00	13.57	6.28	19.85
RSMML	115.00	90.80	106.40	85.55	65.00	462.75
INDSIL HYDRO	3.66	3.09	2.30	7.00	11.16	27.21
RAW EDGE INDUSTRIAL SOLUTIONS LTD	0.00	0.00	0.00	0.00	0.00	0.00
FERRO ALLOYS	15.66	22.09	15.42	31.03	34.86	119.06
SHYAM CENTURIES	0.00	0.00	0.00	0.00	0.00	0.00
SHIRPUR GOLD	0.00	0.00	0.00	0.00	0.00	0.00
NAGPUR POWER	0.00	0.00	0.00	0.00	0.00	0.00
FACOR ALLOYS	0.00	0.00	0.00	0.00	0.00	0.00
IMPEX FERRO TECH	0.00	0.00	0.00	0.00	0.00	0.00
NMDC	2656.79	7703.53	2610.84	1858.01	0.00	14829.17
HINDUSTAN ZINC LTD	1375.00	1751.00	1402.00	1652.00	1962.00	8142.00
CEETA INDUSTRIES LTD	0.00	0.00	0.00	0.00	0.00	0.00
GLITTEK GRANITE LTD	0.00	0.00	0.00	0.00	0.00	0.00
KIOCL LTD	22.44	8.15	2.25	0.98	1.20	35.02
FOUNDRY FUEL PRODUCTS LTD	0.00	0.00	0.00	0.00	0.00	0.00
ASI INDUSTRIES LTD	7.76	3.87	13.88	11.68	5.04	42.23
PACIFIC INDUSTRIES LTD	10.59	21.50	10.62	3.72	0.00	46.43
ORIENTAL TRIMAX LTD	0.00	0.00	0.00	0.00	0.00	0.00
TOTAL	**7081.89**	**13035.54**	**5041.96**	**3933.59**	**3668.34**	**32761.33**

Source: Authors own calculation

Note: Green – highest expenditure, Red – lowest expenditure

Table 11 Detailed assessment of CSR expenditure on environment by the selected companies for the time period 2014–2015 to 2018–2019 (in lakhs)

Company	Environment					
	2014-2015	2015-2016	2016-2017	2017-2018	2018-2019	Overall
COAL INDIA	0.00	531.41	188.68	0.00	26.75	746.84
VEDANTA	49.84	22.18	0.22	0.21	0.55	73.00
MOIL	36.21	13.36	19.56	15.31	0.99	85.43
GUJ MINERALS	30.31	0.00	1.27	18.10	445.64	495.32
MAITHAN ALLOYS	0.00	0.00	0.00	0.00	0.00	0.00
SANDUR MANGANES	0.00	22.85	0.00	0.00	0.00	22.85
INDIA METALS	20.00	2.00	0.00	0.00	0.00	22.00
OMDC	0.00	0.00	0.00	0.00	0.00	0.00
DECCAN GOLD	0.00	0.00	0.00	0.00	0.00	0.00
ASHAPURA MINES	0.00	0.00	0.00	0.00	0.00	0.00
20 MICRONS	0.00	0.00	0.00	0.00	0.00	0.00
RSMML	0.00	0.00	0.00	5.00	1.00	6.00
INDSIL HYDRO	0.00	0.00	0.00	0.00	0.00	0.00
RAW EDGE INDUSTRIAL SOLUTIONS LTD	0.00	0.00	0.00	0.00	0.00	0.00
FERRO ALLOYS	30.21	30.14	28.37	1.09	2.55	92.36
SHYAM CENTURIES	0.00	0.00	0.00	0.00	0.00	0.00
SHIRPUR GOLD	0.00	0.00	0.00	0.00	0.00	0.00
NAGPUR POWER	0.00	0.00	0.00	0.00	0.00	0.00
FACOR ALLOYS	0.00	0.00	0.00	0.00	0.00	0.00
IMPEX FERRO TECH	0.00	0.00	0.00	0.00	0.00	0.00
NMDC	282.96	103.38	2645.02	2790.65	500.00	6322.01
HINDUSTAN ZINC LTD	223.00	234.00	160.00	148.00	258.00	1023.00
CEETA INDUSTRIES LTD	0.00	0.00	0.00	0.00	0.00	0.00
GLITTEK GRANITE LTD	0.00	0.00	0.00	0.00	0.00	0.00
KIOCL LTD	0.00	5.00	0.00	0.00	0.00	5.00
FOUNDRY FUEL PRODUCTS LTD	0.00	0.00	0.00	0.00	0.00	0.00
ASI INDUSTRIES LTD	0.00	0.00	0.00	0.94	0.04	0.98
PACIFIC INDUSTRIES LTD	0.00	0.00	0.00	0.00	0.00	0.00
ORIENTAL TRIMAX LTD	0.00	0.00	0.00	0.00	0.00	0.00
TOTAL	672.53	964.32	3043.12	2979.30	1235.52	8894.79

Source: Authors own calculation
Note: Green – highest expenditure, Red – lowest expenditure

Table 12 Detailed assessment of CSR expenditure on Drinking Water and Sanitation by the selected companies for the time period 2014–2015 to 2018–2019 (in lakhs)

Company	Drinking Water And Sanitation					
	2014-2015	2015-2016	2016-2017	2017-2018	2018-2019	Overall
COAL INDIA	18.04	56.53	0.00	2407.72	198.12	2680.41
VEDANTA	30.47	22.62	8.30	7.27	5.58	74.24
MOIL	41.72	29.44	37.12	101.00	0.00	209.28
GUJ MINERALS	88.23	1.26	673.16	139.00	205.37	1107.02
MAITHAN ALLOYS	0.00	0.00	0.00	0.00	0.00	0.00
SANDUR MANGANES	0.00	0.00	0.00	0.00	51.60	51.60
INDIA METALS	0.00	0.00	0.00	0.00	0.00	0.00
OMDC	13.77	11.47	0.00	0.00	0.00	25.25
DECCAN GOLD	0.00	0.00	0.00	0.00	0.00	0.00
ASHAPURA MINES	0.00	0.00	0.00	0.00	0.00	0.00
20 MICRONS	0.00	0.00	0.00	0.00	0.00	0.00
RSMML	0.00	0.00	0.00	0.00	0.40	0.40
INDSIL HYDRO	4.79	1.10	0.00	0.00	0.00	5.89
RAW EDGE INDUSTRIAL SOLUTIONS LTD	0.00	0.00	0.00	0.00	0.00	0.00
FERRO ALLOYS	0.00	0.00	0.00	0.00	0.00	0.00
SHYAM CENTURIES	0.00	0.00	0.00	0.00	0.00	0.00
SHIRPUR GOLD	0.00	0.00	0.00	0.00	0.00	0.00
NAGPUR POWER	0.00	0.00	0.00	0.00	0.00	0.00
FACOR ALLOYS	0.00	0.00	0.00	0.00	0.00	0.00
IMPEX FERRO TECH	0.00	0.00	0.00	0.00	0.00	0.00
NMDC	769.93	0.00	0.00	0.00	0.00	769.93
HINDUSTAN ZINC LTD	0.00	0.00	0.00	0.00	0.00	0.00
CEETA INDUSTRIES LTD	0.00	0.00	0.00	0.00	0.00	0.00
GLITTEK GRANITE LTD	0.00	0.00	0.00	0.00	0.00	0.00
KIOCL LTD	34.59	29.79	21.90	11.50	0.47	98.25
FOUNDRY FUEL PRODUCTS LTD	0.00	0.00	0.00	0.00	0.00	0.00
ASI INDUSTRIES LTD	5.35	9.58	10.47	13.56	14.16	53.12
PACIFIC INDUSTRIES LTD	0.00	0.00	0.00	0.00	0.00	0.00
ORIENTAL TRIMAX LTD	0.00	0.00	0.00	0.00	0.00	0.00
TOTAL	1006.89	161.79	750.95	2680.05	475.70	5075.38

Source: Authors own calculation

Note: Green – highest expenditure, Red – lowest expenditure

Table 13 Detailed assessment of CSR expenditure on empowerment by the selected companies for the time period 2014–2015 to 2018–2019 (in lakhs)

Company	Empowerment					
	2014-2015	2015-2016	2016-2017	2017-2018	2018-2019	Overall
COAL INDIA	839.42	1691.64	0.00	21.60	384.87	2937.53
VEDANTA	30.00	90.41	5.65	3.45	4.97	134.48
MOIL	0.00	0.00	166.53	309.95	0.00	476.48
GUJ MINERALS	3426.27	0.36	0.70	62.87	123.62	3613.81
MAITHAN ALLOYS	50.00	0.00	0.00	0.00	0.00	50.00
SANDUR MANGANES	0.00	0.00	0.00	0.00	0.00	0.00
INDIA METALS	1.00	1.00	26.00	1.00	2.00	31.00
OMDC	0.10	0.00	0.00	0.00	0.00	0.10
DECCAN GOLD	0.00	0.00	0.00	0.00	0.00	0.00
ASHAPURA MINES	0.00	0.00	0.00	0.00	0.00	0.00
20 MICRONS	0.00	0.00	0.00	5000.00	0.00	5000.00
RSMML	0.00	0.00	0.00	0.00	20.20	20.20
INDSIL HYDRO	0.00	0.00	0.00	0.00	0.00	0.00
RAW EDGE INDUSTRIAL SOLUTIONS LTD	0.00	0.00	0.00	0.00	0.00	0.00
FERRO ALLOYS	1.26	8.45	0.00	0.00	0.00	9.71
SHYAM CENTURIES	0.00	0.00	0.00	0.00	0.00	0.00
SHIRPUR GOLD	0.00	0.00	0.00	0.00	0.00	0.00
NAGPUR POWER	0.00	0.00	0.00	0.00	0.00	0.00
FACOR ALLOYS	0.00	0.00	0.00	0.00	0.00	0.00
IMPEX FERRO TECH	0.00	0.00	0.00	0.00	0.00	0.00
NMDC	188.93	137.74	135.70	137.99	155.00	755.36
HINDUSTAN ZINC LTD	114.00	97.00	152.00	357.00	540.00	1260.00
CEETA INDUSTRIES LTD	0.00	0.00	0.00	0.00	0.00	0.00
GLITTEK GRANITE LTD	0.00	0.00	0.00	0.00	0.00	0.00
KIOCL LTD	0.00	0.00	0.00	0.00	0.00	0.00
FOUNDRY FUEL PRODUCTS LTD	0.00	0.00	0.00	0.00	0.00	0.00
ASI INDUSTRIES LTD	0.30	2.00	1.00	0.00	0.00	3.30
PACIFIC INDUSTRIES LTD	0.00	0.00	0.00	0.00	0.00	0.00
ORIENTAL TRIMAX LTD	0.00	0.00	0.00	0.00	0.00	0.00
TOTAL	4651.28	2028.60	487.58	5893.86	1230.66	14291.98

Source: Authors own calculation
Note: Green – highest expenditure, Red – lowest expenditure

Table 14 Detailed assessment of CSR expenditure on livilihood by the selected companies for the time period 2014–2015 to 2018–2019 (in lakhs)

Company	Livelihood					
	2014-2015	2015-2016	2016-2017	2017-2018	2018-2019	Overall
COAL INDIA	0.00	0.00	0.00	0.00	0.00	0.00
VEDANTA	356.49	82.94	3.05	4.53	3.44	450.45
MOIL	0.00	0.00	0.00	0.00	0.00	0.00
GUJ MINERALS	0.00	0.87	94.00	38.20	0.00	133.07
MAITHAN ALLOYS	0.00	0.00	0.00	0.00	0.00	0.00
SANDUR MANGANES	0.00	0.00	0.00	0.00	0.00	0.00
INDIA METALS	0.00	0.00	0.00	0.00	0.00	0.00
OMDC	0.00	0.00	0.00	0.00	0.00	0.00
DECCAN GOLD	0.00	0.00	0.00	0.00	0.00	0.00
ASHAPURA MINES	0.00	0.00	0.00	0.00	0.00	0.00
20 MICRONS	0.00	0.00	0.00	0.00	0.00	0.00
RSMML	0.00	0.00	0.00	0.00	0.00	0.00
INDSIL HYDRO	0.00	0.00	0.00	0.00	0.00	0.00
RAW EDGE INDUSTRIAL SOLUTIONS LTD	0.00	0.00	0.00	0.00	0.00	0.00
FERRO ALLOYS	0.00	0.00	0.00	0.00	0.00	0.00
SHYAM CENTURIES	0.00	0.00	0.00	0.00	0.00	0.00
SHIRPUR GOLD	0.00	0.00	0.00	0.00	0.00	0.00
NAGPUR POWER	0.00	0.00	0.00	0.00	0.00	0.00
FACOR ALLOYS	0.00	0.00	0.00	0.00	0.00	0.00
IMPEX FERRO TECH	0.00	0.00	0.00	0.00	0.00	0.00
NMDC	158.41	0.00	0.00	0.00	0.00	158.41
HINDUSTAN ZINC LTD	436.00	218.00	221.00	1199.00	1632.00	3706.00
CEETA INDUSTRIES LTD	0.00	0.00	0.00	0.00	0.00	0.00
GLITTEK GRANITE LTD	0.00	0.00	0.00	0.00	0.00	0.00
KIOCL LTD	0.00	0.00	0.00	0.00	0.00	0.00
FOUNDRY FUEL PRODUCTS LTD	0.00	0.00	0.00	0.00	0.00	0.00
ASI INDUSTRIES LTD	0.00	0.00	0.00	0.00	0.00	0.00
PACIFIC INDUSTRIES LTD	0.00	0.00	0.00	0.00	0.00	0.00
ORIENTAL TRIMAX LTD	0.00	0.00	0.00	0.00	0.00	0.00
TOTAL	950.90	301.81	318.05	1241.73	1635.44	4447.93

Source: Authors own calculation
Note: Green – highest expenditure, Red – lowest expenditure

Table 15 Detailed assessment of CSR expenditure on disaster relief by the selected companies for the time period 2014–2015 to 2018–2019 (in lakhs)

Company	Disaster Relief					
	2014-2015	2015-2016	2016-2017	2017-2018	2018-2019	Overall
COAL INDIA	0.00	0.00	0.00	0.00	0.00	0.00
VEDANTA	0.00	0.00	0.00	0.00	0.00	0.00
MOIL	0.00	0.00	0.00	0.00	0.00	0.00
GUJ MINERALS	0.00	0.00	0.00	1000.00	0.00	1000.00
MAITHAN ALLOYS	0.00	0.00	0.00	0.00	0.00	0.00
SANDUR MANGANES	0.00	0.00	0.00	0.00	0.00	0.00
INDIA METALS	0.00	0.00	0.00	11.00	1.00	12.00
OMDC	0.00	0.00	0.00	0.00	0.00	0.00
DECCAN GOLD	0.00	0.00	0.00	0.00	0.00	0.00
ASHAPURA MINES	0.00	0.00	0.00	0.00	0.00	0.00
20 MICRONS	0.00	0.00	0.00	0.00	0.00	0.00
RSMML	0.00	0.00	0.00	0.00	0.00	0.00
INDSIL HYDRO	0.00	0.00	0.00	0.00	0.00	0.00
RAW EDGE INDUSTRIAL SOLUTIONS LTD	0.00	0.00	0.00	0.00	0.00	0.00
FERRO ALLOYS	0.00	0.00	0.00	0.00	0.00	0.00
SHYAM CENTURIES	0.00	0.00	0.00	0.00	0.00	0.00
SHIRPUR GOLD	0.00	0.00	0.00	0.00	0.00	0.00
NAGPUR POWER	0.00	0.00	0.00	0.00	0.00	0.00
FACOR ALLOYS	0.00	0.00	0.00	0.00	0.00	0.00
IMPEX FERRO TECH	0.00	0.00	0.00	0.00	0.00	0.00
NMDC	0.00	0.00	0.00	0.00	0.00	0.00
HINDUSTAN ZINC LTD	0.00	0.00	0.00	0.00	0.00	0.00
CEETA INDUSTRIES LTD	0.00	0.00	0.00	0.00	0.00	0.00
GLITTEK GRANITE LTD	0.00	0.00	0.00	0.00	0.00	0.00
KIOCL LTD	0.00	0.00	0.00	0.00	2.94	2.94
FOUNDRY FUEL PRODUCTS LTD	0.00	0.00	0.00	0.00	0.00	0.00
ASI INDUSTRIES LTD	0.00	0.00	0.00	0.00	0.00	0.00
PACIFIC INDUSTRIES LTD	0.00	0.00	0.00	0.00	0.00	0.00
ORIENTAL TRIMAX LTD	0.00	0.00	0.00	0.00	0.00	0.00
TOTAL	0.00	0.00	0.00	1011.00	3.94	1014.94

Source: Authors own calculation

Note: Green – highest expenditure, Red – lowest expenditure

Table 16 Detailed assessment of CSR expenditure on rural upliftment by the selected companies for the time period 2014–2015 to 2018–2019 (in lakhs)

Company	Rural Upliftment					
	2014-2015	2015-2016	2016-2017	2017-2018	2018-2019	Overall
COAL INDIA	44.99	488.34	199.38	269.11	709.35	1711.17
VEDANTA	591.26	426.56	10.05	4.87	5.82	1038.56
MOIL	4232.84	381.82	556.17	384.92	97.52	5653.27
GUJ MINERALS	561.63	0.05	45.29	57.28	50.17	714.42
MAITHAN ALLOYS	0.00	0.35	0.25	0.54	0.22	1.36
SANDUR MANGANES	0.00	0.00	0.00	0.00	0.00	0.00
INDIA METALS	17.00	9.00	2.00	3.00	40.00	71.00
OMDC	9.04	0.07	2918.00	93789.00	0.00	96716.11
DECCAN GOLD	0.00	0.00	0.00	0.00	0.00	0.00
ASHAPURA MINES	0.00	0.00	0.00	0.00	0.00	0.00
20 MICRONS	0.00	0.00	0.00	30501.00	0.00	30501.00
RSMML	25.00	59.02	163.98	78.00	48.13	374.13
INDSIL HYDRO	0.00	0.00	0.50	1.00	0.00	1.50
RAW EDGE INDUSTRIAL SOLUTIONS LTD	0.00	0.00	0.00	0.00	0.00	0.00
FERRO ALLOYS	0.00	0.00	0.00	1.48	1.20	2.68
SHYAM CENTURIES	0.00	0.00	0.00	0.00	0.00	0.00
SHIRPUR GOLD	0.00	0.00	0.00	0.00	0.00	0.00
NAGPUR POWER	0.00	0.00	0.00	0.00	0.00	0.00
FACOR ALLOYS	0.00	0.00	0.00	0.00	0.00	0.00
IMPEX FERRO TECH	0.00	0.00	0.00	0.00	0.00	0.00
NMDC	4901.35	7204.55	3481.28	4055.99	4907.15	24550.32
HINDUSTAN ZINC LTD	1047.00	1705.00	452.00	789.00	944.00	4937.00
CEETA INDUSTRIES LTD	0.00	0.00	0.00	0.00	0.00	0.00
GLITTEK GRANITE LTD	0.00	0.00	0.00	0.00	0.00	0.00
KIOCL LTD	0.00	9.00	0.00	0.00	5.90	14.90
FOUNDRY FUEL PRODUCTS LTD	0.00	0.00	0.00	0.00	0.00	0.00
ASI INDUSTRIES LTD	0.00	1.50	4.16	2.00	13.45	21.11
PACIFIC INDUSTRIES LTD	0.00	0.00	0.00	0.00	0.00	0.00
ORIENTAL TRIMAX LTD	0.00	0.00	0.00	0.00	0.00	0.00
TOTAL	11430.11	10285.26	7833.06	129937.19	6822.91	166308.54

Source: Authors own calculation

Note: Green – highest expenditure, Red – lowest expenditure

Table 17 Detailed assessment of CSR expenditure on Others by the selected companies for the time period 2014–2015 to 2018–2019 (in lakhs)

Company	Others					
	2014-2015	2015-2016	2016-2017	2017-2018	2018-2019	Overall
COAL INDIA	15.83	1196.33	15.29	1485.11	61.63	2774.19
VEDANTA	127.02	1.51	0.09	7.11	3.29	139.02
MOIL	464.00	778.00	285.00	150.00	0.99	1677.99
GUJ MINERALS	70.61	0.00	114.95	95.35	119.08	399.99
MAITHAN ALLOYS	0.00	0.00	0.51	5.15	20.00	25.66
SANDUR MANGANES	0.00	0.00	0.00	0.00	10.00	10.00
INDIA METALS	29.00	29.00	16.00	17.00	3.00	94.00
OMDC	2.88	0.00	0.00	0.00	0.00	2.88
DECCAN GOLD	0.00	0.00	0.00	0.00	0.00	0.00
ASHAPURA MINES	0.00	0.00	0.00	0.00	0.00	0.00
20 MICRONS	0.00	0.00	0.00	5.00	0.10	5.10
RSMML	0.00	6.61	0.00	0.00	30.00	36.61
INDSIL HYDRO	0.00	0.00	0.00	0.00	0.00	0.00
RAW EDGE INDUSTRIAL SOLUTIONS LTD	0.00	0.00	0.00	0.00	0.00	0.00
FERRO ALLOYS	0.00	0.00	0.00	0.50	0.00	0.50
SHYAM CENTURIES	0.00	0.00	0.00	0.00	10.02	10.02
SHIRPUR GOLD	0.00	0.00	0.00	0.00	0.00	0.00
NAGPUR POWER	0.00	0.00	0.00	0.00	0.00	0.00
FACOR ALLOYS	0.00	0.00	0.00	0.00	0.00	0.00
IMPEX FERRO TECH	0.00	0.00	0.00	0.00	0.00	0.00
NMDC	866.12	615.49	1501.00	1517.64	1141.46	5641.71
HINDUSTAN ZINC LTD	623.00	871.00	492.00	1315.00	1989.00	5290.00
CEETA INDUSTRIES LTD	0.00	0.00	0.00	0.00	0.00	0.00
GLITTEK GRANITE LTD	0.00	0.00	0.00	0.00	0.00	0.00
KIOCL LTD	0.00	0.00	0.00	2.50	7.00	9.50
FOUNDRY FUEL PRODUCTS LTD	0.00	0.00	0.00	0.00	0.00	0.00
ASI INDUSTRIES LTD	1.06	0.73	5.01	2.42	2.49	11.71
PACIFIC INDUSTRIES LTD	2.88	0.00	0.00	0.00	0.00	2.88
ORIENTAL TRIMAX LTD	0.00	0.00	0.00	0.00	0.00	0.00
TOTAL	2202.40	3498.67	2429.85	4602.78	3398.06	16131.77

Source: Authors own calculation
Note: Green – highest expenditure, Red – lowest expenditure

References

Abuya, W. O. (2016). Mining conflicts and corporate social responsibility: Titanium mining in Kwale, Kenya. *The Extractive Industries and Society, 3*(2), 485–493.

Abuya, W. O. (2018). *Mining conflicts and corporate social responsibility in Kenya's Nascent mining industry: A call for legislation, social responsibility, Ingrid Muenstermann*, IntechOpen. Retrieved from https://www.intechopen.com/books/social-responsibility/mining-conflicts-and-corporate-social-responsibility-in-kenya-s-nascent-mining-industry-a-call-for-l

Anguelovski, I. (2011). Understanding the dynamics of community engagement of corporations in communities: The iterative relationship between dialogue processes and local protest at the Tintaya Copper Mine in Peru. *Society & Natural Resources, 24*(4), 384–399.

Arko, B. (2013). Corporate social responsibility in the large scale gold mining industry in Ghana. *Journal of Business and Retail Management Research, 8*(1), 81–90.

Banerjee, S. P. (2004). Social dimensions of mining sector. *IE (I) Journal-MN, 8*, 5–10.

Barnard, C. I. (1938). *The functions of the executive*. Cambridge: Harvard University Press.

Beatty, C., & Fothergill, S. (1996). Labour market adjustment in areas of chronic industrial decline: The case of the UK coalfields. *Regional Studies, 30*, 637–650.

Bice, S. (2013). No more sun shades, please: Experiences of corporate social responsibility in remote Australian mining communities. *Rural Society, 22*(2), 138–152.

Bodruzic, D. (2015). Promoting international development through corporate social responsibility: the Canadian government's partnership with Canadian mining companies. *Canadian Foreign Policy Journal, 21*(2), 129–117.

Boon, J. (2009). *Corporate social responsibility (CSR) in the mineral exploration and mining industry–perspectives on the role of "home" and "host" governments University of Ottawa; M.A. (Dissertation/Thesis), ProQuest Dissertations Publishing, United States. (MR51636).*

Bowen, H. (1953). *Social responsibilities of the businessman*. New York: Harper.

Buenar, P. B., & Hevina, S. D. (2011). Organizational antecedents of a mining firm's efforts to reinvent its CSR: The case of golden star resources in Ghana. *Business and Society Journal, 116* (4), 467–507.

Clark, J. M. (1939). *Social control of business*. New York: McGraw-Hill.

Conway, C. (2003). Tracking health and well being in Goa's mining belt, Case Study5, Ecosystem Approach to Human Health, International Development Research Centre, Canada.

Critcher, C., Schubert, K., & Waddington, D. P. (1995). *Regeneration of the coalfield areas: Anglo-German perspectives, (Anglo-German Foundation for the Study of Industrial Society)*. Pinter.

Diale, A. J. (2014). Corporate social responsibility in South African mining industry: Necessity, conformity or convenience? *International Journal of Business and Economic Development, 2* (1), 1–13.

Dobele, A. R., Westberg, K., Steel, M., & Flowers, K. (2014). An examination of corporate social responsibility implementation and stakeholder engagement: A case study in the Australian mining industry. *Business Strategy and the Environment, 23*(3), 145–159.

Dong, S. D., Burritt, R., & Qian, W. (2014). Salient stakeholders in corporate social responsibility reporting by Chinese mining and minerals companies. *Journal of Cleaner Production, 84*(1), 59–69.

Franks, D. M., Davis, R., Bebbington, A. J., Ali, S. H., Kemp, D., & Scurrah, M. (2014). Conflict translates environmental and social risk into business costs. *Proceedings of the National Academy of Sciences USA, 111*(21), 7576–7581.

Hamann, R. (2003). Mining companies' role in sustainable development: The 'why' and 'how' of corporate social responsibility from a business perspective. *Development Southern Africa, 20* (2), 237–254.

Hamann, R., & Kapelus, P. (2004). Corporate social responsibility in mining in Southern Africa: Fair accountability or just greenwash? *Development, 47*(3), 85–92.

Hilson, G. (2012). Corporate social responsibility in the extractive industries: Experiences from developing countries. *Resources Policy, 37*(2), 131–137.

Imbun, B. Y. (2007). Cannot manage without the 'significant other': Mining, corporate social responsibility and local communities in Papua New Guinea. *Journal of Business Ethics, 73*(2), 177–192.

Ingamells, A. T., Holcombe, S., & Buultjens, J. (2011). Economic development and remote desert settlements. *Community Development Journal, 46*(4), 436–457.

Jamali, D., & Mirshak, R. (2007). Corporate social responsibility (CSR): Theory and practice in a developing country context. *Journal of Business Ethics, 72*(3), 243–262.

Kapelus, P. (2002). Mining, corporate social responsibility and the "Community": The case of Rio Tinto, Richards Bay minerals and the Mbonambi. *Journal of Business Ethics, 39*(3), 275–296.

Kemp, D. (2009). Community relations in the global mining industry: Exploring the internal dimensions of externally orientated work. *Corporate Social Responsibility and Environmental Management., 17*, 1–14.

Kepore, K., Higgins, C., & Goddard, R. (2013). What do indigenous communities think of the CSR practices of mining companies? *Journal of business systems, governance and ethics, 8*(1), 34–50.

Keyes, R. (1992). Mine closures in Canada: Problems, prospects and policies. In C. Neil, M. Tykkyläinen, & J. Bradbury (Eds.), *Coping with closure: An international comparison of mine town experiences* (pp. 27–43). London: Routledge.

Kotilainen, J., Prokhorova, E., Sairinen, R., & Tiainen, H. (2015). Corporate social responsibility of mining companies in Kyrgyzstan and Tajikistan. *Resources Policy, 45*, 202–209.

Kreps, T. J. (1940). *Measurement of the Social Performance of Business: In an Investigation if Concentration of Economic Power for the Temporary National Economic Committee, Monograph NO.7*. Washington: Government Printing Office.

Lahiri-Dutt, K. (2008). Digging to survive: Women's livelihoods in South Asia's small mines and quarries. *South Asian Survey, 15*(2), 217–244.

Lydenberg, S. D. (2005). *Corporations and the public interest: Guiding the invisible hand*. San Francisco: Berrett-Koehler Publishers.

MCA. (2013). *Companies Act, 2013, Ministry of Corporate Affairs, Government of India, New Delhi*. http://www.mca.gov.in/Ministry/pdf/CompaniesAct2013.pdf

Muller, A. (2006). Global versus local CSR strategiesm. *European Management Journal, 24*(2), 189–198.

Mzembe, A. N., & Meaton, J. (2014). Driving corporate social responsibility in the Malawian mining industry: A stakeholder perspective. *Corporate Social Responsibility and Environmental Management, 21*(4), 189–201.

Mzembe, A. N. (2016). Doing stakeholder engagement their own way: Experience from the Malawian mining industry. *Corporate Social Responsibility and Environmental Management, 23*(1), 1–14.

Randall, J. E., & Ironside, G. (1996). Communities on the edge: An economic geography of resource dependent communities in Canada. *Canadian Geographer, 40*(1), 17–35.

Sharma, D., & Bhatnager, P. (2015). Corporate social responsibility of mining industries. *International Journal of Law and Management, 57*(5), 367–372.

Slack, K. (2012). Mission impossible? Adopting a CSR-based business model for extractive industries in developing countries. *Resources Policy, 37*, 179–184.

Storey, K. (2001). Fly-in/Fly-out and Fly-over: mining and regional development in Western Australia. *Australian Geographer, 32*(2), 133–148.

The International Council on Mining and Metals (ICMM). (2015). Annual Review 2015

Viviers, S., & Boulder J. M. (2008). *Corporate social responsibility in the mining sector: Critical issues*. International society of business, economics and ethics: USA.

Waddington, D. (2004). Making the difference in the Warsop Vale: The impact of government regeneration policy and community development on a Nottinghamshire Ex mining community. *Social Policy and Society, 3*, 21–31.

Waddington, D., Critcher, C., Dicks, B., & Parry, D. (2001). *Out of Ashes? The social impact of industrial contraction and regeneration on Britain's Mining Communities, Regions, Cities and Public Policy Series*. London: The Stationary Office.

Welker, M. (2014). *Enacting the corporation: An American mining firm in post authoritarian Indonesia*. Berkeley: University of California Press.

Wheeler, D., Colbert, B., & Freeman, R. E. (2003). Focusing on value: Reconciling corporate social responsibility, sustainability and a stakeholder approach in a network world. *Journal of General Management, 28*(3), 1–28.

World Commission on Environment and Development. (1987). *Our common future*. Oxford: Oxford University Press.
Yakovleva, N. (2005). *Corporate social responsibility in the mining industries*. Hampshire: Ashgate Publishing Limited.
Yakovleva, N., & Vazquez-Brust, D. (2012). Stakeholder perspectives on CSR of mining MNCs in Argentina. *Journal of Business Ethics, 106*(2), 191–211.

Sumona Ghosh has been associated with St. Xavier's College, Kolkata since 2002. She was the head of the Department of Law from 2003 to 2018. Presently, she is the Joint Coordinator of the Foundation Course. After completing her post-graduation in Commerce with rare distinction, she has been conferred with the Degree of Philosophy in Business Management by the University of Calcutta on 31st of July 2014. Her area of research was on Corporate Social Responsibility (CSR). The title of her doctoral dissertation was 'Pattern of Participation of Public and Private Sector Companies in Corporate Social Responsibility Activities'.

She has published in journals of national and international repute. She has been highly acclaimed for her guest lectures on CSR in premier institutes of higher learning including the Indian Institute of Management (Calcutta) and Indian Institute of Management (Shillong). She has taken sessions in Management Development Programmes conducted by premier institutes on CSR. She has presented papers on CSR at various national and international conferences. Her research interest lies in Corporate Social Responsibility, Sustainable Development, Integrated Reporting and Philosophies of Management. She is also a Certified Assessor for Sustainable Organizations (CASO), certification conferred upon her by UBB GmBH Germany.

She is also the recipient of the 'Bharat Jyoti Award' for meritorious services, outstanding performance and remarkable role given by India International Friendship Society for the year 2012 given by Dr Bhishma Narain Singh, Former Governor of Tamil Nadu and Assam, on 20th of December 2012.

Part III
Europe

A Circular Economy Strategy for Sustainable Value Chains: A European Perspective

Mark Anthony Camilleri

Abstract The European Union (EU) institutions are increasingly raising awareness on the circular economy (CE) agenda. They are encouraging marketplace stakeholders to engage in sustainable production and consumption behaviours by urging them to reduce, reuse, restore, refurbish, remanufacture, and recycle resources in all stages of their value chain. Therefore, this chapter presents a cost-benefit analysis of the circular economy strategy. Afterwards it features a critical review of some of the latest European regulatory guidelines, instruments and principles appertaining to the CE agenda. It sheds light on EU's (2020) new circular economy plan for a cleaner and more competitive Europe. Therefore, this research examines the EU's key propositions on the value chains of different products. The findings suggest that the circular economy's sustainable development model and its regenerative systems are increasingly minimising industrial waste, emissions, and energy leakages through the creation of long-lasting designs that can improve resource efficiencies. This contribution implies that successful CE practices are sustainable in the long run as they will ultimately add value to the business as well as to our natural environment. In conclusion, the researcher puts forward his recommendations to policy makers and practitioners.

1 Introduction

The economic models of many countries are mostly built on the premise of "take-make consume and dispose" patterns of growth (Kirchherr et al. 2017; Camilleri 2017a, 2018a; EU 2015a; EMF 2013). The manufacturing of products have customarily followed such a linear model that assumes that resources are abundant, available and cheap to dispose of; as every product is usually bound to reach its 'end of life' at some stage (Camilleri 2018b; Ghisellini et al. 2016; EU 2014). When

---

M. A. Camilleri (✉)
University of Malta, Msida, Malta
e-mail: mark.a.camilleri@um.edu.mt

© Springer Nature Switzerland AG 2021
S. Vertigans, S. O. Idowu (eds.), *Global Challenges to CSR and Sustainable Development*, CSR, Sustainability, Ethics & Governance,
https://doi.org/10.1007/978-3-030-62501-6_7

products are no longer useable or required, they are often discarded as waste that is either incinerated or dumped in landfills (Kirchherr et al. 2017; Murray et al. 2017; Haas et al. 2015). The circular economy proposition differentiates itself from the linear economic systems that rely on resource depleting systems that are usually characterised by high externalities, including emissions and waste generation. The circular economy is intended to reduce waste as extant resources and materials are used more efficiently (EU 2014, 2020a, b; Stahel 2016).

The sustainable consumption of resources was recently listed as one of the priority areas of the European Green Deal as the European Union (EU) has also recognised the importance to reduce resource extraction as the reserves of some of the globe's key elements and minerals shall be depleted within the next years. Extant economic models are still relying too much on resource extraction and depletion. Currently, the national economies are very dependent on the use of natural resources as they provide crucial raw materials for business, industry as well as for individuals. The increase in global extraction of resources is driven by higher living standards and from major infrastructural investments that are happening in developing and transitioning countries (EU 2020b). The rise of the rapid urbanisation within the emerging economies is expected to intensify the competition for certain raw materials and to destroy our natural environment (Stubbs and Cocklin 2008). Eventually, this can (it already had in some parts of the world) a devasting effect on the globe's climate. The projections are that the demand for the world's resources would more than double between 2015 and 2060 (UNEP 2019). The global consumption materials or products including biomass resources (like fruit and vegetables), fossil fuels (like gas, oil or coal), metals (like iron, aluminium and copper that are used in construction and electronics manufacturing) as well as non-metallic minerals (that are used for construction, notably sand, gravel and limestone) are expected to double in the next forty years (OECD 2019a), while annual waste generation is projected to increase by 70% by 2050 (World Bank 2018). Therefore, there is scope for governments and policy makers to incentivise practitioners to use circular economy systems, as better eco-designs, waste prevention as well as the reuse and recycling of materials can result in operational efficiencies and cost savings, whilst reducing waste and emissions (Camilleri 2018a; EEA 2018; Kirchherr et al. 2017; Haas et al. 2015).

In this light, this chapter critically analyses costs and benefits of the circular economy. Afterwards, it sheds light on the latest strategies and policy developments within the European Union (EU) context. Them, it clarifies how the EU is promoting the circular economy approaches in different industry sectors as it encourages the sustainable production and consumption of resources and materials. In conclusion, this contribution puts forward the managerial implications to policy makers and industry practitioners.

2 The Circular Economy Strategy

The term "circular economy" is increasingly being used by politicians, business practitioners and even within the civil society. Many individuals may usually associate it with recycling and/or better waste management. They may perceive that its eco-design will improve the sustainable production and consumption of the businesses and their customers. Therefore, the circular economy approaches are intended to reduce the absolute resource consumption during manufacturing (Stahel 2016; Haas et al. 2015). At the same time, it can also minimise the generation of waste and unwanted externalities.

2.1 The Benefits of the Circular Economy

Hence, the governments ought to monitor and control the consumption of their land and water with a view to setting reduction targets in future (EU 2020a, b). Moreover, it is imperative that they regularly examine the carbon emissions and footprints during the production and consumption of resources in order to bridge further actions between climate change agenda and the circular economy approach. Firms are encouraged to continuously re-examine their extant operations, management systems and production processes as they need to identify value-added practices (Porter and van der Linde 1995). Their industrial operations can be improved through redesigned processes, the elimination of some of them, the modification of certain technologies and/or inducting new technology. Prakash (2002) suggested that the businesses could adopt management systems that create the right conditions to reduce their negative impact on the natural environment. He posited that this could take place in the following ways: (1) repair—extend the life of a product by repairing its parts; (2) recondition—extend the life of a product by significantly overhauling it; (3) remanufacture—the new product is based on old ones; (4) reuse—design a product so that it can be used multiple times; (5) recycle—products can be reprocessed and converted into raw material to be used in another or the same product, and (6) reduce—even though the product uses less raw material or generates less disposable waste, it could still deliver benefits that are comparable to its former version. These preventative and restorative practices are related to the circular economy.

The circular economy's closed loop systems could minimise the cost of dealing with pollution, emissions and environmental degradation (Geissdoerfer et al. 2017; Peeters et al. 2014; Geng et al. 2013; Geng and Doberstein 2008; Stubbs and Cocklin 2008).

2.2 The Costs of the Circular Economy

There may be circular economy processes that may result in negative outcomes. The idea that businesses can design longer-lasting products be considered as advantageous. However, the longevity of product designs may not be possible or feasible. The longevity of the products' life could not be sustainable and/or efficient in ecological terms. The long-lasting products that do not break down quickly may consume more useful energy and release more entropy than those that are designed towards a more natural outcome. For example, a bamboo chopstick would be more sustainable than a plastic fork as it can be recycled and removed from the biosphere. The bamboo chopstick is made of natural nutrients, whilst the plastic fork has technical nutrients. The latter cannot be re-assimilated back into the natural environment.

There are some environmentally friendly technologies, including the wind farms and solar panels that have certain metals and compounds in them that may be difficult to recycle, upcycle or downcycle. Moreover, such structures may require frequent servicing and replacements of their key components. Hence, the prices of these green technologies may not always reflect the real costs of their materials. These structures consist of technological components that will necessitate energy-expensive servicing and/or replacement, as nothing lasts forever in an entropic universe (Murray et al. 2017). Therefore, it may not be sustainable to delay or prolong these green products' lifecycle through regular servicing.

The transition toward a zero-waste model could prove to be a very difficult endeavour for many businesses. The new sustainable technologies could be more expensive than other technologies. The smaller businesses may not have access to adequate and sufficient financial resources to make green investments. Therefore, there may still be a low demand for the circular economy technologies, particularly if the businesses engage in short term planning. Hence, they will not perceive the business case for the long term, sustainable investment (Camilleri 2017a; Moratis et al. 2018). Alternatively, the businesses may not be interested in new technologies that will require them to implement certain behavioural changes (Lieder and Rashid 2016; Porter and van der Linde 1995). Technology is one of the key factors in the development of a circular economy model. However, businesses may not always be in a position to invest in economic and efficient infrastructures.

There may be governments that will probably introduce the regulatory instruments relating to the circular economy. The business practitioners will be expected to comply with the relevant laws if they want to maintain their legitimacy with the governments and marketplace stakeholders. Arguably, while some companies may expect the respective governments to provide a legislative framework, others may resent any mandatory changes that will be imposed on them. In many cases, it is very likely that they would opt to remain in their status quo, where they keep using traditional, linear economic models (Bocken et al. 2016).

There may be other challenges that could slow down or prevent the industry practitioners' engagement in the circular economic approaches. For example, the emerging economies may decide not to follow the international guidelines and recommendations to achieve the sustainable development goals. Their authorities and policy makers may not enforce the recommended practices on social and environmental matters in their respective jurisdictions (Estol et al. 2018). The developing countries' economies will probably support the corporate businesses that are providing employment to their citizens, rather than burdening them with strict regulations (Rasiah et al. 2010; Jensen 2003). Their governments may not introduce hard legislation to trigger the corporations' sustainable production and consumption behaviours as this could impact on the businesses' prospects. Therefore, such businesses are not even expected to engage in corporate responsible and sustainable behaviours (Amin-Chaudhry 2016). As a result, they will not safeguard or protect the environment. They may not mitigate their externalities, including their emissions or unwanted waste, as such actions would require changing or upgrading the extant technologies or practices. The social and environmentally responsible behaviours will probably usurp more corporate resources in terms of time, labour and money. Of course, the businesses that are operating is such contexts may also have to face contingent issues like weak economic incentives; access to finance; shortage of advanced, green technologies; and a lack of appropriate performance standards in their workplace environments, among other issues. The businesses may see little economic incentives to save energy, material and water. The producers can easily transfer the manufacturing costs to their customers in the form of higher sale prices.

The financial services institutions, including banks may decide not to invest in environmental-friendly technologies as they are indifferent to CSR investments (Goss and Roberts 2011). Similarly, several governments may not be in a position to provide economic instruments, such as grants or tax incentives to support the businesses and clusters to implement closed loop or product service systems (Camilleri 2018a; Tukker 2015; Wang et al. 2008; Tukker and Tischner 2006; Mont 2002). Conversely, other governments from the advanced economies possess the financial resources to incentivise the industry practitioners' circular economic practices. For instance, most EU countries have introduced intelligent, substantive and reflexive regulations for the performance assessment of government agencies, corporations and large undertakings (Camilleri 2015b). These jurisdictions are following the EU (2014) directive on non-financial reporting, among others, as they disclose material information on their environmental, social and governance (Venturelli et al. 2019; Camilleri 2015a, 2018a).

3 The European Union's Policies on the Circular Economy

In the past few years, the EU has raised awareness on the importance of the circular economy agenda. It has published various action plans relating to this topic, including; "Innovation for a sustainable future—The Eco-innovation Action Plan"; "Building the Single Market for Green Products"; "Facilitating better information on the environmental performance of products and organisations"; "Green Action Plan for SMEs: enabling SMEs to turn environmental challenges into business opportunities"; "Closing the loop—An EU action plan for the Circular Economy"and "Investing in a smart, innovative and sustainable Industry—A renewed EU Industrial Policy Strategy", among others (see EU 2017). EU (2015a) anticipated that new business models, eco-designs and industrial symbiosis can move the community towards zero-waste; reduce greenhouse emissions and environmental impacts. Its 'Resource Efficient Europe' initiative involved the coordination of cross-national action plans and policies on the formulation of sustainable growth. The circular economy proposition was intended to bring positive environmental impacts, real cost savings, and greater profits. EU (2015a, b) indicated that its envisaged improvements in waste prevention and eco-designs, the use and reuse of resources, and similar measures could translate to the net savings of € 600 billion, or 8% of annual turnover (for European businesses); while reducing total annual greenhouse gas emissions by 2–4%. This EU (2015a) communication anticipated that the markets for eco-industries will double between 2010 and 2020. It also posited that internationally, resource-efficiency improvements are in demand across a wide range of industrial sectors.

Moreover, in 2015, the European Fund for Strategic Investments (EFSI) has also announced a new financing avenue for future investments in infrastructure and innovation, that could be relevant for circular economy projects and closed loop systems. In December of the same year, the EU launched its Circular Economy Package, which included revised legislative proposals to "close the loop" of product lifecycles through greater recycling and re-use. Its action plan specified that the inefficient use of resources in production processes can lead to lost business opportunities and significant waste generation throughout a product's life. It suggested that product designs could make products more durable or easier to repair, upgrade or remanufacture. Operational improvements would enable recyclers to disassemble products in order to recover valuable materials and components. Therefore, the EU legislative proposals on waste management included long-term targets for the sustainable consumption and production of resources. They also encouraged the reuse and recycling of materials, including plastics, food waste; critical raw materials from electronic devices; construction and demolition resources; as well as from biomass and bio-based products, among other items. Moreover, the EU's action plan recommended further innovative investments in fields such as waste prevention and management, food waste, remanufacturing, sustainable process industry, industrial symbiosis, and in the bioeconomy to support the circular economy and global supply chains (EU 2015a). Eventually, in March 2017, the EU Commission and the European Economic and Social Committee organised a Circular Economy Stakeholder Conference, where it reported on the delivery and progress of some of its Action Plan. It also established a Finance Support Platform with the European Investment Bank (EIB) and issued important guidance documents to Member States on the conversion of waste to energy.

The latest European Union (EU) Commission's (2020) Circular Economy Action Plan emphasises that it relies on the stakeholders' active engagement to achieve the

European Green Deal2 for a climate-neutral, resource-efficient, circular economy. The EU has reiterated its commitment to implement the 2030 Agenda for Sustainable Development as well as its Sustainable Development Goals on resource efficiency and decoupling; on sustainable management and efficient use of natural resources; on land-degradation neutrality; and on halting biodiversity loss (EU 2020a, b). Its recent working document provides a comprehensive account about the current situation. It discusses about the opportunities and challenges, for various actors across the globe, to engage in sustainable production and consumption behaviours. Hence, it puts forward its key recommendations for the circular economy agenda.

The latest circular economy (CE) strategy is even more ambitious. EU (2020) is planning to increase the bloc's GDP by an additional 0.5%, it envisages that it would create more than 700,000 jobs by 2030 (this was announced a few days before the outbreak of the 2019–2020 Coronavirus—COVID19). The EU's (2020) strategy is aimed at making the European economy even more sustainable, as the Commission is pushing the member states to exploit the untapped potential of greener product life cycles. Another European sustainability policy (that is another component of the European Green Deal) is entitled; 'Farm to Fork' (F2F), is focused on the agriculture, aquaculture and sustainable food production. Specifically, this policy covers the entire food supply chain. Its underlying objective is to reduce the usage of resources and unnecessary waste by promoting the sustainable production and consumption of food. It also supports consumers choose healthy and sustainable diets as they are given relevant information on the foods' nutritional value and environmental footprint. One of its key objectives is to contribute towards achieving a circular economy. It aims to reduce the environmental impact of the food processing and retail sectors by taking action on transport, storage, packaging and food waste (EU 2019a).

The objectives of both the CE and F2F strategies are to reduce waste including food waste. The measures from the circular economy plan refer to the reduction on the use of packaging and over-packaging. At the same time, it encourages the businesses and their customers to use and re-use or recycled materials, where they can. Very often, the industry practitioners are finding ways to use fewer resources in the manufacturing and packaging processes (EU 2019a). Of course, it is in their interest to seek new solutions to enhance their operational efficiency and to reduce their manufacturing costs. The EU Commission is engaging with key stakeholders operating in different value chains to identify the opportunities and challenges for the sustainable production and consumption of the following products.

3.1 Batteries and Vehicles

The EU has introduced a batteries directive as from 2006. Since them it has been subject to a number of revisions. Its latest legislative proposal builds on the previous directive (EU 2006) and on the work of the Batteries Alliance. In a nutshell, the EU's action plan is proposing specific rules and measures to improve the collection and recycling rates of all batteries (EU 2020a, b). It also emphasised the importance to phase out non-rechargeable batteries where alternatives exist. This EU document also referred to sustainability and transparency requirements for the production of batteries and has taken into account the carbon footprint from the sourcing of their raw materials through their production and consumption. The EU is pushing the automotive industry to comply with its stricter CO_2 emission standards. It is encouraging the European vehicle owners to purchase Lithium-ion powered, electric vehicles (EV). The Commission has recognised that the use of lithium plays a strategic role in achieving zero-emission mobility, climate neutrality and technological leadership. At the same time, it contributes to enhance the sustainability of its economy.

Lithium is an important raw material for the production ofEV batteries, one of the six strategic value chains for the European economy. This metal is not featured in the EU's list of critical raw materials as it is not scarce. Its supply is not considered to be at risk. The EU has its own lithium reserves that are available for industrial-scale extraction. It very likely, that this soft, silvery-white alkali metal will be included in the EU's list of critical raw materials, as Europe will require higher volumes of lithium to mass-produce rechargeable batteries. Therefore, the EU together with the European battery manufacturing industry, is backing new initiatives to develop lithium mining and recycling activities as part of its concerted efforts to increase its production. In this light the EU has launched the European Battery Alliance in 2017. It brought together automakers, chemical and engineering executives in a bid to compete with Asian and American manufacturers. The EU wants to develop a strategic value chain for the manufacturing of electric car batteries within its territory.

However, the Europeans extant reliance on imported oil and gas should not be replaced by another dependency on lithium, cobalt, copper, and/or other raw materials that industries need for its green transition. One of the potential challenges is to recycle the used batteries after about 10 years. The used batteries can possibly play a much bigger role in secondary raw materials. Hence, the EU should ensure that there will be a business case for the stakeholders in the automotive industry to operate sustainably by recycling lithium from waste batteries in their manufacturing processes. Recently, COVID19 has led to a drop in the price of lithium, due to oversupply concerns. This issue may result in more usage of this white metal.

3.2 Construction and Buildings

The Commission's European Green Deal had put forward its recommendations for energy efficiency in the EU in line with its circular economy principles, including the longer life expectancy of the buildings and reasonable recovery targets for construction and demolition waste, among other proposal. The EU's (2020) action plan is promoting the circularity principles to improve the durability and adaptability of buildings throughout their lifecycle. This document is addressing the safety and functionality of material recovery as it also makes reference to the sustainability performance of construction products. For example, it promotes initiatives to reduce soil sealing, to rehabilitate abandoned or contaminated brownfields and to increase the safe, sustainable and circular use of excavated soils.

One quarter of all waste that is generated in the EU comes from the construction and buildings industry (EU 2019b). Such waste can be prevented by extending the lifetime of extant buildings, by reducing the living space per capita of new dwellings and by reusing building components for the same purpose for which they were conceived (Tam and Lu 2016). In practice, it may prove very difficult to reuse the building and construction materials like scrap metal, used cement or wood products, for the very same purpose for which they were conceived (Pomponi and Moncaster 2017; Smol et al. 2015). For the time being, the used materials from demolition and renovation works are not suitable for reuse or for high-grade recycling (EEA 2020). Therefore, the construction industry cannot achieve the EU's waste policy objectives, in terms of waste prevention. Currently, they are not in a position to use the recycled resources from the building sites, and/or to reduce the hazardous materials from construction waste (EEA 2020; EU 2019b)).

Nevertheless, this sector is still considered a priority sector according to the latest European CE action plan (EU 2020a, b). In fact, the EU countries are scrutinising their waste management policies and practices. Several member states are recovering the construction waste for backfilling operations. Such waste and rubble are used to fill holes in construction sites. Alternatively, they may be using low-grade recovery waste such crushed cement or stones in road works. The cost to re-use construction material can be affected by the national and local circumstances; the mismatch of supply and demand; as well as by the logistical issues, like moving materials over long distances (Pomponi and Moncaster 2017; Smol et al. 2015). Moreover, such practices may also have an impact on the nearby communities and/or the natural environment. The construction and the development of the properties that use recycled materials may still require considerable scarce resources, in terms of time and money to complete them. Notwithstanding, the builders may be reluctant to re-use construction material that lacks an adequate certification of tested performance from a recognised authority. The testing of the building materials can be expensive as it involves the thorough analyses of the samples to mitigate the risks of further use. These costs will be added to the material costs for the builders and may possibly override any savings from the reuse of extant material. Moreover, the latest construction techniques are increasingly combining traditional and novel building materials. This practice could impede the future builders to deconstruct and reuse the construction material. Notwithstanding, the reuse of some of the demolition waste, including furniture and equipment, may require requires selective demolition. In this case, the building contractor will have to sift through the construction waste to sort, clean and repair possible resources. Currently, there is an increase in demand for used construction and demolition materials. Such materials are usually given away free of charge or may be easily obtained through online marketplaces.

Better waste prevention as well as higher quality recycling can be achieved if certain measures are taken to improve the information on the construction materials that are being used in new buildings. This way, the building contractors would perceive the quality and value of the construction industry's secondary materials. Circular economy-inspired actions, like improved information sharing on material properties, and the optimal re-utilisation of secondary raw materials can go a long way in fostering the circular economy within the construction sector.

3.3 Electronics and ICT

Electrical and electronic equipment continues to be one of the fastest growing waste streams in the EU, with current annual growth rates of 2%. It is estimated that less than 40% of this waste is recycled (EU 2020a). About two out of three Europeans would like to keep using their technological devices for longer, if their performance is not significantly affected (Eurobarometer 2020).

The waste of electrical and electronic equipment (WEEE) may include computers, printers, televisions, refrigerators, mobile phones, tablets and laptops. These electrical and electronic devices are the fastest growing waste streams within the EU. Their materials and components are made from scarce, hazardous resources. For example, the smartphones' components may include very precious and rare metals like gold, silver, copper, platinum and palladium, among others. The mining and extraction of these metals has an impact on the natural environmental at local, regional and global scales, as ecosystems are destroyed through waste spills and pollution (Liu et al. 2009). As a result, there will be negative effects on the bio-diversity and on environmental sustainability. In this light, the EU's Directives on "waste electrical and electronic equipment" (WEEE) and on the "restriction of the use of hazardous substances" (RoHS) were intended to improve the collection, treatment and recycling of electronics at the end of their life. The electrical and electronic materials can cause major environmental and health problems if they are not collected, treated and/or recycled in an appropriate manner.

The first EU's WEEE Directive entered into force in February 2003 (WEEE 2002). It provided information to the European citizens about how to dispose of their used electrical and electronic devices. In the same year, another EU legislation was enacted to restrict the use of hazardous substances in electrical and electronic equipment (RoHS 2002). This legislation restricted the use of hazardous substances including lead, mercury, cadmium, and hexavalent chromium and flame retardants such as polybrominated biphenyls (PBB) or polybrominated diphenyl ethers (PBDE) to safeguard the health and safety of individuals and protect the environment from heavy metals. Eventually, in 2012, the Commission revised this Directive that became effective on the third January 2013 (RoHS 2017). This revised legislation provided useful guideline for the creation of collection schemes where business, industry or consumers could return their hazardous substances free of charge. In a similar vein, the EU has introduced another new WEEE directive to tackle the increased waste stream of EEEs (WEEE 2012). This directive also stipulates that the EU member states are expected to collect and report data on the electrical and electronic products that are sold, collected, recycled and recovered. Various electrical and electronic materials may have different hazardous components in them. Therefore, the EU's WEEE (2012) directive specified that there are different targets for the recovery, reuse and recycling of different categories of electrical and electronic resources. A few EU countries are also disclosing other information on the reuse of whole appliances (on a voluntary basis). Currently, other devices like smart phones and tablets are being designed to be energy efficient, durable, repairable and/or upgradable, as their components can be reused or recycled (EU 2019c, 2020c). In addition, the upcoming Eco-design Working Plan will set out further regulatory measures on chargers for such devices, as it suggests the introducing a common charger for different devices, increasing the durability of charging cables, and providing incentives to decouple the purchase of chargers from the purchase of new devices (EU 2019c).

3.4 Food, Water and Packaging

An estimated 20% of the total food produced is lost or wasted within the EU. Therefore, the food value chain and its generated waste is also increasing the pressure on the natural environment. In this light, EU (2008) has proposed a target to reduce food waste in its Farm-to-Fork Strategy (see EU 2019a, d). EU (2020a) has reaffirmed its commitment to increase the sustainability of food distribution and consumption. It indicated that it may amend its Drinking Water Directive (EU 2019d) to make tap water drinkable and accessible in public places. This way, there will be less dependence on the bottled water and its packaging waste.

This document also raised awareness about its previous directive on reusable and recyclable packaging of materials and products (See EU 1994). In 2017 the waste from packaging materials accounted to 173 kg per inhabitant. This figure is poised to increase year after year. It specified that a sustainable products initiative will include details on its proposed legislation to direct businesses and consumers to substitute single-use packaging, tableware and cutlery by reusable products in food services. The EU could restrict some packaging materials for certain applications, in particular where there are alternative, reusable products. These initiatives complement extant collection systems where there the packaging waste is separated at source. Currently, it is planning to establish specific rules for the safe recycling of food contact materials. The Commission may consider other systems where the consumer goods could be handled safely without packaging (EU 2020a).

3.5 Nutrients

EU (2020a) announced that the Commission will develop an Integrated Nutrient Management Plan, with a view to ensuring that the member states will ensure the long-term sustainability of their nutrient resources and to stimulate the markets for recovered nutrients. For instance, this document refers to the new Water Reuse Regulation that encourages circular approaches to reuse water for agricultural purposes. The Commission also proposed that the secondary water could be reused by other industries (EU 2019e). Currently, is assessing the possibility to use wastewater treatment and sewage sludge as a natural means of nutrient removal like algae.

Recently, the Farm Sustainability Tool (FaST) was developed to help farmers manage the use of nutrients in their farm. This tool (which is free of charge) was proposed in the framework of the Good Agricultural and Environmental Conditions (GAECs) and is part of the new common agricultural policy (CAP) proposals for 2021–2027. In a nutshell, it aims to facilitate the sustainable use of fertilisers while boosting the use of the digital technologies in the agricultural sector. These latest CAP proposals are aimed to improve the competitiveness of the European farms, and to raise awareness about their environmental responsibility and on matters relating to the climate change. This tool can be accessed through a personal computer and via smart devices including smart phones and tablets. It provides useful data about the farms' resources, including crops, soil, the proximity of protected areas, the legal limits on the use of nutrients, animals, as well as the manure generated by them, among other issues. It may provide the farmers a nutrient management plan, including customised recommendations. For example, it includes information on; how to improve crop fertilisation, how to reduce nutrient leakages in ground water or rivers; how to increase soil quality and how to reduce greenhouse gas emissions. The tool will also make sense economically as it will help to decrease the use of nutrients and/or to increase crop yield. In both cases, this will lead to enhance the farmers' revenues and operational efficiencies. A few EU countries are already customising the functions and services of FaST to ensure that it is adapted to the local conditions, whilst taking advantage of the extant knowledge (EU 2019e).

3.6 Plastics

The production of virgin plastic has increased 200-fold since 1950. It has grown at the rate of 4% a year since 2000, even though it takes more than 400 years to degrade (WEF 2019). To date, 12% of this plastic has been incinerated and just 9% has been recycled (Geyer et al. 2017; National Geographic 2018). This material carries a carbon footprint when burned as waste. However, most plastics still exist in landfills or in our natural environments. Last year the EU passed ambitious new laws to reduce the consumption of single-use plastics. Currently, its research agenda is focused to investigate the risk and occurrence of microplastics in our food and drinking water. In this light, many European governments have imposed restrictions on the use of plastic bags. These measures were followed by considerable reductions of such waste in the citizens' litter. As a result, there was also less plastic in marine environments (Earthwatch Institute 2019).

The European Parliament had emphasised that the prevention of plastic waste should be one of the Commission's first priorities (EP 2018). Other European entities and stakeholders have pointed out that reducing the use of this material is required to improve our natural environment. Notwithstanding, the manufacture of these materials are increasing the demand for petrochemicals as they make up 99% of all plastics. For decades, oil and gas companies have created markets for their by-products or manage them as waste streams (OECD 2019b; Wong et al. 2015).

EU's (2020a) clarified that it is planning to introduce more stringent legislative instruments to eliminate the single-use food packaging, tableware and cutlery and replace them with reusable products by 2021. It has renewed its focus on tackling plastic pollution as it is expecting the manufacturers to reduce its use in foreseeable future. The commission recommended that they utilise biodegradable and recyclable packaging for their products. However, despite the EU (2019f) specified that the practitioners ought to reduce the consumption of the single-use plastic products and packaging, the industry is still producing and utilising plastic materials. At this stage, any efforts to reduce their leakage and/or to recycle them are inadequate and insufficient, as its consumption is still on the rise. The high rates of the consumption of plastics appear to be incompatible with addressing the environmental damage that is associated with their leakage (Wong et al. 2015). There are several stakeholders, including governments, intergovernmental organisations (EU 2019f, 2020a; EASAC 2020; UNEP 2018, 2020; OECD 2019b) and non-governmental organisations in different contexts, that are increasingly calling for the reduction of plastic products. The industry practitioners are discouraged to continue the non-essential production as well as the unnecessary consumption of these materials.

3.7 Textiles

The production of textiles is highly globalised as there are billions of consumers that demand clothing, footwear and household products (Hu et al. 2011). On average the Europeans consume about 26 kg of textiles per person, per year. In the EU, there are around 171,000 companies that are providing employment to more than 1.7 million people within the textile and apparel industries. However, the EU is still a net importer of textiles and its exports comprise intermediate textile products, such as technical fibres and high-quality fabrics (EEA 2020).

Globally, the polyester is the most used fibre. It is produced from carbon-intensive processes requiring more than 70 million barrels of oil each year. The remaining fibres are mainly derived from cotton that necessitate vast stretches of land and water. The emerging economies are experiencing the major pressures to use their land to cultivate cotton.

From an EU consumption perspective, the clothing, footwear and household textiles are the fourth highest pressure category after food, housing and transport, as their production involves primary raw materials and water. Specifically, the textile industry is ranked as the second highest in terms of land use and it is fifth with regards to greenhouse gas emissions. It is estimated that the global manufacturing of textiles generates around 15–35 tonnes of CO_2 per tonne of produced textiles. Within the EU, the production and handling of clothing, footwear and household textiles have generated emissions of 654 kg CO_2 per person (in 2017). Only 25% of this production and handling took place in Europe (see EEA 2020). Moreover, the processes to produce textiles involve substantial amounts and a variety of chemicals. There are about 3500 substances that are used in textile production. Of these, 750 have been classified as hazardous for human health and 440 as hazardous for the environment. It is estimated that about 20% of global water pollution is caused by dyeing and finishing textile products, affecting the health of workers and local communities (EEA 2020). Furthermore, the washing of textiles also releases chemicals and microplastics into household wastewater.

Therefore, the production of textiles is impacting on our natural eco-systems and is putting considerable pressures on our climate (Hu et al. 2011). This is one of the industries that is affecting our climate in every phase of its production and consumption processes: from the manufacturing of fibres and textile products to distribution and retail, use of textiles, collection, sorting and recycling, and final waste management. Currently, the EU consumers are discarding about 11 kg of textiles per person, per year. A significant proportion of these material are exported to the developing countries, incinerated, or landfilled, as their recycling remains very low (EEA 2020). It is estimated that less than 1% of all textiles worldwide are recycled into new textiles (EMF 2017).

The European Commission has recognised that the textile industry as a priority area in one of its latest documents, entitled; "Towards an EU product policy framework contributing to the circular economy" (see EU 2015b). Furthermore, the new European Commission President Ursula von der Leyen has also pledged her commitment to implement the new circular economy action plan. In her own words, she specified that she would focus on "sustainable resource use, especially in resource intensive and high-impact sectors such as textiles and construction" (EUWID 2019). In fact, this year, the Commission is proposing a comprehensive EU Strategy for Textiles (EU 2020a). The aim of this strategy is to bolster the competitiveness and sustainability innovation in the textiles sector. In sum, the EU's (2020a) document specified that this can be achieved by employing its new sustainable product framework. This framework suggests that the manufacturers of textile materials ought to follow eco-designs and sustainability measures to ensure that they are engaging in closed loop systems. Such systems would require that they use and reuse secondary raw materials, tackle the presence of hazardous chemicals, empower the businesses and their consumers to choose sustainable textiles and have easy access to re-use and repair services (EU 2020a). Moreover, the EU and the respective governments of its member states needs to provide incentives and support to facilitate closed loop and product-service systems, by raising awareness on such sustainable production and consumption processes, by increasing their engagement with different marketplace stakeholders in the value chain, and by enabling industry practitioners to access and to avail themselves of secondary resources (EU 2020a; Camilleri 2018b).

4 Discussion

In the past decade, the EU Commission and its agencies have been instrumental in raising more awareness on sustainable innovation and on circular economic models. This contribution suggests that there is scope for the EU and the respective governments of its member states to forge relationships with marketplace stakeholders, including suppliers and distributors in order to implement the circular economy. Currently, the public regulatory authorities as well as the industry practitioners are increasingly seeing the potential economic, social, environmental and climate benefits of engaging in the circular economy's closed loop and product service systems (Camilleri 2018b; Lieder and Rashid 2016; Ghisellini et al. 2016; Tukker 2015; Tukker and Tischner 2006). Many manufacturers are already perceiving the business case for the circular economy as they are using and reusing secondary resources. There are various businesses and non profit organisations that are engaged in repairing, refurbishing, restoring and/or recycling materials.

On the other hand, there are other practitioners that are opting to remain in their status quo as they rely on traditional, linear economic models (Bocken et al. 2016). This chapter referred to some other challenges that could slow down or prevent the businesses' engagement in the circular economy's sustainable approaches. It presented a cost-benefit analysis of advancing the circular economy strategy. Afterwards, it featured a critical review of the latest European Circular Economy Action Plan that comprises a sustainable product policy framework (EU 2020a, b). This research evaluated the Commission future-oriented agenda that is intended to foster a cleaner and more competitive Europe. In a nutshell, it shed light on the value chains of different products and resources, including batteries and vehicles, electronics/electrical products and components, packaging, food, water and nutrients, textiles, construction and building materials and plastics, among others. Arguably, it may appear that this plan has paved the way for more interrelated circular economy initiatives in different industry sectors. The EU is encouraging the businesses as well as their consumers to engage in sustainable production and consumption behaviours and to use and reuse products, materials and resources. This way they will minimise their impact on the natural environment, in terms of the generation of waste and emissions. Indeed, the EU's latest policy proposals as well as its action plans are a step in the right direction as they include key recommendations on our way forward towards a regenerative circular economy model.

5 Key Recommendations

The transition to the circular economy ought to be systemic, deep and transformative, within the EU context and beyond. It is very likely that it will be disruptive as it requires cooperation of all stakeholders including the policy makers, industry practitioners, consumers and non-government organisations, among others. The transition towards the circular economy can be facilitated if the governments would create a favourable climate for responsible, positive impact investing, by providing technical assistance and mobilising financial resources for sound investments in sustainable innovations (Camilleri 2020; 2015b). For instance, the European Green Deal Investment Plan (EIP) is currently supporting the sectors relating to the provision of sustainable energy, energy efficiency, sustainable cities and sustainable agricultural practices, among other areas.

The transition towards the circular economy is dependent on the stakeholders' willingness and capacity to collaborate and forge long term relationships with one another. There are many lessons to be learned from good practice as well as on failures. Success stories point out on the need to promote the economic return on investment, process improvements and product benefits to motivate businesses and investors to shift from the linear economy to the circular economy. The case studies of the responsible businesses that resorted to closed loop systems are indicating that there is market for used materials and resources. Moreover, many countries are providing the appropriate conditions for the development of clusters at supply chain level, as they enable businesses to exchange waste resources (EU 2015b; Geng et al. 2009). Any reductions in waste could translate to lower costs for the businesses themselves, as they do not need to handle, transport and store waste prior to its disposal. As a result, businesses will consume lower energy in their production processes. The main challenge is to create the right environment where businesses collaborate in supply chains. The development of clusters and eco-industrial parks would help them to turn their unwanted externalities into useable materials for others. Some industries lend themselves to circular initiatives more than others as the recirculation of their resources may be straightforward. For example, primary industries such as iron, steel and aluminium may need to be incentivised to minimise their waste in closed loop systems. Other industries, including solar and wind energy technologies, battery production and biotech materials will inevitably have to be re-assessed on the basis of their recirculation potential and performance over their whole life cycles.

The circular economy concept has the potential to maximise the functioning of global ecosystems Camilleri 2018a; Kirchherr et al. 2017; Murray et al. 2017; Stubbs and Cocklin 2008). Everyone has a responsibility to bear for the products' disposal at their end of the life. The businesses are encouraged to measure their direct and indirect environmental impacts. Circular-economy measures and eco labels may have to be developed for the big businesses and large undertakings. At present, there is no way of measuring how the companies are effectively circulating materials and resources in their manufacturing processes. The circular economy metrics can assess the ecological footprint of the businesses' sustainable production and consumption of materials and products during their operational processes. These metrics may also measure the economic benefits that can accrue from the adoption of closed loop systems. There is also scope for many corporations to legitimate and consolidate their social licence to operate. They are expected to comply (if they are still not complying) with the EU's non-financial reporting directive (Venturelli et al. 2019; Camilleri 2015a) and provide adequate and sufficient information on how they are transiting towards the circular economy strategy in their corporate disclosures (Camilleri 2017b, 2018b).

Acknowledgements The author thanks the editors and their reviewers for their constructive remarks and suggestions.

References

Amin-Chaudhry, A. (2016). Corporate social responsibility–from a mere concept to an expected business practice. *Social Responsibility Journal, 12*(1), 190–207.
Bocken, N. M., de Pauw, I., Bakker, C., & van der Grinten, B. (2016). Product design and business model strategies for a circular economy. *Journal of Industrial and Production Engineering, 33*(5), 308–320.
Camilleri, M. A. (2015a). Environmental, social and governance disclosures in Europe. *Sustainability Accounting, Management and Policy Journal, 6*(2), 224–242.
Camilleri, M. A. (2015b). Valuing stakeholder engagement and sustainability reporting. *Corporate Reputation Review, 18*(3), 210–222.
Camilleri, M. A. (2017a). Case study 5: Closing the loop of the circular economy for corporate sustainability and responsibility. In *Corporate sustainability, social responsibility and environmental management* (pp. 175–190). Cham: Springer.
Camilleri, M. A. (2017b). The integrated reporting of financial, social and sustainability capitals: A critical review and appraisal. *International Journal of Sustainable Society, 9*(4), 311–326.
Camilleri, M. A. (2018a). The circular economy's closed loop and product service systems for sustainable development: A review and appraisal. *Sustainable Development, 27*(3), 530–536.
Camilleri, M. A. (2018b). Theoretical insights on integrated reporting: The inclusion of non-financial capitals in corporate disclosures. *Corporate Communications: An International Journal, 23*(4), 567–581.
Camilleri, M. A. (2020). The market for socially responsible investing: A review of the developments. Social Responsibility Journal, Forthcoming https://doi.org/10.1108/SRJ-06-2019-0194.
Earthwatch. (2019). *Plastic rivers: Tackling the pollution on our doorsteps.* Retrieved April 13, 2020, from https://earthwatch.org.uk/get-involved/plastic-rivers
EASAC. (2020). *Packaging plastics in the circular.* European Academies Science Advisory Council, Halle Germany. Retrieved April 13, 2020, from https://easac.eu/fileadmin/PDF_s/reports_statements/Plastics/EASAC_Plastics_complete_Web_PDF.pdf
EEA. (2018). *Waste prevention in Europe — Policies, status and trends in reuse in 2017.* Luxembourg: European Environment Agency.
EEA. (2020). *Improving circular economy practices in the construction sector key to increasing material reuse, high quality recycling.* Retrieved April 13, 2020, from https://www.eea.europa.eu/highlights/improving-circular-economy-practices-in
EMF. (2013). *Towards the circular economy, Ellen MacArthur foundation rethinking the future.* Retrieved April 13, 2020, from http://www.ellenmacarthurfoundation.org/assets/downloads/publications/TCE_Report-2013.pdf
EMF. (2017). *A new textiles economy: Redesigning fashion's future.* Retrieved April 13, 2020, from https://www.ellenmacarthurfoundation.org/publications/a-new-textiles-economy-redesigning-fashions-future
EP. (2018). *A European strategy for plastics in a circular economy.* Retrieved April 13, 2020, from https://www.europarl.europa.eu/doceo/document/TA-8-2018-0352_EN.html
Estol, J., Camilleri, M. A., & Font, X. (2018). European Union tourism policy: An institutional theory critical discourse analysis. *Tourism Review, 73*(3), 421–431.

EU. (1994). *European Parliament and Council Directive 94/62/EC of 20 December 1994 on packaging and packaging waste, OJ L 365 31.12.1994, p. 10*. Retrieved April 13, 2020, from https://eur-lex.europa.eu/legal-content/en/TXT/?uri=CELEX:31994L0062

EU. (2006). *Directive 2006/66/EC of the European Parliament and of the Council of 6 September 2006 on batteries and accumulators and waste batteries and accumulators and repealing Directive 91/157/EEC, OJ L 266, 26.9.2006, p. 1*. Retrieved April 13, 2020

EU. (2008). *Directive 2008/98/EC of the European Parliament and of the Council of 19 November 2008 on waste and repealing certain Directives (Text with EEA relevance)*. Retrieved April 13, 2020, from https://eur-lex.europa.eu/legal-content/EN/TXT/?uri=celex%3A32008L0098

EU. (2014). *Non-financial reporting*. European Commission, Brussel, Belgium. Retrieved April 13, 2020, from http://ec.europa.eu/internal_market/accounting/non-financial_reporting/index_en.htm

EU. (2015a). *Closing the loop - An EU action plan for the Circular Economy, Communication from the Commission to the European Parliament, the Council, The European Economic and Social Committee and the Committee of the Regions*. Retrieved April 13, 2020, from http://eur-lex.europa.eu/legalcontent/EN/TXT/HTML/?uri=CELEX:52015DC0614andfrom=EN

EU. (2015b). *Towards an EU product policy framework contributing to the circular economy*. Retrieved April 13, 2020, from https://ec.europa.eu/info/law/better-regulation/have-your-say/initiatives/1740-Towards-an-EU-Product-Policy-Framework-contributing-to-the-Circular-Economy

EU. (2017). *Council conclusions on eco-innovation: enabling the transition towards a circular Economy*. European Council of the European Union. Retrieved April 13, 2020, from http://www.consilium.europa.eu/en/press/press-releases/2017/12/18/council-conclusions-oneco-innovation-transition-towards-a-circular-economy/#

EU. (2019a). *Farm to fork strategy for sustainable food*. Retrieved April 13, 2020, from https://ec.europa.eu/food/farm2fork_en

EU. (2019b). *Construction and demolition waste*. European Commission, Brussel, Belgium. Retrieved March 5, 2020, from https://ec.europa.eu/environment/waste/construction_demolition.htm

EU. (2019c). *Preparatory study for the ecodesign working plan 2020-2024*. Retrieved March 31, 2020, from https://ec.europa.eu/growth/content/preparatory-study-ecodesign-working-plan-2020-2024_en

EU. (2019d). *The drinking water directive*. Retrieved March 31, 2020, from https://ec.europa.eu/environment/water/water-drink/legislation_en.html

EU. (2019e). *A new tool to increase the sustainable use of nutrients across the EU*. Retrieved March 31, 2020, from https://ec.europa.eu/info/news/new-tool-increase-sustainable-use-nutrients-across-eu-2019-feb-19_en

EU. (2019f). *EU Directive 2019/904 Of the European Parliament and of the Council of 5 June 2019 on the reduction of the impact of certain plastic products on the environment*. Retrieved April 13, 2020, from https://eur-lex.europa.eu/legal-content/EN/TXT/PDF/?uri=CELEX:32019L0904

EU. (2020a). *Commission Staff Working Document. Leading the way to a global circular economy: state of play and outlook*. European Commission, Brussel, Belgium. Retrieved March 31, 2020, from https://ec.europa.eu/environment/circular-economy/pdf/leading_way_global_circular_economy.pdf

EU. (2020b). *A new circular economy plan for a cleaner and more competitive Europe*. European Commission, Brussels, Belgium. Retrieved March 31, 2020, from https://eur-lex.europa.eu/legal-content/EN/TXT/?qid=1583933814386&uri=COM:2020:98:FIN

EU. (2020c). *Guidance for the assessment of material efficiency: Application to smartphones*. JRC Technical Report. Luxembourg. Retrieved April 13, 2020, from https://publications.jrc.ec.europa.eu/repository/bitstream/JRC116106/jrc116106_jrc_e4c_task2_smartphones_final_publ_id.pdf

Eurobarometer. (2020). *Special Eurobarometer 503: Attitudes towards the impact of digitalisation on daily lives*. *EU Open Data Portal*. Retrieved April 13, 2020, from https://data.europa.eu/euodp/en/data/dataset/S2228_92_4_503_ENG

EUWID. (2019). *New European action plan for the circular economy*. EUWID Recycling and Waste Management. Retrieved April 13, 2020, from https://www.euwid-recycling.com/news/policy/single/Artikel/new-european-action-plan-for-the-circular-economy.html

Geissdoerfer, M., Savaget, P., Bocken, N. M., & Hultink, E. J. (2017). The circular economy–a new sustainability paradigm? *Journal of Cleaner Production, 143*, 757–768.

Geng, Y., & Doberstein, B. (2008). Developing the circular economy in China: Challenges and opportunities for achieving leapfrog development. *The International Journal of Sustainable Development and World Ecology., 15*(3), 231–239.

Geng, Y., Sarkis, J., Ulgiati, S., & Zhang, P. (2013). Measuring China's circular economy. *Science, 339*(6127), 1526–1527.

Geng, Y., Zhang, P., Côté, R. P., & Fujita, T. (2009). Assessment of the National eco-Industrial Park Standard for promoting industrial Symbiosis in China. *Journal of Industrial Ecology, 13*(1), 15–26.

Geyer, R., Jambeck, J. R., & Law, K. L. (2017). Production, use, and fate of all plastics ever made. *Science Advances, 3*(7), 1–5.

Ghisellini, P., Cialani, C., & Ulgiati, S. (2016). A review on circular economy: The expected transition to a balanced interplay of environmental and economic systems. *Journal of Cleaner Production, 114*, 11–32.

Goss, A., & Roberts, G. S. (2011). The impact of corporate social responsibility on the cost of bank loans. *Journal of Banking & Finance, 35*(7), 1794–1810.

Haas, W., Krausmann, F., Wiedenhofer, D., & Heinz, M. (2015). How circular is the global economy? An assessment of material flows, waste production, and recycling in the European Union and the world in 2005. *Journal of Industrial Ecology, 19*(5), 765–777.

Hu, J., Xiao, Z., Zhou, R., Deng, W., Wang, M., & Ma, S. (2011). Ecological utilization of leather tannery waste with circular economy model. *Journal of Cleaner Production, 19*(2), 221–228.

Jensen, N. M. (2003). Democratic governance and multinational corporations: Political regimes and inflows of foreign direct investment. *International Organization, 57*(3), 587–616.

Kirchherr, J., Reike, D., & Hekkert, M. (2017). Conceptualizing the circular economy: An analysis of 114 definitions. *Resources, Conservation and Recycling, 127*, 221–232.

Lieder, M., & Rashid, A. (2016). Towards circular economy implementation: A comprehensive review in context of manufacturing industry. *Journal of Cleaner Production, 115*, 36–51.

Liu, Q., Li, H.-M., Zuo, X.-L., Zhang, F.-F., & Wang, L. (2009). A survey and analysis on public awareness and performance for promoting circular economy in China: A case study from Tianjin. *Journal of Cleaner Production, 17*, 265–270.

Mont, O. K. (2002). Clarifying the concept of product–service system. *Journal of Cleaner Production, 10*(3), 237–245.

Moratis, L., Melissen, F., & Idowu, S. O. (2018). *Sustainable Business Models*. Cham: Springer.

Murray, A., Skene, K., & Haynes, K. (2017). The circular economy: An interdisciplinary exploration of the concept and application in a global context. *Journal of Business Ethics, 140*(3), 369–380.

National Geographic. (2018). *A whopping 91% of plastic isn't recycled*. Retrieved April 13, 2020, from https://www.nationalgeographic.com/news/2017/07/plastic-produced-recycling-waste-ocean-trash-debris-environment/

OECD. (2019a). *Global Material Resources Outlook to 2060*. Retrieved April 13, 2020, from https://www.oecd.org/environment/global-material-resources-outlook-to-2060-9789264307452-en.htm

OECD. (2019b). *Policy approaches to incentivize sustainable plastic design*. Environment Working Paper No. 149. Retrieved April 13, 2020, from https://www.oecd.org/officialdocuments/publicdisplaydocumentpdf/?cote=ENV/WKP(2019)8&docLanguage=En

Peeters, J. R., Vanegas, P., Tange, L., Van Houwelingen, J., & Duflou, J. R. (2014). Closed loop recycling of plastics containing flame retardants. *Resources, Conservation and Recycling, 84*, 35–43.

Pomponi, F., & Moncaster, A. (2017). Circular economy for the built environment: A research framework. *Journal of Cleaner Production, 143*, 710–718.

Porter, M. E., & Van der Linde, C. (1995). Green and competitive: Ending the stalemate. *Harvard Business Review, 73*(5), 120–134.

Prakash, A. (2002). Green marketing, public policy and managerial strategies. *Business Strategy and the Environment, 11*(5), 285–297.

Rasiah, R., Gammeltoft, P., & Jiang, Y. (2010). Home government policies for outward FDI from emerging economies: Lessons from Asia. *International Journal of Emerging Markets, 5*(3-4), 333–357.

RoHS. (2002). *Directive 2002/95/EC of the European Parliament and of the Council of 27 January 2003 on the restriction of the use of certain hazardous substances in electrical and electronic equipment.* Retrieved March 15, 2020, from https://eur-lex.europa.eu/legal-content/EN/TXT/HTML/?uri=CELEX:32002L0095&from=EN

RoHS. (2017). *EU Directive 2017/2012 of the European Parliament and of the Council of 15 November 2017 amending Directive 2011/65/EU on the restriction of the use of certain hazardous substances in electrical and electronic equipment.* EU Commission, Brussels, Belgium. https://eur-lex.europa.eu/legal-content/EN/TXT/HTML/?uri=CELEX:32017L2102&from=EN

Smol, M., Kulczycka, J., Henclik, A., Gorazda, K., & Wzorek, Z. (2015). The possible use of sewage sludge ash (SSA) in the construction industry as a way towards a circular economy. *Journal of Cleaner Production, 95*, 45–54.

Stahel, W. R. (2016). Circular economy: A new relationship with our goods and materials would save resources and energy and create local jobs. *Nature, 531*(7595), 435–439.

Stubbs, W., & Cocklin, C. (2008). Conceptualizing a sustainability business model. *Organization and Environment, 21*(2), 103–127.

Tam, V. W. Y., & Lu, W. (2016). Construction waste management profiles, practices, and performance: A cross-jurisdictional analysis in four countries. *Sustainability, 8*(2), 190–206.

Tukker, A. (2015). Product services for a resource-efficient and circular economy–a review. *Journal of Cleaner Production, 97*, 76–91.

Tukker, A., & Tischner, U. (2006). Product-services as a research field: Past, present and future. Reflections from a decade of research. *Journal of Cleaner Production, 14*(17), 1552–1556.

UNEP. (2018). *Single use plastic.* Retrieved April 13, 2020, from https://wedocs.unep.org/bitstream/handle/20.500.11822/25496/singleUsePlastic_sustainability.pdf

UNEP. (2019). *Global resources outlook. Natural Resources for the future we want.* Retrieved April 13, 2020, from https://www.resourcepanel.org/file/1191/download?token=oxkXHwCD

UNEP. (2020). *Stocktaking of initiative on plastic waste.* Retrieved April 13, 2020, from http://www.basel.int/Implementation/Plasticwaste/PlasticWastePartnership/Meetings/PWPWG1/tabid/8305/ctl/Download/mid/23074/Default.aspx?id=9&ObjID=22867

Venturelli, A., Caputo, F., Leopizzi, R., & Pizzi, S. (2019). The state of art of corporate social disclosure before the introduction of non-financial reporting directive: A cross country analysis. *Social Responsibility Journal., 15*(4), 409–423.

Wang, G. H., Wang, Y. N., & Zhao, T. (2008). Analysis of interactions among barriers to energy saving in China. *Energy Policy, 36*, 1879–1889.

WEEE. (2002)**.** *Directive 2002/96/EC of the European Parliament and of the Council of 27 January 2003 on waste electrical and electronic equipment (WEEE) - Joint declaration of the European Parliament, the Council and the Commission relating to Article 9.* Retrieved April 13, 2020, from https://eur-lex.europa.eu/legal-content/EN/TXT/?uri=CELEX:32002L0096

WEEE. (2012). *Directive 2012/19/EU of the European Parliament and of the Council f 4 July 2012 on waste electrical and electronic equipment (WEEE).* Retrieved April 13, 2020, from https://eur-lex.europa.eu/legal-content/EN/TXT/?uri=CELEX:32012L0019

WEF. (2019). *5 steps that could end the plastic pollution crisis – and save our ocean.* Retrieved April 13, 2020, from https://www.weforum.org/agenda/2019/03/5-steps-that-could-end-the-plastic-pollution-crisis-and-save-our-oceans-eb7d4caf24/

Wong, S. L., Ngadi, N., Abdullah, T. A. T., & Inuwa, I. M. (2015). Current state and future prospects of plastic waste as source of fuel: A review. *Renewable and Sustainable Energy Reviews, 50*, 1167–1180.

World Bank. (2018). *What a waste 2.0: A global snapshot of solid waste management to 2050.* Retrieved April 13, 2020, from https://www.worldbank.org/en/news/infographic/2018/09/20/what-a-waste-20-a-global-snapshot-of-solid-waste-management-to-2050

Mark Anthony Camilleri is an Associate Professor in the Department of Corporate Communication at the University of Malta. He successfully finalised his full-time PhD (Management) in three years' time at the University of Edinburgh in Scotland—where he was also nominated for his "Excellence in Teaching". He also holds an MBA from the University of Leicester and an MSc from the University of Portsmouth, among other qualifications. During the past years, Mark taught business subjects at under-graduate, vocational and post-graduate levels in Hong Kong, Malta, UAE and the UK.

Prof. Camilleri is a member in the Global Reporting Initiative (GRI)'s Stakeholder Council, where he is representing the European civil society. He is a scientific expert in research for the Ministero dell' Istruzione, dell' Universita e della Ricerca (in Italy) and a reviewer for the Austrian Science Fund (FWF). He is an editorial board member in a number of academic journals, conferences and committees. Mark has authored and edited seven books for Emerald, IGI Global Springer, among others. He published more than 100 contributions in high-impact, peer-reviewed journals, chapters and conferences.

Recycling Initiatives in Romania and Reluctance to Change

Silvia Puiu

Abstract Romania is one of the countries that received in 2018 an early warning report from European Union, for the risk of not reaching the target of 50% reuse/recycling of municipal waste until 2020. The report highlighted that the landfilling rate was one of the highest in Europe (69%) and the recycling rate was only 13% in 2016. These numbers are worrying, reflecting a lack of administrative measures but also a reluctance to change from the population, mainly because of the lack of education regarding recycling and the negative consequences of waste for the environment. The main objectives of the paper are to analyze statistics in Romania and other EU countries on waste and recycling, the causes of this phenomenon and to establish the necessary measures that could be implemented in order to raise the recycling rate and the awareness of the population regarding environmental issues. The research methodology consists in applying a survey to analyze the reluctance to change of the population and their perception on recycling and reuse but also in conducting interviews with specialists from public authorities and NGOs regarding this issue. The results will consist in offering solutions for raising the awareness on these problems, developing the infrastructure for recycling waste, highlighting good practice examples from other EU countries.

1 Introduction

Climate change became a reality and countries should seriously take into account the 13th Goal of the 2030 Agenda for Sustainable Development adopted by United Nations—Climate Action. This goal can be achieved by implementing environmental policies. Thus, the European Union proposed such a policy in order to reduce gas emissions until 2050 comparing with the level registered in 1990 (40% reduction till 2030, 60% till 2040 and 80-95% till 2050). These objectives will keep the increase of the global temperature below 2 degrees Celsius (European Commission 2011).

S. Puiu (✉)
University of Craiova, Faculty of Economics and Business Administration, Craiova, Romania

According to Eurostat (2020a), waste is responsible for 3.1% of the gas emissions, being the fourth source, after fuel combustion (76.7%), agriculture (9.8%) and industry (8.4%). From the level in 1990, there is a 42% reduction in the share of this sector in the total gas emissions, but still the smallest one comparative with the other sectors (Eurostat 2020a). The European statistics analyze the three types of waste management: incineration, landfill and recycling. Landfill waste was reduced by more than 50% between 1995 and 2017. This was possible by raising the recycling rate and realizing that some raw materials used in the industry are limited, thus recycling being seen as a solution.

At the level of European Union, the changes regarding waste management were possible with the help of specific directives such as: Waste Framework Directive (Directive 2008/98/EC), Landfill Directive (Council Directive 1999/31/EC), Waste Electrical and Electronic Equipment—WEEE (Directive 2002/96/EC, Directive 2012/19/EU), Directive 94/62/EC on packaging and packaging waste, Directive 2000/53/EC on end of life vehicles and Directive 2006/66/EC on batteries and accumulators and waste batteries and accumulators. All these directives became a part of the Circular Economy Package that was published in 2015. Stahel (2019) consider the circular economy as the path to a more efficient economy and also to a more sustainable one. It is in fact the opposite of the old model of a linear economy where limited resources are wasted. The package published by European Commission was based mainly on the waste management concept.

Circular economy does not waste resources, the ideal being not to throw anything away and reintroduce resources in the economy. The advantages are not only for the environment but also for the health of the population. Waste has a negative impact on the economy, the environment and the health of people. Alam and Ahmade (2013), p. 168 consider waste "a serious health hazard" and should be managed accordingly. "Profound health repercussions" are also highlighted as a consequence of a poor waste management by Ijjasz-Vasquez (2018), Senior Director of The World Bank.

A briefing of European Parliamentary Research Service (2016) mentions the important gap between the EU member states regarding waste management, offering as an example Germany with a share of recycling and composting of 65% and the case of Romania with only 3%. Thus, the target for 2020 (reducing household waste by 50%) was in danger of not being accomplished in many EU countries. At the same time, the share of landfill had important variations between countries (less than 5% in some of them and in others, more than 70%, the average being 31%).

There are various waste streams related to either materials (paper, glass, metal, textile, wood, plastics, bio-waste) or products (electronics, batteries and accumulators, end of life vehicles, construction and demolition waste). Municipal waste is represented by the waste produced by the public and this is difficult to manage because there are numerous producers and the responsibility is shared among them (European Commission 2018). Ferronato and Torretta (2019) appreciate that a solid waste management is necessary for reducing the environmental footprint and there should be treatment solutions for each type of waste.

European Commission (2018) revised the recycling targets for municipal waste—55% by 2025, 60% by 2030 and 65% by 2035, for packaging waste—65% by 2025 and 70% by 2030 and for landfilling—10% by 2035.

2 Recycling in Europe

The reasoning behind the need to reach the recycling targets is represented by the need to use the limited resources of the planet more efficiently and reintroduce them in the economy that will thus become a circular economy with an increased productivity. According to European Environment Agency (2019a), recycling rates for each category of waste stream (municipal waste, packaging waste, electrical and electronic equipment waste) improved considerably between 2004 and 2017 in Europe. Thus, municipal waste recycling rate improved from 30% in 2004 to 46.4% in 2017, packaging waste recycling rate increased from 54.6% in 2005 to 67.1% in 2016 and that of electrical and electronic equipment increased from 27.8% in 2010 to 41.2% in 2016. The share of municipal waste in the total waste of European Union member states is 10%, with important gaps between the European countries as we can notice in Table 1. This can be recycled or composted, incinerated or landfilled. The aim is to increase the recycling rate and thus introducing materials into the circular economy and at the same time reducing the waste deposited by landfilling or incinerated that can pose serious threats for environment and health.

As we can notice in Table 1, Germany has the highest recycling rate for municipal waste (68% in 2017), followed by five countries that have a rate of more than 50% (Austria, Slovenia, Belgium, Netherlands, Switzerland) and other five countries with a rate of more than 45% (Italy, Lithuania, Luxembourg, Sweden, Denmark). From the EU countries, ten have a higher recycling rate comparing with the EU-28 average. Greece, Cyprus, Romania, Turkey and Malta have the lowest recycling

Table 1 Municipal waste recycling rate in Europe between 2004 and 2017

	2004	2017		2004	2017		2004	2017
Germany	56%	68%	EU-28	30%	46%	Spain	31%	33%
Austria	57%	58%	United Kingdom	23%	44%	Slovakia	6%	30%
Slovenia	20%	58%	France	29%	43%	Estonia	25%	28%
Belgium	52%	54%	Finland	34%	41%	Portugal	13%	28%
Netherlands	47%	54%	Ireland	29%	41%	Croatia	3%	24%
Switzerland	49%	53%	Norway	37%	39%	Latvia	5%	23%
Italy	18%	48%	Czechia	12%	38%	Greece	10%	19%
Lithuania	2%	48%	Bulgaria	17%	35%	Cyprus	3%	16%
Luxembourg	41%	48%	Hungary	12%	35%	Romania	1%	14%
Sweden	44%	47%	Poland	6%	34%	Turkey	1%	9%
Denmark	41%	46%	Iceland	18%	33%	Malta	6%	6%

Source: table created using data from European Environment Agency (2019a)

Table 2 Packaging waste recycling rate in Europe between 2005 and 2016

	2005	2016		2005	2016		2005	2016
Belgium	76.8%	81.9%	Italy	53.7%	66.9%	Romania	23%	60.4%
Denmark	52.5%	79%	Austria	66.9%	66.8%	Cyprus	11.1%	59.8%
Czechia	59%	75.3%	Greece	41.8%	66.1%	Poland	29.5%	58%
Netherlands	59.4%	72.6%	France	53.3%	66%	Latvia	47%	57.7%
Germany	68.2%	70.7%	Slovakia	29.8%	65.8%	Norway	–	57.2%
Spain	50.4%	70.3%	United Kingdom	54.4%	64.7%	Estonia	40.3%	56%
Lithuania	32.5%	69.5%	Finland	43.2%	64.7%	Croatia	–	54.7%
Slovenia	45.3%	69.4%	Bulgaria	30.8%	63.8%	Iceland	–	51.6%
Sweden	48.2%	68.2%	Luxembourg	62.6%	61.5%	Hungary	45.9%	49.7%
Ireland	55.6%	67%	Portugal	44.3%	60.9%	Malta	8.1%	39.7%

Source: table created using data from European Environment Agency (2019a)

rates. The most important evolution between 2004 and 2017 was registered by Lithuania (an increase from 2% to 48%). Thus, only six countries exceeded the target for 2020 (50% recycling rate). The gap between the European countries is important and there should be implemented more measures by the governments to increase this rate and create the foundation for a circular economy that on the long term will create more benefits and lead to sustainable development.

Regarding the packaging waste recycling rate, the situation is presented in Table 2 where we can see that the gaps between European countries are not so significant as in the situation of municipal waste. Many countries are close to reaching the 2025 target (65%).

From the 30 European countries that were analyzed, 15 exceeded the recycling target for 2025 for packaging waste. At the level of EU-28, between 2005 and 2016 this rate increased with 13%, from 54% to 67%. From the other 15, 13 registered a recycling rate higher than 50% and only two countries had a rate below 50% (Hungary and Malta, the former being very close to 50% and the latter under 40%). According to European Environment Agency (2019a), the increase at the level of the EU slowed after 2013, the difference between 2013 and 2016 being only 2%. The differences between countries are more important in the case of municipal waste but an explanation could be the fact that the first target for this type of waste was established in 2008 for 2020, meanwhile for the packaging waste, the targets were established in 1994. The packaging waste refer to waste generated by companies meanwhile municipal waste refer to the one generated by households (including packaging waste), so it is much easier to determine companies to comply to some regulations and also much easier to control them.

If the recycling rate has to be higher, the landfilling rate has to decrease until 2035 to 10%. Landfilling has important negative consequences on environment, health and the economy and the tendency should be to reduce it as much as possible. This should be correlated with education of the population regarding recycling because not all materials are recyclable and improper recycling could do more harm than good. We notice from Table 3 that ten European countries (Switzerland, Sweden, Denmark, Germany, Belgium, Finland, Netherlands, Austria, Norway, Luxembourg) succeeded to reach the target of 10% since 2017, having rates below this level.

The landfilling rate is very high in countries like Slovakia, Bulgaria, Romania, Croatia, Greece, Cyprus and Malta that have a landfilling rate of more than 60% in 2017. The average rate in European Union for landfilling is 23.5% in 2017, 16 of the 28 countries (at the time of the latest data provided by European Environment Agency) have a rate above the average in European Union. What is more interesting is the case of Switzerland that has no landfilled waste, the country having a landfilling tax since 2001 (Herczeg 2013) and also successfully managing to reuse, recycle or producing energy from its waste. Landfilling should be the last solution for waste management, the other more desirable ones are recycling, composting and reusing the waste. This should be seen as a valuable resource that can be reintegrated in the economy, creating the circular economy that brings numerous benefits for all actors involved. And reducing unnecessary consumption can also lead to less waste in general.

Table 1 shows the municipal waste recycling rate in Europe but the quantity of municipal waste generated by a person in each of these countries is also important. From Table 4 we notice that there are also serious differences between the countries. The ideal would be to generate less waste or if the quantity is higher to at least recycle more.

According to Eurostat (2020b), the recycling of materials (in kg per person) in European Union increased from 54 kg in 1995 to 150 kg in 2018 and the composting increased from 33 kg to 84 kg during these years. This reveals an improved situation and what Table 4 shows is that developed countries in Europe generate more waste because people there consume more and this is normal. Romania is the European country with the least amount of municipal waste generated per person (only 272 kg). Further, we will present the recycling situation in Romania taking into account Eurostat data and also conducting a qualitative research.

3 Recycling in Romania

European Commission (2019), p. 3 appreciates that Romania should take more measures in order to reach the targets for recycling and landfilling, considering that waste management in this country "remains a key challenge" and that the progress was purely formal, referring to the National Plan for Waste Management adopted in 2017. We see from Table 1 that Romania had a municipal waste recycling rate of only 14% in 2017, far from the 50% target established for 2020. Another aspect revealed by the report of European Commission referred to the lack of awareness on environmental issues and the lack of private initiatives in this area, most companies being preoccupied by the costs and not by green innovations in their sectors. At the same time, we notice from Table 3 that the landfilling rate is very high

Table 3 Municipal waste landfill rates in Europe between 2006 and 2017

	2006	2017		2006	2017		2006	2017
Malta	86.56%	93.49%	Czechia	76.83%	48.45%	Luxembourg	17.34%	6.91%
Cyprus	95.78%	81.98%	Poland	90.99%	41.77%	Norway	18.23%	3.47%
Greece	87.17%	80.06%	Lithuania	97.9%	33.02%	Austria	10.03%	2.08%
Croatia	100%	75.38%	Latvia	94.5%	31.34%	Netherlands	2.88%	1.41%
Romania	99.35%	71.05%	Ireland	63.9%	26.16%	Finland	57.85%	0.92%
Bulgaria	76.63%	61.97%	Italy	62.75%	25.71%	Belgium	9.86%	0.89%
Slovakia	81.12%	60.57%	EU-28	44.6%	23.5%	Germany	0.66%	0.88%
Iceland	90%	57.27%	France	36.24%	21.59%	Denmark	5.05%	0.84%
Spain	59.74%	53.59%	Estonia	79.87%	19.92%	Sweden	5.08%	0.44%
Portugal	64.17%	49.58%	United Kingdom	60.32%	16.87%	Switzerland	0%	0%
Hungary	81.18%	48.64%	Slovenia	81.83%	12.81%			

Source: table created using data from European Environment Agency (2019b)

Table 4 Municipal waste generated per person in Europe in 2018 (in kg)

	Quantity		Quantity		Quantity
Denmark	766	Greece	504	Latvia	407
Malta	640	Italy	499	Estonia	405
Cyprus	637	European Union	492	Hungary	381
Germany	615	Slovenia	486	Czechia	351
Luxembourg	610	Spain	475	Poland	329
Austria	579	Lithuania	464	Romania	272
Ireland	576	Sweden	434	United Kingdom	463
Finland	551	Croatia	432	Norway	739
France	527	Bulgaria	423	Switzerland	703
Netherlands	511	Slovakia	414	Iceland	656
Portugal	508	Belgium	411		

Source: table created using data from Eurostat (2020b)

(71.5%) comparing with the EU average (44.6%). In these conditions, Romania should increase recycling and composting (thus reducing its landfilling rate). The municipal waste per person is the lowest in Romania (272 kg) as we see in Table 4, this at least being a good aspect for the moment, taking into account that the recycling rate is low and the landfilling rate is high.

According to the Romanian Government (2017) that adopted the National Plan for Waste Management (NPWM), waste management integrated systems (WMIS) are in course of implementation in most counties. These systems target investments for creating a proper infrastructure, acquiring equipment needed for separate collection, waste treatment, composting and having legal deposits for waste. Separate collection in Romania is made on three fractions (paper, plastic/metal, glass) in 24 of the 34 counties where WMIS are implemented. As the Environmental Implementation Review of European Commission (2019) states, the most important problems of the country are the noncompliant deposits, the high rate of landfilling, the lack of infrastructure for separate collection and low collection of bio-waste. As solutions, the report highlights that there could be implemented measures like: a landfilling tax that could lead to an increase of the municipal waste recycling rate; closing improper and illegal deposits; to develop proper ones; treat the waste before disposal on landfills; establish fines for those local authorities that do not succeed to reach the recycling and landfilling rates; and to make separate collection accessible for people. We can add to these the need to inform citizens regarding separate collection, the importance of recycling and the environmental and health consequences of having a high landfilling rate.

Accessing European funds for the implementation of WMIS, there were created 39 Intercommunity Development Associations in order to properly manage the waste in Romania (Romanian Government 2017). There are some discrepancies between the report of European Commission (2019) and the NPWM of the Romanian government (2017). The former mentions the many problems of Romania regarding waste management and the formal NPWM that was adopted, the dependency on EU funds and bad management of environmental aspects. The latter has a more optimistic tone, appreciating that most objectives included in the previous NPWM were reached between 2003 and 2013 but still mentioning a few problems: low level of composting, of recycling, of separate collection, landfilling without a previous waste treatment, some lacks in the national legislation and administrative problems.

4 Research Methodology

The methodology takes into account the objectives of the research: identify the main differences between Romania and the other European countries and create a context for the present situation regarding recycling and waste management in general in this country; establish the causes for having a low recycling rate, a high landfilling rate and the lowest waste quantity per capita; highlight solutions that can be implemented by the government in order to reach the EU targets on waste and reduce the gaps.

In order to reach these objectives, there were formulated two main research questions:

RQ1: What are the main obstacles that prevent Romania from reaching the targets on recycling and landfilling, from the perspective of public authorities and NGOs and also that of individuals?

RQ2: What is the level of knowledge and awareness regarding environmental issues (separate collection, recycling, reducing, reusing, waste management) and the population willingness to change their habits in order to increase the recycling rate and help authorities in the process of waste management?

To answer these questions, the following tools were used as research methodology: comparison of the data on waste management in Romania and the other European countries; a documentary analysis using the latest NPWM published in Romania and the report of European Commission (2019) on this country's situation; an analysis using the survey applied to individuals in order to identify the obstacles to recycling, if there is a reluctance to change or not, what are the solutions for a better waste management in the population's perspective and also to see the level of knowledge and awareness among citizens in matters related to recycling, separate collection and waste management; a qualitative analysis consisting in conducting interviews with public authorities and NGOs dealing with waste management responsibilities.

5 Analysis of Individuals' Perspective on Recycling

For this analysis, we prepared a survey that was conducted in March 2020. The online questionnaire was created with Google Forms and distributed on social media channels. The questionnaire was distributed in various online groups to an audience of 8000 members. There were only 243 people that submitted their answers; the demographic characteristics of the members of the social media groups reveal a person with the age between 25 and 45, mostly from urban areas. Unfortunately, the response rate is low (3.04%) and the limitations of this research tool are important. We cannot extrapolate the answers to the entire Romanian population because the sample is not representative. We can still use the data gathered to identify problems and obstacles and offer an idea about the recycling behaviour of these individuals, their willingness to accept changes and their knowledge and awareness on issues related to waste management. To create an equilibrium and present the situation in Romania more objectively, we conducted a qualitative analysis consisting in interviews with public authorities and NGOs responsible of waste management. All these data are also correlated with those provided by Eurostat. The questionnaire included also open questions where respondents expressed their opinions, thus offering qualitative data that can help us to create a better image on the perspectives of these people.

Asked if they are informed about waste management, separate collection and recycling, 69.1% of the respondents said they are informed and very well informed, followed by those who mentioned they are less informed (27.6%). Only 3.3% of them said they are not at all informed. Correlating this with the low recycling rates in Romania, we can say that there is either an important difference between real knowledge and the perceived knowledge on these issues or a lack of motivation to change their behaviour. The increasing rates during previous years (we see in Table 1 that from 2004 to 2017 there is a progress from 1% to 14% for the municipal waste recycling rate) show us that the situation improved but not so much. Therefore, the real knowledge is definitely lower than the perceived knowledge taking into account the tendency of individuals to over evaluate. 94.7% of the respondents said they want to know more about these issues (92.31% of those very well informed, 93.8% of those well informed and 97.3% of those less informed), thus strengthening our conclusion that their perceived knowledge is higher than the real knowledge. Otherwise those who said they are very well informed would not need to learn the basics of separate collection and recycling. Learning is a lifelong process so no matter if the perceived knowledge is higher or not than the real one, the fact that most of the respondents answered they are willing to learn more is a good thing creating the premises for a higher recycling rate in the future. The same people that said they want to learn more, appreciate the importance of separate collection and recycling for the environment. The fact they consider these issues important show their awareness regarding waste management.

Another set of questions targeted the recycling habits of people and their willingness to change some of these habits in order to contribute to a higher recycling rate in Romania. Most of the respondents (63.8%) mentioned they recycle even if this requires an effort from their part. In the colloquial language, people mostly use the term recycle and not separate collection. Another term that entered the colloquial language is selective collection even if the correct term defined in Law 211/2011 republished in 2014 (Romanian Parliament 2014) is separate collection. 26.3% of the respondents said they recycle only if they do not have to make an important effort, in other words only if they have accessibility. This share is rather high and could lead to an increase of the recycling rate if authorities would implement more measures and create the proper infrastructure for separate collection and recycling of waste. People are willing to change their habits but they need the same willingness from the authorities responsible with waste management. This question is related with the one where respondents mentioned if they separately collect waste at home, going afterwards to the containers nearby. Most of them (43.6%) said they do this all the time, followed by 29.6% of them that said they do this very rarely because the containers are far from where they live. Also, 25.1% mentioned they do not recycle because the authorities do not offer the adequate infrastructure. From these motives, we can presume the number of people collecting separately would be higher if they would have the proper infrastructure accessible to them, making it easier for them to have a behaviour that protects the environment. The respondents mentioned they recycle mostly plastics (77.4%), paper (69.1%), glass (65.4%), batteries (60.5%), aluminium (45.3%).

We also wanted to address the motives for both recycling and not recycling and thus help us identify the obstacles, aspects that we included in the interviews conducted with local authorities and NGOs. So, with all the limits of this analysis, we succeeded to establish some reasons for which people do not recycle in order for authorities to be able to implement more measures in this direction and thus achieving the recycling targets. As reasons for recycling, most of the respondents mentioned the environment protection (87.9%), followed by protecting our health (59.3%), bringing benefits for the economy (32.5%), feeling useful (29%). As reasons for not recycling, most of them mentioned the lack of infrastructure for separate collection and recycling (78%), followed by not knowing what to recycle and how (10%). All these could be corrected by authorities investing more in infrastructure and in campaigns to raise the awareness and the level of knowledge regarding waste management, recycling, separate collection.

Asked about the accessibility of the recycling and separate collection in their town, most of them (42%) said the accessibility is low, followed by those who mentioned the lack of accessibility (12.8%). Only 24.7% of the respondents appreciated that the separate collection is accessible to them. The answers to this question correlated with the reasons for not recycling can offer valuable insights for the authorities in order to better understand the way people make their decisions regarding separate collection and contribute to a change in their behaviour with appropriate measures. In correlation with these questions, we asked the respondents about the actual distance between their homes and the containers for separate collection, this being one of the indicators for the accessibility. If the distance is higher, the accessibility decreases and with it also the recycling rate in Romania.

Thus, most of the respondents (51.4%) said these containers are at more than 500 m away, followed by those who said they are at less than 500 m (27.6%). The distance that someone is willing to go for collecting separately depends on the effort put in this activity. Only 37.4% of them said they would go far than 500 m so there is an important loss of waste that can be recycled because these containers are too far from people's homes; 26.7% of the respondents said they would collect separately only if these containers are at less than 250 m away and 34.6% of them would be willing to go to containers situated between 250 and 500 m from their home. These data cannot be extrapolated to the entire population due to the limitations mentioned at the beginning but they can be used as a starting point for drawing some directions for the authorities. The fact that containers are too far from someone's home is a reality and it is difficult for people to carry bags with plastics, paper, glass walking more than 500 m to collect separately. It is the authorities' responsibility to create a proper infrastructure and ensure accessibility for everyone.

In order to find if people are aware of what is happening to the waste that is not recycled in Romania, most of them (51.9%) answered they do not know. This is worrying because Romania has a very landfilling rate (71.05% in 2017). In order to reach the EU target for landfilling (10% in 2035), Romania would have to take more efficient measures and at the same time organizing campaigns to inform citizens about all these issues that affect everyone. Most respondents said they would recycle more if there would be more containers for separate collection close to them (97.53%), if there would be more campaigns for raising awareness and inform citizens about waste management (92.18%) and at the same time giving more fines for those who do not collect separately (88.48%). Asked about what should authorities do to raise the recycling rate, they mentioned mostly the same measures that they appreciated as convincing them to recycle more: placing the containers close to where people live (84%), creating more campaigns for raising the knowledge and awareness related to waste management and climate change (69.1%), giving fines for those who do not collect separately (68.7%) and offer bags for each household in order to increase the accessibility of separate collection and make it easier for people to change their behaviour (60.1%). The last measure would also facilitate the way to identify people who do not collect separately.

Most of the respondents (73.7%) would agree to pay an additional tax of 1 euro/month or 10 euros/year if that will help local authorities to have a more efficient waste management. They had to justify their choice, this open question being optional. 60.9% of the respondents chose to express their opinion thus showing they are interested on these issues ("recycling is important", "it is also our duty to protect the environment", "paying the tax would motivate me more to collect separately", "if you want quality, you have to pay more"). Those who will not accept to pay the tax offered as arguments their low income and the lack of trust in the authorities ("the problem is not money but the way the authorities manage the budget", "there is corruption", "the authorities are interested in making profits not in protect the environment", "the money will be used for something else"). According to Transparency International (2020), the Corruption Perception Index in Romania in 2019 was 44 on a scale from 0 (very corrupt) to 100 (very clean). Thus, it is easy to understand why some people would not be so confident to pay another tax to the local authorities. From those who would accept to pay the tax, there were many who also expressed their scepticism but still understood the importance of recycling. These reasons are helpful not as a prediction for collecting a dedicated tax but as an indicator of the way people perceive the situation and also the relationship with the authorities. In Romania, there are WMIS in most counties created with the help of European funds. So, there are financial resources but the problem is with the capacity to manage these systems, as European Commission (2019) also mentioned in its last report on the progress made by Romania. The answers given to the open question can serve authorities at creating better campaigns that target all the concerns people have regarding waste management and thus changing the general perception and increasing the level of trust in them.

The analysis offers answers to the two RQ formulated for this research, thus the obstacles to recycling from individuals' perspective are: the inappropriate infrastructure, the low level of accessibility to the infrastructure, the lack of awareness and knowledge regarding the importance of recycling and waste management. In order to change their habits, people want more accessibility to the containers for separate collection, more fines for those who do not separately collect and also more campaigns for raising the level of knowledge and awareness regarding environmental issues, thus revealing there is a need to know more about recycling, separate collection and waste management in general.

6 Qualitative Analysis of Public Authorities and NGOs' Perspectives on Recycling

In order to raise the level of objectivity of the analysis and extend the research beyond the individuals' perspective, we conducted a qualitative analysis during March 2020 consisting in interviews with representatives of public authorities and NGOs with activities and responsibilities in the domain of waste management or having as a main objective protection of the environment. Using this tool, we addressed also the problems highlighted by individuals in the questionnaire discussed above. Anticipating a high rate of non-responses, we sent the interview by e-mail (for public authorities and NGOs) and social media channels (for NGOs in addition to the e-mail). We sent 117 e-mails and seven reminders on social media for NGOs. From these, we received answers from seven NGOs developing activities for the protection of the environment, activities related to separate collection, recycling and educational activities for raising the awareness on the importance of these issues. Among public authorities, we received responses from the Ministry of Environment, Water and Forests, Oradea City Hall and from two Intercommunity Development Associations (IDA). IDAs have as a main objective public utility services (Romanian Parliament 2006) and were created by local authorities in a county in order to access European funds for waste management. So, they are NGOs but strongly connected with local authorities. Thus, local administration assign IDAs to act in their name on issues related to waste management (separate collection, recycling).

The interview consisted in six questions and the respondents expressed the willingness to be part of the research revealing the organizational name. They also had the possibility to be included in the paper using an identification number but all of them agreed that the name of the organization be fully disclosed in the present research. The answers to these questions are centralised in Table 5 for NGOs and in Table 6 for public authorities and IDAs.

Q1: Which are the main measures initiated by you (as an NGO or public authority) for raising separate collection in the community?
Q2: What are your main objectives regarding separate collection and recycling?
Q3: Do you think separate collection and recycling rates in Romania/your region are improving or the opposite?
Q4: Which are the main obstacles or limits to separate collection in Romania/your region?
Q5: What measures could be taken to raise the level of separate collection?
Q6: What is the level of knowledge and awareness among citizens regarding separate collection/recycling and what do you do for raising this level?

From the answers received from the NGOs (Table 5), we notice that they are very active in the community either by information and awareness campaigns or by developing the infrastructure for separate collection. Regarding their objectives, with one exception—Kogayon (because is not directly involved in separate collection but promotes ecotourism and environment protection), the NGOs aim to raise the awareness of people on environmental issues, contribute to the increase of the recycling rate of municipal waste or WEEE, advocating for legislation or public policies changes (Viitor Plus), acting as a communication platform for all actors involved (Romanian Association of Waste Management—RAWM), implementing good practice examples. Q3 is subjective because it required their perspective on how things evolved in the last years. From the seven NGOs, five perceive things as regressing taking into account the gap between the recycling rates in Romania and the targets established at EU level (Ecotic is the exception, considering that recycling of WEEE is improving and also RAWM who keeps a neutral tonality). Still, some of them (Environ for WEEE collection, Mai Mult Verde for their project on clean waters) appreciated the increased openness of people regarding separate collection and recycling. Most of the obstacles identified by the NGOs are: the inefficient management, the lack of infrastructure for separate collection or an inappropriate one (as RAWM states, the system door-to-door would be more useful than the one based on bell containers), the lack of knowledge, awareness and responsibility among citizens, improper implementation of the legislation, lack of control from authorities and not applying punitive measures. The measures for raising separate collection are those which solve the problems: developing the infrastructure and making it more available for households, apply the legislation and take punitive measures against those breaking the regulations, organize more information

Table 5 NGOs' perspective on separate collection and recycling

	Q1	Q2	Q3	Q4	Q5	Q6
Environ (NGO for WEEE collection)	Awareness campaigns, education activities in schools, trainings, developing a national network of 4500 points for collection of batteries and electrical waste, a partnership with 150 local authorities	Campaigns for promoting separate collection and raising the awareness and responsibility, correlated with the development of more collection centres and extending the partnership with local authorities to 1000 entities	Far from reaching the recycling target in 2020, there are still many gaps in legislation but people are more aware and more informed about the need to collect separately and recycle	The indifference of authorities, almost half of Romanian live in a rural area and there the infrastructure is not so accessible or does not exist, the indifference of people who do not use appropriately the containers for separate collection	Implement the measures that were successful in other European countries (developing the infrastructure, a plan for separate collection and rigorous controls to make it happen, information campaigns for the public, taxation in accordance with the waste generated and more fines for those who do not respect the law)	There is some progress in the last 10 years but not so consistent and Environ is actively involved in information campaigns addressed to the community, public authorities and companies
Ecotic (NGO for WEEE collection)	Partnership with public authorities, development of 8100 collection points in the country, information and collection campaigns	Raising the awareness through information campaigns, developing the infrastructure for collecting WEEE and reaching the target of 45% for WEEE collection	The evolution of WEEE collection is good (85,000 tons collected in 2019 compared to 52,000 tons in 2017, a third of this being collected through Ecotic)	Insufficient infrastructure and support from public authorities, a low level of awareness from the population	More awareness and information campaigns, an increased support from authorities, involvement of all actors (retailers, consumers, NGOs, authorities), development of the infrastructure for WEEE collection	Ecotic Caravan received Best LIFE Environment Award from European Commission (more than 400,000 people informed), local campaigns, newsletters, guides
Kogayon	Focused on environment protection,	Not the case	There is an important regress in the last	Bad management; lack of control, gaps	To have better laws for making it	Citizens are informed and

		ecotourism, education on environmental issues, not being responsible of separate collection		10 years, a bad management and sceptical about public statistics, considering recycling at a lower rate than it is declared	in legislation and poor implementation	mandatory for citizens to have a contract for waste collection, including for demolition/construction, to have more controls in the territory and punish those who do not respect the law, giving more fines	willing to collect separately, the problem being the lack of infrastructure. Still, there is a need for information campaigns especially in poor regions and this is part of our policy
Viitor Plus – association for sustainable development	Launching 5 programs: Recicleta, Harta Reciclarii, Ecoprovocarea, BiroulECO	Raising the level of knowledge, advocating for changes in legislation or to properly implement it, offering eco-friendly alternatives for separate collection, being a model for authorities or companies to develop similar initiatives	No, the direction is not good taking into account the low rate of recycling compared to the target	Lack of infrastructure for separate collection, lack of formal education on these issues, lack of knowledge and awareness, not enough controls and fines	Developing infrastructure for separate collection and making it accessible directly to the waste producers, placing special containers close to the place of waste generation, information campaigns, dedicated courses in schools, more fines and controls, making people accountable for the way they collect	People are not well informed, authorities are indifferent, few fines applied. We have numerous campaigns, support circular economy and believe all of us should reuse and reduce as much as possible

(continued)

Table 5 (continued)

	Q1	Q2	Q3	Q4	Q5	Q6
Mai Mult Verde	Developed the program *Cu apele curate* (*With clean waters*), we created a model of good practice for cleaning the Danube of waste, correlated with information campaigns and workshops	Identifying the obstacles to collecting plastics especially near rivers in order to find solutions to reduce the waste and increase the recycling in these regions	The target for 2020 is far from reaching, population is not incentivized to collect separately and landfilling is a very high rate. But at a lower level, we met numerous people open to our program	Authorities' lack of involvement, not applying the legislation, lack of knowledge among citizens correlated with their lack of trust in authorities regarding separate collection and recycling	Developing the infrastructure, informing citizens and applying fines for local authorities who do not implement the appropriate measures in this area	There are no national campaigns in schools, so the level of knowledge is rather low. NGOs organize most of these campaigns. We have an online presence and inform people on plastic pollution, forestation, food waste
Foundation Terra Mileniul III	Developing numerous projects to increase the level of knowledge and awareness, contributing to the development of public policies regarding waste management, monitoring waste management	Raising the recycling rate, diminishing the generated waste, reducing the landfilling. Other operational objectives: raising awareness, advocating for the principle *Pay as you throw*	A wrong direction: informal collectors were eliminated, citizens are not forced by law to collect separately, the containers are not accessible to population, there are no projects for developing circular economy, lack of control and monitoring from authorities, illegal imports of waste, no	Inadequate legislation that encourages monopoly of sanitation companies, a poor application of penalties and fines, authorities delay applying European Directives on waste management, lack of infrastructure	Applying the European Directives, encouraging competition on the waste management market, giving more fines for breaking the law, developing the infrastructure	People in big cities are well informed but the barriers to separate collection create serious problems. There is a need for information campaigns in rural areas. We have sporadic campaigns for raising awareness but unfortunately there are no grants for environmental NGOs from public

				consistent penalties applied		authorities, most funds coming from EU, private companies, Norway and Switzerland
Romanian Association of Waste Management (RAWM) - a communication platform for all actors with responsibilities and an interest in waste management	Promoting the interests of the industry, having a proactive role in the legislative process, organizing workshops, information campaigns, trainings, monitoring waste management and offering suggestions for improvement		Raising the recycling rate by analysing the reports of separate collection from all actors and offering recommendations for improvement to public authorities	The existing system consists in placing islands of containers in different regions but it is difficult for people especially in winter to go there. The door-to-door system would be more useful, people receiving bags for separate collection. This system started in some cities but requires time	Inefficiency of present systems used for separate collection. A door-to-door system will mean additional costs for the population and an analysis should be made to see where to keep islands of containers and where to use the door-to-door system. Inefficiency of sorting and recycling plants	Information campaigns, education in schools from early ages, penalties and fines, technology development to create eco-friendly products, legislation to discourage production with materials that are not recyclable. Many producers intentionally create products with limited life and difficult to repair. The increased consumption is also a factor to the increased waste	Regular events with its members for informing people about the benefits of separate collection and recycling, publishing a periodic journal (Waste Management), information campaigns organized by RAWM members addressed to the population

Table 6 Public authorities' and IDAs' perspective on separate collection and recycling

	Q1	Q2	Q3	Q4	Q5	Q6
Ministry of Environment, Water and Forests	Measures: The National Strategy for Waste Management (2013) and the National Plan for Waste Management (2017), creating the legislative background for environment and individuals' health protection	Reaching a target of 50% for the municipal waste recycling rate. The responsibility for the appropriate infrastructure belongs to local authorities	Romania will not reach its target of 50% recycling rate for municipal waste	Inefficient management, laws that are not applied, delays in implementing waste management systems, insufficient waste sorting, lack of infrastructure, lack of knowledge among the community regarding waste management	Application of the laws on environment protection, the waste management law (Law no. 211/2011) by all the actors responsible with waste management. The Ministry supports local authorities in their efforts for an efficient waste management by: promoting education on environmental issues, environment protection, forbidding investments in protected areas, changing consumption habits of people organizing awareness and information campaigns, making the infrastructure for separate collection available to all citizens, implementing projects promoting sustainable development	There is a need for more awareness and information campaigns and the Ministry supports local authorities with the measures they implement

Recycling Initiatives in Romania and Reluctance to Change   181

| Oradea City Hall | Measures: 2 main points for separate collection, containers for two fractions (dry and wet) for homeowners' associations (those living in block of flats), yellow bags for houses to recycle the dry steam, subway containers for separate collection, the public sanitation company collects and transports the waste separately to the sorting plant and the biowaste to the ecological plant | Reaching the targets for the separate collection of municipal waste (50% for 2020, 60% for 2021 and 70% from 2022). The quantity of the waste collected separately from the total municipal waste is considered 33% if the sanitation company responsible of the transportation of the sorted waste is not able to establish the share and composition of waste | There is a good direction | People do not collect separately and throw the waste mixed | Applying the principle *Pay as you throw* included in OUG no. 74/2018, implementing the door-to-door system (households receive bags for each type of waste, with a bar code), personalising the frequency for collecting waste and the volume of generated waste depending on the needs of the individual, measuring the quantity of the generated waste (with bar codes on the bags and weighing them) | The public sanitation company is responsible of the awareness and information campaigns of the public. The objective is to raise the level of knowledge and making people more responsible and aware of the consequences on the environment and their health if they do not collect separately |

(continued)

	Q1	Q2	Q3	Q4	Q5	Q6
ADIS Iasi (the IDA in Iasi county)	Monitoring the sanitation service in Iasi county, making recommendations for improving the separate collection system, supporting the efforts for adequate infrastructure, being an active part in the information campaigns on environment protection and waste management	To raise the amount of waste that is collected separately and reach the targets established at the EU level	Things are improving if all actors do their part, there are information campaigns, an adequate infrastructure, punitive measures	Lack of infrastructure and knowledge and indifference from people who do not collect separately	By information campaigns, developing the infrastructure for separate collection and applying punitive measures for breaking the law	We live in a digital age so only those who do not want to be informed are not informed. We have frequent information campaigns on radio and TV
ADI Ecodolj (the IDA in Dolj county)	Using European funds within the Waste Management Integrated System, we created an adequate infrastructure in urban and rural areas for separate collection, sorting and transfer plants, vehicles equipped for separate collection of waste, applying the system door-to-door in some rural areas and collecting the bags once in a fortnight, in the other regions the system used is with bell type containers	Reaching the target of a recycling rate of 50% from the total waste till the end of 2020, applying the principle *Pay as you throw*, based on one of these tools – volume, weight, frequency, personalised bags, establishing and applying tariffs for separate collection for the waste generator and fines for not collecting separately, a separate collection rate of 50% in 2020, 60% in 2021 and 70% in 2022 for paper, glass, metals, plastics	There is a positive trend especially since we started applying the system door-to-door in 15 rural communities in Dolj county	Lack of information, responsibility and awareness, the reluctance to change coming from the consumption habits, the way they dispose the waste, the difficulty to accept additional taxes for separate collection or paying fines for breaking the laws on environment protection	Information campaigns especially addressed to rural areas and children because they are more receptive to the process of separate collection in the system door-to-door	There is a need for information campaigns, but with the ones we contribute to we hope for a change of mentality in the future and expect better results regarding separate collection in the community

campaigns, education in schools from early ages and one that was mentioned only by RAWM—a legislation that discourages producers from creating products with a short life, because high consumption leads to a high amount of waste and the problem in Romania should be looked also through these lens. Regarding the level of knowledge and awareness, all NGOs are involved in information campaigns, but they recognise the need for more because there is room for improvement.

Public authorities and IDAs have different perspectives because only a few of the public entities in Romania responded to the e-mail we sent, these being centralized in Table 6. Thus, Ministry of Environment, Water and Forests reflects the national perspective - neutral responses, more objectives, presents only the facts and is not emotional like we noticed in the case of the NGOs. The Ministry launched The National Strategy for Waste Management (2013) and the National Plan for Waste Management (2017), aims for reaching the target of 50% for the municipal waste recycling rate (even if this is not attainable till 2020 – Q3) and supports local authorities in their information campaigns (Q1, Q2 and Q6). As measures for raising the level of recycling rates, the Ministry mentions: applying the laws, organizing information campaigns, making the infrastructure available to all citizens, changing the consumption habits.

Oradea City Hall reflects the local perspective (they are the only local authority that responded) which is responsible of separate collection: they invested in the infrastructure (containers for dry and wet steam for blocks of flats and bags for houses, subway containers, sorting and ecological plants), aiming for reaching the targets for separate collection (50% for 2020, 60% for 2021 and 70% for 2020); information campaigns are organized by the public sanitation company. They appreciate things are improving in Romania (Q3), but as the main problem for the low recycling rates they consider people are those to blame because they do not collect separately (Q4). As measures to fight against this obstacle, Oradea City Hall mentions the principle *pay as you throw* that would be more efficient (Q5).

The IDAs that responded are two, one from Dolj county (in the south-west of the country) - ADI Ecodolj and the other from Iasi county (in the north-east of the country) - ADIS Iasi. ADI Ecodolj has a more optimistic perspective, appreciating the improvements in collection since they implemented the door-to-door system in 15 rural communities meanwhile ADIS Iasi maintains a neutral tonality (things are improving if all actors do their part). The main obstacles (Q4) for not having a higher recycling rate are similar in their opinion: lack of infrastructure, lack of knowledge and indifference of people (mentioned by ADIS Iasi) and lack of knowledge, awareness, reluctance to change (mentioned by ADI Ecodolj). We notice that ADI Ecodolj is the only one which did not mention infrastructure as a problem and they appreciate that campaigns especially in rural areas are measures that can raise the level of knowledge and awareness. Their objectives are: reaching the target of 50% recycling rate, applying the principle *Pay as you throw*, organizing information campaigns.

ADIS Iasi mentions some measures for reducing the obstacles: developing the infrastructure, punitive measures for breaking the law and information campaigns, considering that only those who do not want or are not interested lack the knowledge because we live in a digital age. Comparing the two IDAs, we notice that ADI Ecodolj mentioned the rural areas where there is higher need for information, meanwhile ADIS Iasi appreciates that we can be easily informed because we have access to technology. Almost half of Romanians (46.4%) live in rural areas (INS 2018, p. 10) and 81% of households have access to internet (Eurostat 2019), so also those in rural areas can be informed, but the problem in these regions is with being poor and more concerned about day-do-day living.

The qualitative analysis consisting in the 11 interviews with specialists in public authorities, IDAs and NGOs reveals the answers to the RQs established at the beginning of this study. Among the obstacles they mentioned (RQ1), there are: the lack of infrastructure and its accessibility for the population, the low level of knowledge and awareness on environmental issues, the lack of consistent punitive measure for violating the law. There are similarities between the perspective of NGOs and that of public authorities and IDAs but there are also differences, most of the NGOs mentioning also the inefficient waste management for which public authorities are responsible. Regarding RQ2, the level of knowledge and awareness regarding separate collection, recycling, waste management is rather low but things are improving and some people became aware of these issues. Still, the improvement is not so significant because the population is reluctant to change especially in poor regions and rural areas, Romania having almost half of the population in rural communities.

7 Conclusions

In conclusion, Romania registered between 2004 and 2017 some progress regarding the municipal waste recycling rate (Table 1), the packaging waste recycling rate (Table 2) and landfilling rate (Table 3), but still below the targets established at the EU level. Thing are improving but not significantly. The present research is mostly a qualitative analysis on two levels, that of individuals who were questioned using an online survey and that of public authorities, IDAs and NGOs responsible of or focused on environmental protection and waste management in Romania and which have a wider perspective. By the questions we addressed, we intended to identify the main obstacles that prevent Romania from reaching its targets and also if and how the level of knowledge and awareness on environmental issues impacts these figures. The obstacles identified are: the lack of infrastructure or accessibility for all citizens, an inefficient management, there are no consistent fines for those breaking the law, inefficient systems for separate collection (bell type containers are not so efficient and accessible as the door-to-door system), the principle *pay as you throw* is not applied and people are not incentivized to collect separately because there is no consequence if they do not do it. NGOs, IDAs and public authorities mentioned the low level of knowledge and awareness combined with a lack of responsibility among citizens meanwhile individuals consider they are informed. This gap can be explained by the difference between the perceived knowledge and real knowledge and the fact we, as individuals, tend to overestimate what we know. Another explanation is given by the fact that public authorities and NGOs might have a wider perspective on this than individuals in our analysis. There is no national study on the level of knowledge but the low rate of separate collection and not enough social pressure from individuals to authorities regarding these issues might suggest there is less awareness on environmental issues. So many obstacles we identified show that there is not only the fault of individual and their reluctance to change but also a consequence of the way authorities manage the problem. The solutions for the problems identified in the research are: raise the level of knowledge and awareness with information campaigns and education in schools starting from early ages; apply the principle *pay as you throw* to make people more responsible and incentivized to collect separately and correctly; apply the laws on environment protection and give fines for those breaking the law; developing the infrastructure for separate collection, sorting and recycling and making it more accessible for people; encouraging production of innovative products which have a longer life and can be repaired. Individuals' reluctance to change comes from the difficulty to change their consumption behavior and the way they dispose the waste. Additionally, many people live in rural and poor areas where access to information is low and they are not so preoccupied to collect separately or use bell type containers. Probably this is why ADI Ecodolj stated there they implemented the system door-to-door for 15 rural communities and they noticed an improvement of separate collection rate afterwards. Even if Romania has the least amount of waste generated per capita in European Union, an increased consumption means also more waste. In a country with underdeveloped infrastructure for separate collection and recycling, people have to understand that the 3 Rs of waste management (reduce, reuse, recycle) are all equally important, idea also expressed in the interview of Romanian Association of Waste Management.

The present study is useful to all actors involved (individuals, companies, public authorities, NGOs, IDAs) in order to understand the perspective of the other part and have a wider view on how things are and how they can be improved for the benefit of both environment and our health because we are all responsible for environment protection.

References

Alam, P., & Ahmade, K. (2013). Impact of solid waste on health and the environment. *Special Issue of International Journal of Sustainable Development and Green Economics, 2,* 1.

European Commission. (2011). *A Roadmap for Moving to a Competitive Low Carbon Economy in 2050. COM(2011) 112 final*. Retrieved from https://eur-lex.europa.eu/legal-content/EN/TXT/?uri=celex:52011DC0112

European Commission. (2018). *Report from the Commission to the European Parliament, the Council, the European Economic and Social Committee and the Committee of the Regions on the Implementation of EU Waste Legislation, Including the Early Warning Report for Member States at Risk of Missing the 2020 Preparation for Re-use/Recycling Target on Municipal*

Waste. COM(2018) 656 final. Retrieved from https://ec.europa.eu/environment/waste/pdf/waste_legislation_implementation_report.pdf

European Commission. (2019). *The EU Environmental Implementation Review 2019. Country Report – Romania*. SWD(2019) 130 final. Brussels 4.4.2019.

European Environment Agency. (2019a). *Waste Recycling. Indicator Assessment*. Retrieved from www.eea.europa.eu/data-and-maps/indicators/waste-recycling-1/assessment-1

European Environment Agency. (2019b). *Diversion of Waste from Landfill. Indicator Assessment*. Retrieved from www.eea.europa.eu/data-and-maps/indicators/diversion-from-landfill/assessment

European Parliamentary Service. (2016). *Circular economy package*. Briefing. PE 573.936. Retrieved from www.europarl.europa.eu/EPRS/EPRS-Briefing-573936-Circular-economy-package-FINAL.pdf

Eurostat. (2019). *Digital economy and society statistics – household and individuals*. Retrieved from https://ec.europa.eu/eurostat/statistics-explained/index.php/Digital_economy_and_society_statistics_-_households_and_individuals

Eurostat. (2020a). *Climate change – driving force*. Retrieved from https://ec.europa.eu/eurostat/statistics-explained/index.php?title=Climate_change_-_driving_forces

Eurostat. (2020b). *492 Kg of municipal waste generated per person in the EU*. Retrieved from https://ec.europa.eu/eurostat/en/web/products-eurostat-news/-/DDN-20200318-1?inheritRedirect=true&redirect=/eurostat/en/news/whats-new

Ferronato, N., & Torretta, V. (2019). Waste mismanagement in developing countries: A review of global issues. *International Journal of Environmental Research and Public Health, 16*(6), 1060.

Herczeg, M. (2013). Municipal waste management in Switzerland. Working paper for European Environment Agency.

Ijjasz-Vasquez, E. (2018). *Foreword to What a Waste 2.0: A Global Snapshot of Solid Waste Management to 2050* by Kaza, Silpa, Lisa Yao, Perinaz Bhada-Tata, and Frank Van Woerden. Overview booklet. World Bank, Washington, DC. License: Creative Commons Attribution CC BY 3.0 IGO. Retrieved from https://openknowledge.worldbank.org/bitstream/handle/10986/30317/211329ov.pdf

INS. (2018). *Romania in cifre. Breviar statistic*. Retrieved from https://insse.ro/cms/files/publicatii/Romania_in_cifre_breviar_statistic_2018.pdf

Romanian Government. (2017). *The national plan for waste management*. Retrieved from www.mmediu.ro/app/webroot/uploads/files/2018-01-10_MO_11_bis.pdf

Romanian Parliament. (2006). *Law 51/2006 on public utility community services, republished in 2013*. Retrieved from http://legislatie.just.ro/Public/DetaliiDocument/70015

Romanian Parliament. (2014). *Law 211/2011 on waste management, republished in 2014*. Retrieved from http://legislatie.just.ro/Public/DetaliiDocument/133184

Stahel, W. R. (2019). *The circular economy: A user's guide*. New York: Routledge.

Transparency International. (2020). *Corruption perception index 2019*. Retrieved from www.transparency.org/cpi2019#report

Silvia Puiu is a PhD Lecturer in the Department of Management, Marketing and Business Administration at the Faculty of Economics and Business Administration, University of Craiova, Romania. She has a PhD in Management since 2012 and she teaches Management, Ethics Management in Business, Marketing, Public Marketing, Creative Writing in Marketing, Marketing in NGOs and Marketing of SMEs. In 2015, Silvia Puiu graduated postdoctoral studies after a 16-month period in which she made a research on *Ethics Management in the Public Sector of Romania*. During the scholarship, she spent 2 months in Italy at University of Milano – Bicocca conducting a comparative research between Romania and Italy regarding Corruption Perception Index, ethics management in the public sector and regulations on whistle blowing. During the last years, Silvia Puiu published more than 40 articles in national and international journals and the proceedings of international conferences. Her research covers topics from corporate social responsibility, strategic management, ethics management, public marketing and management. She is a reviewer for journals indexed in DOAJ, Cabell's, REPEC, EBSCO, Copernicus or Web of Science (Sustainability, The Young Economists Journal and Annals of Eftimie Murgu University). Since 2014, Silvia Puiu is a member of Eurasia Business and Economics Society. Some of the relevant works: *Strategies on Education about Standardization in Romania*, chapter in the book *Sustainable Development*, Springer, 2019; *NEETs – a Human Resource with a High Potential for the Sustainable Development of the European Union*, chapter in the book *Future of the UN Sustainable Development Goals*, Springer, 2019; *Entrepreneurial Initiatives of Immigrants in European Union*, The Young Economists Journal, no. 31, 2018.

Part IV
Strategic

Catalyst, Not Hindrance: How Strategic Approaches to CSR and Sustainable Development Can Deliver Effective Solutions for Society's Most Pressing Issues

Veronica Broomes

Abstract Corporate Social Responsibility (CSR) has evolved from businesses having activities rooted primary in philanthropy to a stage where approaches to Corporate Social Responsibility have become strategic. This evolution has led to multinational enterprises being pivotal in solving a selection of pressing societal issues in health, agriculture and the creation of sustainable livelihoods.

Despite the rapid expansion in the number of businesses publishing their CSR and Sustainability reports, progress has been slow and sporadic in businesses across several industries. However, businesses that integrate the three key pillars of sustainable development—social, environmental and economic—provide a foundation on which to build a strategic approach to CSR. Taking an approach informed by sustainability helps businesses have clarity about their purpose. This enables the core business to be in sync with values of the company, internal stakeholders as well as align with expectations of the wider society in which those businesses operate. This shift albeit a gradual one for at least a decade demonstrates how businesses and their stakeholders can drive innovation, future-proof investments and create an enabling environment for enhanced benefits to shareholders as well as other stakeholders.

Drawing on examples from the extractive industry, manufacturing and supply chains for consumer goods, this chapter highlights how strategic approaches to CSR and sustainable development can be effective tools in solving pressing issues in society as well as serve as a catalyst to achieve specific Sustainable Development Goals. Moreover, embedding CSR in core business operations improves quality of life and livelihoods in communities as well as strengthens the capability of businesses to apply pragmatic solutions in addressing pressing societal issues.

V. Broomes (✉)
School of Applied Social Studies, Robert Gordon University, Aberdeen, UK

1 Introduction

Over the past two decades, there has been growing acknowledgement that businesses have responsibilities to their shareholders as well as other stakeholders in society. This is highlighted in several publications, for example, Elkington (1997), WBCSD (2006), Latapí Agudelo et al. (2019), AB InBev (2012). In addition, Kols (2016) in considering the social responsibility of international business observed how Corporate Social Responsibility (CSR) has been shaped by businesses taking action on ethical, environmental and social dimensions of their operations.

The literature on CSR is replete with reports of businesses showcasing their various CSR activities (Schwartz and Carroll 2003; Harris 2011; CSR Wire undated; Triple Pundit undated; Kumar 2017). Furthermore, the work of Davidson *et al.* (2019) quantified monetary spend on corporate philanthropy/CSR by Fortune 500 firms and assessed the influence of materialism by Chief Executive Officers (CEOs) on CSR. Barnett et al. 2019 argued that despite much research since the late 1940s about CSR as practised by companies, there was a paucity of literature about the social impacts of CSR until about 2000. Other studies examined the performance of firms using criteria such as effect of CSR on attraction and retention of employees (WBCSD 2006; Buciuniene and Kazlauskaite 2012; CIPD 2013), explored design and methods in measuring CSR and impacts (Crane *et al.* 2017) as well as highlighted the shift made by businesses to be more responsive to societal issues. The latter was in contrast to their previous focus of merely debating the ethics of various degrees of social responsibility (Frederick 1978; Cochran 2007) or that the responsibility of businesses to society was to focus on making profits (Friedman 1970). Contemporary society expects businesses to do more than make profits and obey laws. To this end, CSR initatives by private sector businesses have evolved through voluntary action rather than legislation (Carroll 2015, EC 2011, Newell 2012).

Mattan and Moon (2008) considered implicit and explicit approaches to CSR in their work to develop a framework for comparative analysis of CSR as practised in the United States of America (USA) relative to that practised in Europe. The authors observed that businesses in the USA seemed willing to embrace CSR, a direct contrast to what was perceived as reluctance on the part of businesses in Europe. However since the article by Mattan and Moon in 2008, much has changed and there is evidence of an explosion in the growth of CSR reports by businesses in Europe as well as in other regions of the world. Indeed, many are published on sites such as Greenbiz (undated), CSR wire (undated), Triple Pundit (undated) and Reuters Events (undated). At end of June 2020, over 71,000 reports (from 10,000 plus businesses) were registered on the UN Global Compact's portal (UN Global Compact 2020).

In this chapter, the author uses examples from the extractive industry manufacturing/food processing and supply chains to demonstrate how strategic approaches to CSR and sustainable development evolved effective tools to solve pressing issues in society. In addition, this article will highlight the considerable potential businesses have, through their activities, goods or services supplied, to serve also as a catalyst in achieving several of the goals set in the United Nations (UN) Sustainable Development Goals (SDGs) by 2030 (UN 2015).

2 Exploring the Connection Between CSR, Sustainability and Sustainability Development Goals

The connection between CSR and Sustainability is evident through changes made by businesses as they evolved from having CSR with a focus on philanthropy to engagement with CSR initiatives that were congruent with their brand and core business. In part, the shift was driven by considerations arising from adopting frameworks such as a triple bottom line reporting (Elkington 1997), shared-value approach (Porter and Kramer 2011), requirements for supply chain auditing or influenced by perspectives of boards, employees and consumers.

In developing the Sustainable Development Goals agreed in 2015 (UN 2015), consideration was given to the role of business in influencing society and how business actions can have direct positive impacts on progress in achieving given targets under each SDG. This led to businesses being included as stakeholders whose participation is viewed as integral in achieving specific SDGs. This allows business to identify specific SDG(s) aligned with priorities of the business. Such recognition creates opportunities for initiatives made by business at the local level to contribute, as appropriate, in meeting part of larger global development objectives.

As suggested by Witte and Dilyard (2017), the inclusion of business in the SDGs could be viewed as acknowledging their importance in society. Moreover, in a report of the late 2019 simulation event, the Centre for Health Security reiterated that whilst authority needed in a global health pandemic is vested in governments, the resources needed in any such pandemic are held by businesses (Centre for Health Security 2020). This led to a recommendation for the creation of a new international organisation in which governments, businesses, as well as charities, banks and academics can work together to solve the challenges arising from a pandemic (Centre for Health Security 2020).

2.1 CSR

In the early stages of CSR evolving as a concept, the term was used by businesses to describe philanthropic support of 'worthy causes', be these donations to charities and communities groups or employees volunteering time, skills and expertise at local schools or senior citizens groups. In the main CSR has been used to refer to non-core activities of a businesses beyond what are their legal obligations. However, because many philanthropic initiatives are unrelated to core business, it is not unsurprising there is public perception of CSR by some businesses as thinly disguised marketing and public relations exercises. For example, in considering 'conditioning branding' by pharmaceutical firms, Hall and Jones (2008) observed that public health education and awareness raising are reasons given by pharmaceutical companies for discussions about specific pharmaceutical. However, the authors argued that ethical issues arising from such branding can result in the public awareness by pharmaceutical firms being perceived as designed to sell more products.

Another example that seemed to be more about raising brand awareness and increasing sales rather than support for a 'worthy cause' is drawn from the global beverage manufacturer Cocoa Cola™ (Digital Synopsis 2014). Combining traditional and innovative digital marketing strategies the company designed a telephone booth ('Hello Happiness') to enable migrant workers in United Arab Emirates (UAE) to make telephone calls. Migrant workers could make free three-minute calls to their home countries, using Cocoa Cola™ bottle caps as tokens. As a result of using Coca Cola™ bottle caps in exchange for time on overseas telephone calls to their families, the beverage manufacturer saw a rapid rise in number of new customers in the UAE as well as neigbbouring countries (Digital Synopsis 2014).

Increasingly, firms have attributed activities apart from those with philanthropic or social intent to their CSR programme. This has led to a broad selection of issues being grouped under the banner of CSR. These range from the protection of human rights in supply chains so as to prevent exploitation of workers and child labour, environmental stewardship, sustainable use of resources, including biodiversity, carbon off-setting, corporate governance, business ethics through to Human Resource and talent management activities such as inclusion and diversity (Broomes 2014, 2020).

In general, decisions about initiatives selected for inclusion in a CSR programme often reflect priorities and values of businesses. The views of CEOs and boards of directors as well as interests of employees and/or shareholders can be instrumental as well in shaping the final result.

A sample of definitions of CSR from Carroll (1979), ECRC (undated), EC (2011) and ISO (2012) is shown in Box 1.

2.1.1 Drivers of CSR

In addition to internal stakeholders, demands for CSR activities to be more than philanthropy originated from consumers and campaigning groups concerned about issues such as governance and accountability in global corporations and reports of human rights violations (including child labour in factories, people trafficking, modern slavery).

Concerns about issues around ethical sourcing in supply chains have been drivers for CSR and led many to choose to include aspects of this in decisions on procurement. In addition, boards of directors and chief executives were propelled into facilitating the creation of policy statements on CSR or sustainability, formulating guidelines on minimum standards for factories and commissioning supply chain audits to ensure consistency between policy statements and tangible actions to improve supply chains.

> **Box 1 Contemporary definitions of CSR (adapted from Broomes 2014)**
>
> *"CSR is a means of discussing the extent of any obligations a business has to its immediate society; a way of proposing policy ideas on how those obligations can be met; as well as a tool by which the benefits to a business for meeting those obligations can be identified."*
> **Ref:** *ECRC (undated). A guide to Corporate Social Responsibility, 12pp.*
>
> *"The social responsibility of business encompasses the economic, legal, ethical, and discretionary expectations that society has of organizations at a given point in time."*
> **Ref:** *Carroll, Archie B. (1979). 'A Three-Dimensional Conceptual Model of Corporate Performance,' Academy of Management. Review, 1979, 4:4, p. 500.*
>
> *"CSR is the responsibility of enterprises for their impacts on society."To fully meet their corporate social responsibility, enterprises should have in place a process to integrate social, environmental, ethical, human rights and consumer concerns into their business operations and core strategy in close collaboration with their stakeholders, with the aim of: i) maximising the creation of shared value for their owners/shareholders and for their other stakeholders and society at large and ii) identifying, preventing and mitigating their possible adverse impacts.*
> **Ref:** *EC. (2011). Corporate Social Responsibility and responsible business conduct.*
>
> *"Social Responsibility refers to the responsibility of an organisation for the impacts of its decisions and activities on society and the environment that, through transparent and ethical behaviour, is integrated throughout the organisation and practised in its relationships, is in compliance with law and consistent with international norms of behaviour, takes into account the expectations of stakeholders and contributes to sustainable development, including health and the welfare of society. Activities include products, services and processes. Relationships refer to an organisation's activities within its sphere of influence."*
> **Ref:** *ISO. (2012). Guidelines on Social Responsibility. International Standard Organisation.*

2.2 Sustainability

Sustainability is built on the three key pillars of sustainable development -social, economic and environment indicators. Businesses can demonstrated the three key pillars of sustainability through taking a Triple Bottom Line (TBL) approach in the reporting of business performance (Elkington 1997).

Elkington proposed that businesses should apply social, economic and environmental indicators to measure and monitor success, instead of the traditional single bottom line of profits only. Moreover, opting for a TBL approach in how a business operates enables business impacts to be monitored and measured on people, planet and profits. The TBL approach forms the basis for businesses to report on their relationship with people (internal and external stakeholders), the planet (biodiversity and environmental matters) and profits (measuring economic success in use of production factors).

Embedding sustainability and taking strategic approaches to CSR have helped companies find purpose, align core business operations with values of the company, employees as well as expectations of the wider societies in which those businesses operate. In making the change from philanthropic-driven CSR to one in which sustainability is at its core provides numerous opportunities for businesses to foster innovation, future-proof investments create an enabling environment for enhanced benefits to shareholders and other stakeholders as well as be a catalyst to enable attainment of the Sustainable Development Goals.

2.3 Sustainability Development Goals

The United Nations SDGs constitute 17 interconnected goals with associated tasks and activities through which governments and the international community commit to specific actions to address inequalities and in support of development which improves lives and livelihoods of citizens around the world (UN 2015).

SDGs are the successor to the Millennium Development Goals (MDGs), launched in 2001 (UN 2000). Over a 15-year period, the MDGs sought to achieve eight goals, including reduce poverty along with improvements in global indicators for health, education among other areas (UN 2015, 2017).

Although critised for not having been developed through taking an inclusive bottom-up approach, a multi-disciplinary review of 90 publications about the MDGs led to the observation that the MDGs lacked meaningful involvement of representatives from developing countries, did not specify accountable parties for the various targets underpinning each of the eight MDGs, were not always adapted to national needs and ignored development objectives previously-agreed by the countries (Fehling *et al.* 2013). Nonetheless, despite shortcomings, the MDGs can be credited with providing a framework that enabled countries to make inroads in reducing poverty and hunger by over 50% (as measured by number of people living on less than 1US$ per day), decreasing by 100 million the number of slum dwellers (Sachs 2012) and improving live chances through access to basic health care (Overseas Development Institute 2010) and to primary education (UN 2017).

The SDGs have been described as a shared blueprint for developing and developed countries to achieve peace, prosperity and justice for their peoples through taking urgent action to eradicate poverty, reduce inequalities, improve health and education while simultaneously spurring economic growth and addressing climate change and protecting biodiversity (UN 2015). In addition the SDGs make specific reference to businesses in the agenda for 2030. Indeed, even a cursory appraisal of the aims and objectives behind the SDGs reveal how firms and industries can readily align their vision and mission with specific SDGs such as goals addressing inequalities, climate change and environmental degradation.

2.3.1 SDG and Effect of Post Covid-19 Global Pandemic

In early March 2020. the World Health Organisation declared a global pandemic (WHO 2020) caused by Covid-19 coronavirus. By November 2020, covid-19 had infected 53 million plus people and killed over 1.3 million across the world (John Hopkins 2020).

In view of the extraordinary and widely-publicised measures taken by governments in countries around the world because of the global pandemic that started of late 2019/early 2020, it would be instructive to see how strategies to achieve the SDGs are reviewed and refined. Moreover, communites, businesses and governments around the world need to assess, pivot and adapt as a 'new normal' emerges following the devastating impact of covid-19 on people's lives, their communities, livelihoods and the economies of countries. Doubtless some will see it as an opportunity to develop greater resilience to sudden shocks in economies and foster innovation that can deliver urgent solutions to societal challenges. Others in contrast may consider it a time to retreat from partnership working as their focus is on how to survive in the short-term, not address longer-term considerations.

During the annual World Earth Day (21 April 2020), a time when many were working from their homes because of 'lock-down' in countries still battling with the Covid-19 pandemic, the responses from a small group business leaders suggested that businesses were already integrating social and environmental issues to inform decision-making (Clancy 2020). Simultaneously, there was optimism among stakeholders groups that Covid-19 could accelerate initiatives by brands to become more aligned with a sustainability agenda (Clancy 2020).

While it as yet unknown, and too soon to predict how many will take the approach of 'building back better', it cannot be denied that in the aftermath of the global Covid-19 pandemic, businesses that survive will be reviewing how they do business and identifying ways to do so in a more cost effective, competitive and sustainable way. Simultaneously, more attention may also be directed to raising 'business profile' and expanding their reach though cooperation with other stakeholders in working towards achieving specific targets under one or more of the SDGs.

3 CSR: from Philanthropy to Integrating CSR for Sustainability

Over the past 10–15 years, a steadily growing number of businesses have moved away from philanthropic 'add-on' type of CSR initiatives to approaches that are closely aligned with sustainability and core business. This shift however was neither rapid nor widespread. This section discusses the evolution of CSR over four phases (described as four generations).

3.1 CSR 1.0 to CSR 4.0: Key Attributes of the CSR Generations

Broomes (2014) proposed CSR as having evolved through four distinct generations or phases.

Beginning with philanthropy (CSR 1.0), then progressing to employee volunteering and community engagement (CSR 2.0), followed by greater emphasis on reporting and start of triple bottom line reporting (CSR 3.0).

In the fourth and final phase, CSR can be characterised as having sustainability at the heart of CSR alongside a focus on purpose-driven CSR and partnership working. In the fourth phase (CSR 4.0), the process creates 'triple wins' for business, policymakers and communities (people). What this means is that businesses through their operations have positive impacts on a wide range of stakeholders, not only shareholders. Such businesses can benefit also from business-supportive enabling policies created by governments (national and regional/state) and contribute to improved lives and livelihoods in the communities where those businesses operate.

Communities (geographic locations as well as specific interest groups, such are youth, women) that have secured 'wins' through improved livelihoods are more likely to accord social approval to businesses when they contribute to strengthening supply chains and improving small and medium sized enterprises in local communities. In addition, the third set of 'wins' will be by policymakers from their creation of the enabling environments in which businesses can thrive and contribute directly by way of tax revenues and in the process facilitate achieving of national and regional/local development priorities as well as specific SDGs.

Characteristics of each of the four phases of CSR, the CSR generations from periphery to mainstream, are summarised in Table 1.

As part of the shift in how CSR is implemented by businesses, there was growing popularity in creating foundations through which businesses could channel finance for the funding of initiatives that were not core to their operations and were considered as a way of 'giving back' to society through the framework of CSR. Examples of multi-national businesses with foundations include telecommunication enterprises Vodafone (Vodafone Group plc 2020) and Safaricom (Safaricom Foundation 2019); food processors Nestle (Nestlé Foundation 2019) and Kellogg

Table 1 Periphery to mainstream: evolving CSR from philanthropy to 'Triple Wins'

CSR phase	Descriptions of CSR phases—from periphery to mainstream
CSR 1.0	*Philanthropy: monetary donations, creation of foundations
CSR 2.0	*Employee volunteering, community engagement
CSR 3.0	*Inclusive business approaches, growth in CSR Reporting (Triple Bottom Line, Global Reporting Index, United Nations Global Compact), Accountability standard (***AA1000AS) for sustainability reporting***, ISO 26000 guidelines for Social Responsibility standard.
CSR 4.0	*Sustainability embedded in core business: strategic CSR, Inclusive Business, Shared Value (business and society with common purpose); leveraging investment and taxation frameworks; aligning with development goals (local, national, international), creating 'Triple Win CSR' (this delivers simultaneous benefits to businesses, governments, communities)

(WK Kellogg Foundation 2019); oil/petroleum companies Shell (Shell Foundation 2018) and ExxonMobil (ExxonMobil Foundation 2018); technology brands Microsoft (Microsoft 2019) and Cisco (Cisco 2019) and the global news and information services company Thompson Reuters (Thompson Reuters Foundation 2019). The latter has three areas of focus, human rights, media freedom and inclusive economics.

In general, foundations are funded from a percentage or fixed amount of profits generated by the business. The funding received is then used to advance and in support of agreed areas of work. Type of projects funded is wide-ranging. This may include support for environmental groups in local communities, funding scholarships, studying problems of nutrition, increasing access to the internet, use of technology and more widespread availability of computers in schools (Triple Pundit undated; Green Biz undated; Cisco 2020 and UN Global Compact 2020).

As CSR becomes integrated in core activities, it is more aligned with the business. This approach is in marked contrast to philanthropic focus of the 1980s and 1990 as Foundations was the preferred channel for businesses to engage with local communities and the wider society.

The generation of CSR 3.0. is associated with a rapid escalation in reporting on CSR int the annual reports of companies. Indeed, thousands of firms report on their CSR and/or sustainability initiatives and are sharing this information online. This approach is in marked contrast to philanthropic focus of the 1980s and 1990 as Foundations was the preferred channel for businesses to engage with local communities and the wider society. Among popular websites where reports are published include CSR Wire (www.csrwire.com); Triple Pundit (https://bit.ly/2XSM8Nx), Green Biz (www.greenbiz.com/) and the UN Global Compact (http://bit.ly/1N9PRM5).

CSR reports prepared by marketing or public relations teams could result in questions about independence and objectivity of business. This is one reason why businesses should choose to move beyond CSR 3.0 and foster the integration of CSR and sustainability into their operations, instead of treating their CSR initiative as an 'add-on' to core operations of the business.

Fourth generation CSR enables businesses to deliver longer-lasting, more substantial and higher value impact that include Elkington's Triple Bottom Line of taking account of profits, people and planet (Elkington 1997). Inclusive business as advocated by the World Business Council for Sustainable Development (WBCSD 2006) has at its core the integration of sustainability into standard business operations.

Another example of how CSR 4.0 could be achieved is by way of the Shared Value construct of Porter and Kramer (2011). The authors assume there is a common purpose and shared value that drive businesses and society. However, although Porter and Kramer's Shared Value approach to CSR has been adopted by global brands such as Nestlé, the construct of Shared Value has been criticised. For example, Crane *et al.* (2014) suggested the approach reflected a naivety about business compliance. In addition, the authors argued that its simplistic framework did not recognise tensions inherent in responsible business activity nor the role of corporations in society. Nonetheless, the Shared Value concept can be credited with contributing to corporations reflecting on how best to ensure CSR conveyed the purpose of business and serving as a catalyst for innovations that will offer solutions to pressing societal challenges.

Embodied in the strategic and development-linked approach of CSR 4.0 are opportunities for businesses to leverage their CSR strategies during investment negotiations if business goals are aligned with relevant developmental objectives. CSR 4.0 strategies are capable of linking business priorities with relevant national and local government objectives, and promoting meaningful stakeholder engagement in local communities where these businesses operate. Such businesses are more likely to have capability to integrate key features of CSR 4.0, the so-called 'Triple Win' approach to CSR, in a way that will deliver higher returns for shareholders and other key stakeholders.

Application of CSR at the level of CSR 4.0 through a sustainability framework provides a mechanism for partnership working. This in turn can enable businesses to find solutions to societal challenges through engagement with policymakers at national, regional and international levels as well as communities with specific needs and demographics. Such partnership working by businesses, governments and civil society/communities is fundamental to creating 'Triple Wins from CSR' (Broomes 2009, 2011). Moreover, this partnership working fosters consideration of how relevant policies can be harnessed by businesses to make meaningful contributions to specific SDGs. This in turn will feed into solutions for some societal challenges in health, education and manufacturing. In the process CSR can be a catalyst in achieving national developmental priorities.

3.2 Integrating CSR Into Business Operations

Businesses in which CSR is focused on philanthropy and employee volunteering (CSR 1.0 or 2.0, as outlined in Table 1) often allocate CSR to Human Resources Management (HRM) and the marketing functions in a business. This could be a contributory factor to the scope of CSR in the businesses being broadened to cover diversity and inclusion, increasing access to employment and employability of disadvantaged and less socially mobile groups in society through internships and similar work placements opportunities.

Notable examples of brands that have diverted from a philanthropic focus to integrating sustainability across the business are Marks & Spencer (M & S) with its Plan A (M & S 2020) which focuses on its supply chain, and Unilever with its sustainable living plan (Unilever undated) and innovative computer games (Lampikoski et al. 2014). In the case of M & S, the business has set significant targets to become more efficient in use of natural resources and energy by 2025. In addition, savings of over £700 million have been made already by the business could be attributed to the approach of integrating sustainability principles into its decision making (M & S 2020).

It is only when the actions of businesses go beyond their legal obligation that businesses can be considered as truly adding value to communities and societies where they work though initiatives which extend into the social and environmental spheres. Moreover, there is greater likelihood of longer-lasting benefits to shareholders as well as other stakeholders. when such actions are not merely short-term initiatives but integrated into operations of the business and, encompass the three pillars of sustainable development and sustainability.

4 Integrating CSR and Sustainability to Solve Pressing Societal Issues

This section presents an overview of three societal issues to which strategic approaches to CSR and integrating sustainability into business operations resulted in solutions that delivered benefits to businesses, communities and policymakers. The societal solutions addressed health, agriculture and labour exploitation in supply and value chains for fast moving consumer goods and commodities.

4.1 Mining Companies Acting to Reduce Mortality and Morbidity: Benefits for Workers and Families

During the 1990s when there were high numbers of infections from Human Immunodeficiencey Virus (HIV) and Acquired Immuno Deficiency Syndrome

(AIDS). The high level of infections of HIV, deaths from AIDS and resultant devastation in communities in Sub-Saharan Africa led to a societal problem of loss of bread winners in households and many children with life changing experiences following the loss of one or both parents.. The societal problem affected multiple generations and not only economically active adults.

Against this backdrop, mining companies in countries whose employees and their families were directly affected by HIV infections and ensuing Aids epidemic were forced to act. Action taken by mining companies, for example, Ango American plc, resulted in benefits to then current employees, staff as well as families and communities in which the miners lived (Cauvin 2002; George 2006). Often these communities were not in the immediate vicinity of the mine but in remote locations to which employees travelled to when returning to their home village and visiting their families (Ellis 2007; Meyer-Rath *et al.* 2015).

4.1.1 Access to Anti-Retroviral Drugs and Global Funds for AIDS, Malaria and Tuberculosis

The solution found and decision taken by mining companies to stem the loss of lives from AIDS was to fund and facilitate the supply of anti-retro viral drugs to employees, their families and communities in countries where the company operated. In many countries, the decision of mining companies resulted to access to anti-retroviral drugs well in advance of governments responding in a tangible way to arresting the public health crisis (Cauvin 2002; Ellis 2007; Hoen *et al.* 2011).

It can be argued that mining firms had a vested interest and had taken action because they needed to ensure availability of a ready pool of economically active miners for their operations so as to generate profits from which dividends can be paid to shareholders. What cannot be disputed however is that the collective action taken in providing access to anti-retroviral drugs saved tens of thousands of lives (Ellis 2007; Carter 2002). Moreover, their actions paved the way and laid the foundation for establishment in 2002 of the Global Fund for AIDS, malaria and tuberculosis as well as opened the doors to giving focused attention to diseases such as tuberculosis and malaria (Global Fund undated). The Global Fund is now an international partnership designed to accelerate the end of AIDS, tuberculosis and malaria as epidemics.

Actions taken by mining companies to reduce devastation from AIDS led not only to fewer AIDS-related deaths in Sub-Saharan Africa but improved the life chances of people infected with HIV in countries around the world (Carter 2002; Hoen *et al.* 2011). In addition, subsequent partnership working at a global level led to more resources, expertise and funding being used in research on diseases such as malaria and tuberculosis which caused mortality and morbidity in infected people. Previously these diseases were largely neglected and treated as low priority in attracting funding for research (Global Fund undated).

4.2 Business, Human Rights and Modern Slavery in Global Supply Chains

Global businesses operating across multiple jurisdictions employ millions of workers in their supply chains. This means these businesses have a duty of care to employees and a social obligation to ensure suppliers are not operating in a way that causes harm through poor health and safety practices, use of forced labour or employing children (UNHCR 1966, article 7).

Inhumane treatment of workers has been highlighted in reports by media organisations as well as non government organisations. For example, in the 1970s, supply chains linked with Nike™ became infamous for inhumane conditions in factories in Asia that used migrant workers to manufacture high-priced goods that were sold in advanced economies (Bhatnagar *et al.* undated). The lessons learnt by Nike™ in Vietnam were applied subsequently in the creation of a code of practice to improve conditions of factories that were contracted as external suppliers to Nike™ (Lund-Thomsen and Coe 2015). However, as observed by Voiculescu (2011), the role of global companies as guardians of human rights has been controversial.

4.2.1 Exploitation in Global Supply Chains

Since the mid-1970s, global brands manufacturing consumer goods in substandard 'factories' in developing countries have been the subject of public outcry and negative publicity because of exploitative practices in supply chains in Asia, Latin America and Africa (Lund-Thomsen and Coe 2015; Picciotto 2003; Adams *et al.* 2004). Brands have disassociated their businesses from suppliers whose factories are substandard with sweat shop conditions and where workers are subject to abuse and low wages, be they children, women or men. However, despite decades of human rights abuse in supply chains, it was only relatively recently that CSR or sustainability strategies have emerged as having potential to be applied in systematic ways when tackling the scourge of human rights abuses in global supply chains.

4.2.2 *Indicators of Effectiveness of CSR in Supply Chains*

In a review of CSR performance in sustainable supply chain management in Europe, van Opijnen and Oldenziel (2011) identified five key indicators to measure and monitor effectiveness of CSR in the supply chains of 22 companies manufacturing cotton garments, sugar (from sugar cane) and mobile phones.

These indicators were: child labour, freedom of association and collective bargaining, adequate standard of living, unfair price levels and biodiversity. However, in none of the supply chains was there evidence that CSR had been effective in reducing malpractices in areas such as child labour and freedom of association. Moreover, as highlighted recently by Le Pors (2020), major global brands continue to have reports of human rights abuse in their supply chains. Brands such as Carlsberg (breweries), fashion brand Ralph Lauren (fashion) Starbucks (coffee chain) were among those rebuked by 176 institutional investors about failures to respect human rights in the brands' supply chains (Le Pors 2020). These findings suggest that reliance should not be placed only on voluntary meaures in seeking to remove human rights abuse from supply chains.

Nonetheless, improvements have been made since the publication of van Opijnen and Oldenziel (2011). *This includes* addressing better working conditions in factories, removal of sweatshops from supply chains of major brands, reducing instances of child exploitation in supply chains and auditing of first and second tier suppliers (Anon 2020). However, more remains to be done to eradicate exploitation in supply chains.

4.2.3 Initiatives to Reduce Exploitation in Supply Chains

Measures to prevent exploitation in supply chains is an example of a societal issue to which businesses have found solutions and governments created the enabling environment. Solutions include the creation and sharing of best practice guides in supply chains for agricultural commodities, fashion industry and other consumer goods. Moreover, there is heightened awareness about modern slavery in work situations in developing countries as well as in western economies. In the UK, concerns about modern slavery led to enactment of legislation in 2015 (UK Home Office 2015). The act which deals with slavery, servitude and forced or compulsory labour, human trafficking also makes provision for protection of victims.

In the UK, under the Modern Slavery Act (UK Home Office 2015) and accommpanying regulation (UK Crown 2015) it is mandatory for businesses with revenues in excess of £36 million to report of performance of their supply chains. In 2019 almost three-quarter of companies had complied with submitting their report (Transparency in Supply Chains Platform 2019). However, a review of those statements indicated that 90.7% of modern slavery statements failed to meet the minimum requirments of the UK's Modern Slavery Act.

Under the Modern Slavery act, there are four parts to the reporting process that businesses need to follow in reporting how their supply chains are performing. The UK Home Office's document provides specific guidance on each of the four parts, namely how to: (a) writing a slavery and human trafficking statement, (b) structure of the statement, (c) process for approving a statement and (d) publishing that statement as part of the requirement for transparency in performance of supply chains (UK Home Office 2015).

4.2.4 Measures to Reduce Exploitation in Supply Chains: Voluntary and Mandatory

Auditing of supply chains, implementation of best practice for ethical trading and financial penalties are among the measures used to reduce abuses in global supply chains. In addition, because of reputation risk and brand damage, Harris (2011) observed that businesses choose to improve conditions in supply chains rather than dealing with adverse publicity and subsequent brand damage as a consequence of reports of human rights abuses.

Reuters Events: Sustainable Business, known previously as the Ethical Trade Initiative (ETI) and Sedex are service providers working in cooperation with their members to raise standards and improve best practice for working conditions in global supply chains (Reuters Events undated, Anon 2020). In early 2020, for example, Sedex had a membership of over 60,000 organisations from 35 industry sectors, and spread across 180 countries (Anon 2020). Sedex, ETI and similar organisations can play pivotal roles in helping businesses to eliminate human rights abuses in their supply chains.

In addition to voluntary sector supply chain audits, making it mandatory for businesses to assess social impacts as a condition of accessing investment finance is another way in which corporate behaviour can be influenced in the way their supply chains are operated. For example, the International Finance Corporation (IFC) requires businesses to assess social impacts, and as necessary, undertake Human Rights Due Diligence to guard against human trafficking, forced labour, child labour by borrowers and their primary suppliers (IFC 2012).

4.2.5 United Nations Framework on Business and Human Rights

It can be argued that social auditing and applying basic principles of ethical trading in supply chains have been effective in reducing the scourge of human rights abuses. These approaches have contributed also to circumventing the activities of groups involved in modern slavery across supply/value chains of commodities and consumer products destined for sale to customers in western countries.

Of great significance however in raising awareness and deepening understanding of the importance of human rights considerations in global supply chains has been the publication in 2011 of the United Nations Framework on Business and Human Rights (Ruggie 2011). Authored by Professor John Ruggie, this framework sets out key obligations of businesses as well as outlines a process to reduce opportunities for human rights violations in supply chains. The landmark Business and Human Rights Framework document of 2011 sought also to foster greater scrutiny of supply chains. Moreover, with establishment in 2011 of the Working group on issue of human rights and transnational corporations and other business enterprises, the United Nations through its Human Rights Council provided a specific point of focus to promote and disseminate information about the Guiding Principles of Business and Human Rights (OHCHR 2020).

The principles embedded in the three broad pillars underpinning the framework are: (i) protect, (ii) respect and (iii) remedy. 'Protect' and 'Respect' relate to human rights and 'Remedy' becomes necessary when measures to protect fail (Ruggie 2011). Businesses can choose to embrace the framework and voluntarily apply its principles to their operations. This in turn created a forum for the identification and promotion of best practice and sharing of lessons learnt.

Not only does the framework provide guidelines applicable to the process of auditing of supply chains, it also facilitated reporting supply chain performance by businesses as well as contributed to the the emergence of policies and statements about modern slavery in supply chains. In addition, policy makers have since strengthened legislation to signal intolerance for human rights abuses in business. For example, in the UK, legislation on modern slavery was enacted to enable prosecution of modern day slave traders (Home Office 2015).

5 Improved Livelihoods Through Processing Local Raw Materials

In this section, a case study is used to demonstrate how the actions of a business contributed to improving livelihoods in faming communities and in the process contributed to reducing poverty (SDG 1).

This case study is based on actions and approaches taken by the South Africa global brewery (known then as SAB Miller). Over the years, the business has committed to source half of the raw materials used in beer making from over 21,000 small holder farmers across Uganda, South Sudan, Tanzania and Zambia (SAB Miller 2011).

SAB Miller's activities in Uganda had a direct impact on improving income in farming households and reducing the extent of poverty in farming communities. This was achieved through creation of a ready local market for sorghum, barley and maize. The company, now part of ABIn-Bev (2018) since 2016 was significant in finding local solutions to the societal issue of poverty and food insecurity among subsistence farmers in Uganda.

Through facilitating the use of local raw materials in its brewery (manufacturing beers), farmers benefited from SAB Miller's support to make improvements to agronomic practices and strengthen record keeping skills in farming households. In addition, indirect positive impacts were had from the creation of off-farm jobs as reported in the study by INSEAD Business School (SAB Miller 2011). In Uganda, the brewery supported a further 44,000 off-farm jobs in the wider economy (SAB Miller 2011). Furthermore, , through purchase of local crops for commercial beer making in Rwanda and other African countries, SAB Miller created 100,000 farming jobs in Africa (CSR Wire undated).

It could be argued that SAB Miller actions and the decision to use local raw materials in its breweries were driven by enlightened self-interest. That may well be the case. What is clear, however, is that the business advanced not only the interests of its shareholders, but other key stakeholders in host countries. In so doing, the business demonstrated it had positive impacts on poverty reduction and improving the livelihoods of its small holder farming partners and their households. There was indirect positive impact on communities in which members of those farming households lived.

5.1 Creating 'Triple Wins' Through CSR

Through engagement with local suppliers in countries where its subsidiary breweries were located, SAB Miller earned social approval and the licence to operate in those countries. A situation described as demonstrating the institutionalisation of CSR (Carroll 2015). Moreover, as well as integrating of CSR and sustainability in core operations of the business, the firm was able to generate 'triple wins'.

In the example of SA B Miller in Uganda, the three groups of 'winners' were: (i) investor business that secured the supply of locally-grown raw material (sorghum) for the operation of its brewery, deliver greater value to shareholders with the fiscal incentive for using local raw materials as well as earning social licence to operate because of jobs created, (ii) policy makers (government) for creating the policy that offered a fiscal incentive for use of local raw materials and iii) farmers (their business and household) benefiting from having a guaranteed market for their sorghum, improving crop husbandary practices on farms, capacity building to improve on-farm record keeping and higher income for farming households.

In solving the societal issue of poverty in communities with households reliant on subsistence farming, SAB Miller contributed to improved lives (social) and livelihoods (economic) in local communities and benefited from the policy of fiscal incentive for use of local raw materials, instead of imported ones (Broomes 2010, 2014).

6 Business, CSR and SDGs: Catalyst to Find Solutions to Societal Issues

CSR can be used to demonstrate how businessses are contributing to society, beyond what is required for compliance. As reiterated by others, meaningful engagement by businesses is pivotal in making progress towards achieving several of the SDGs (Zagelmeyer and Sinkovics 2019; Witte and Dilyard 2017; Schönherr et al. 2017). Indeed, SDG 12, which focuses on responsible consumption and production requires consorted actions from business. Moreover, best practice will emerge as more firms, including multinational enterprises, adopt and integrate sustainable practices in their operations, including the supply chains.

Using current SDG goals as indicators, the case study given earlier of the SAB Miller brewery in Uganda (and its successor holdings AB InBev) would have contributed to four of the SDGs, 1, 8, 10, 11 and 17. Through improving livelihoods and higher household incomes, SAB Miller contributed to SDG1 (poverty eradication).

Development of vibrant farming businesses for small holding farmers, instead of subsistence farming would have been a direct contribution to SDG8 (decent work and economic growth) and to SDG 10 by way of reducing inequalities between households in sorghum farming communities, as well as off-farm, through creation of sustainable cities and communities (SDG 11). Furthermore partnership working (SDG 17) was evident in the way business engaged with communities and benefited from relevant policies. Furthermore, it is highly likely that the partnership would have contributed to SDG 3 as a result of improved health and well being with greater food security and SDG 3 through improved/better access to education.

In suggesting new areas for academic research to examine how firms can contribute to SDGs, Witte and Dilyard (2017) considered types of policies that would be supportive of business innovation. The authors implied that SDGs will be viewed as stand alone goals that are then introduced to the operations of the business. In practice, however, the initial step should be that of businesses identifying which SDGs are relevant to what they are doing. In this way metaphorical doors can be opened through which businesses can identify opportunities for their products or services to be effecting in solving pressing societal issues.

The absence of suitable products or services to solve societal issues that is highlighted by any of the 17 SDGs represent opportunities for firms to innovate in their respective industry and bring to the market products or services to solve selected societal challenges. Furthermore, it can be argued that it is the need for solutions to problems that have driven innovation rather than specific stand alone policies.

Human rights abuses in global supply chains, including modern slavery, is one of the more pressing societal issues of the twenty-first century. If businesses, law makers and civil society organisations work in partnership (SDG 17), they can cooperate to implement solutions that will eradicate the scourge of human rights abuses in supply chains through improved best practice and enforcement of penalties for infringements. Businesses would then be well placed to make significant contributions to more than half of the SDGs. Examples of relevant targets include ensuring responsible production and consumption (SDG 12), promoting industry, innovation and infrastructure (SDG 9) as well as make direct contributions to poverty alleviation (SDG1) through opportunities for decent work and enhanced economic growth (SDG8). Moreover, this would result in the development of sustainable communities and cities (SDG 11), foster good health and well-being (SDG 3), enhance gender equality (SDG 5) and reducing inequalities (SDG 10).

7 Closing Summary

This chapter shared case studies of three societal problems to which solutions were found through strategic approaches to CSR and integrating sustainability into business operations. The solutions found either created benefits or reduced harm caused to individuals and groups in society. The societal solutions addressed health, agriculture and labour exploitation in supply and value chains for fast moving consumer goods and commodities.

Significant gains can be made when interventions of business are guided by approaches grounded in CSR or sustainability. This in turn could result in finding solutions to societal challenges such as those identified in the SDGs. Nonetheless, much remains to be done. These include measures addressing *societal issues arising from i) health access inequalities and fragility of systems for health care delivery and research, ii) inefficiencies in use of energy resources, iii) lack of opportunities to create sustainable livelihoods because current ones are at risk or have been damaged as a result of the adverse impacts of global climate change as well as iv) eradicating human rights abuses in global supply chains.*

With changes to the perception of multi-national enterprises and their expected role in operating as responsible corporate citizens, businesses have countless opportunities to promote innovation. This can be done through finding solutions for issues of profound importance that threaten the societies in which they operate while ensuring such innovation does not adversely affect profitability and payment of dividends to shareholders. It can be argued that social or societal obligations of businesses are pivotal to ensuring their long-term survival as well as in gaining social licence to operate and trade in a globalised world. Irrespective of jurisdiction, companies are required to comply with laws on governance, financial reporting, duty of care, labour laws, health and safety and environmental protection.

*Formulation of relevant d*eveloping policies and strengthening partnership working will enable businesses to make greater contributions through their goods and services that solve pressing issues. To achieve this outcome will require a paradigm shift in how businesses treat with CSR and sustainability. Is it viewed as a necessary add-on in order to be seen as responsible corporate citizen or a vehicle to raising brand awareness? Or, alternatively, is it a genuine effort to foster a culture of decision-making informed by assessments of value, measured not only in terms of profits for the business, but effects on the environment (planet) and society (people)? Moreover, each business should identify which of society's most pressing needs can be solved through use of its own products or services, including those not yet developed.

In addition, creating opportunities for policy changes supportive of innovation in businesses and simultaneously finding solutions to pressing societal issues could include the:

- Fostering strategic and innovative approaches when revising investment frameworks and negotiation agreements.

- Applying principle-based frameworks such as the UN Global Compact to show how businesses have integrated sustainability into core operations and which specific society issues have been or are being addressed.
- Mapping core operations and activities of business to specific SDGs and set measurable targets to monitor and review progress.

Creating relevant policies and incentives to promote and reward innovation beyond standard business practice will open opportunities for business to be catalysts in finding solutions to some of society's most pressing issues. In this way, policies and incentives can linkinnovative actions with specific national development priorities and objectives and/or targets to achieve specific SDGs. Therein lies the potential for voluntary actions from businesses which take account of social and/or environmental issues beyond those mandated by the laws in each country. Building on a voluntary-led CSR platform can result in wins for businesses, governments and communities.

References

AB InBev. (2012). *SAB Miller plc -Sustainable Development summary report 2012* (24 pp). AB InBev. Accessed from: https://bit.ly/2Sgq8c7

AB InBev. (2018). *Shaping the future -building a better world*. Annual Report 2018. Available at: https://bit.ly/2KmyYjQ

Adams, C., Grost, G., & Webber, W. (2004). Triple Bottom Line: A review of the literature. In A. Hendriques & J. Richardson (Eds.), *The Triple Bottom Line –does it all add up?* (pp. 17–25). Earthscan.

Anon. (2020). *Sedex -About us*. Accessed from: https://www.sedex.com/about-us/

Barnett, M., Henriques, I., Husted, B., & Layrisse, F. (2019). Beyond good intentions: How much does CSR really help society? *Academy of Management Annual Meeting Proceedings, 2019*(1), 17580. Accessed from: https://bit.ly/3brMIpr.

Bhatnagar, D., Rathore, A. R., Moreno Torres, M., & Kanungo, P. (undated). *Nike in Vietnam: The Tae Kwang Vina Factory*. World Bank Case Study 51433. Accessed at: https://bit.ly/2Pc1JlT

Broomes, V. (2009). *'Triple wins' from Foreign Direct Investment. Potential for Commonwealth countries to maximise economic and community benefits from inward investment negotiations – case studies of Belize and Botswana*. Commonwealth Policy Studies Unit and © Commonwealth Secretariat. ISBN: 978-0-9551095-5-3.

Broomes, V. (2010). Shifting paradigms in investment negotiations: Maximising economic and community benefits. In *Commonwealth Finance Ministers Report 2010* (pp. 82–84). Henley Media Group Ltd in Association with The Commonwealth Secretariat. ISBN: 978-0-9563722-5-3.

Broomes, V. (2011). Economic progress in developing countries–opportunities from healthy long-term partnerships. In *Commonwealth Finance Ministers Report 2011* (pp. 14–16). Henley Media Group Ltd in Association with The Commonwealth Secretariat. ISBN: 978-0-9563722-7-7.

Broomes, V. (2014). *Who invests wins -Leveraging Corporate Social Responsibility for CEOs, Investors and Policy Makers to create 'Triple Wins'*. ISBN: 10: 1494272822 ISBN-13: 978-1494272821

Broomes, V. (2020). *Six steps to more purpose-driven CSR beyond the covid-19 pandemic: An integrated approach to CSR*. Article for HR Zone, online publication for HR professionals. Pub. Sift (online business-to-business publisher). Accessed from: https://bit.ly/3founeK

Buciuniene, I., & Kazlauskaite, R. (2012). The linkage between HRM, CSR and performance outcomes. *Baltic Journal of Management, 7*(1), 5–24. Accessed at: https://bit.ly/2BBLgUn.

Carroll, A. B. (1979). A three-dimensional conceptual model of corporate performance. *Academy of Management Review, 4*(4), 500.

Carroll, A. B. (2015). *Corporate social responsibility: The centrepiece of competing and complimentary frameworks*. [Online]. Accessed from: https://bit.ly/2XTABxv

Carter, M. (2002). *Mining company to provide HAART*. Accessed from: https://bit.ly/2UKkIY4

Cauvin, H. E (2002). *Mining company to offer HIV. Drugs to employees*. New York Times. August 7, 2002. Accessed from: https://nyti.ms/2UP2OmK

Centre for Health Security. (2020). *Public-private cooperation for pandemic preparedness and response -a call to action*. John Hopkins Bloomberg School of Public Health, Centre for Health Security. Accessed from: https://bit.ly/2C3KuQL.

CIPD. (2013). *The role of HR in corporate responsibility. Sustainable Organisational Performance Research Programme, Research Report* (February 2013). Published by CIPD. London. United Kingdom. Available at: https://bit.ly/3dpkoUX

Cisco. (2019). *Cisco 2019 CSR report*. Accessed at: https://bit.ly/31obJQd

Clancy, H. (2020). *What happens to corporate sustainability amid the COVID-19 crisis? Some reflections*. Accessed from: https://bit.ly/3auZsdM

Cochran, P. L. (2007). The evolution of corporate social responsibility. *Business Horizons, 50*, 449–454. Accessed from: https://bit.ly/2Ng606P.

Crane, A., Hendriques, I., Husted, B. W., & Matten, D. (2017). Measuring corporate social responsibility and impact: Enhancing quantitative research design and methods in business and society research. *Journal of Business and Society, 56*(6), 787–795. Accessed from: https://bit.ly/2zllvH0.

Crane, A., Palazzo, G., Spence, L. J., & Matten, D. (2014). Contesting the value of "creating shared value". *California Management Review, 56*(2), 130–153. Accessed from: https://bit.ly/2VKR3h3.

CSR Wire. (undated). *3BL Report Alert. The corporate social responsibility Newswire*. Accessed from: https://www.csrwire.com/reports

Davidson, R. H., Dey, A., & Smith, A. J. (*2019*). CEO materialism and corporate social responsibility. *The Accounting Review, 94*(1) (Abstract). Accessed from: https://aaapubs.org/doi/abs/10.2308/accr-52079.

Digital Synopsis Digital Synopsis. (2014). *Firm creates phone booth that accepts bottle caps, instead of coins, for EAU workers to call home*. Digital Synopsis: Accessed from: https://bit.ly/37B58Tw

EC. (2011). *Corporate social responsibility and responsible business conduct*. Accessed from: https://bit.ly/2Vwf78o

ECRC. (undated). *A guide to Corporate Social Responsibility* (12 pp).

Elkington, J. (1997). *Cannibals with forks –the triple bottom line of 21st century business*. London: Capstone. (this edition 1999). ISBN 1-84112-084-7.

Ellis, L. L. (2007). Studies in economics and econometrics - The impact of HIV / AIDS on selected business sectors in South Africa. *Bureau for Economic Research, 31*(1), 29–52. Accessed from: https://bit.ly/3fmccGm.

ExxonMobil Foundation. (2018). *2018 Sustainability Report Highlights*. Accessed at: https://exxonmobil.co/2XZO9Hr

Fehling, M., Nelson, B. D. and Venkatapuram, S. (2013). Limitations of the millennium development goals: A literature review. *Global Public Health, 8*(10), 1109–1122. Accessed from: https://bit.ly/2B4vykX.

Frederick, W. C. (1978). *From CSR1 to CSR2: The maturing of business and society thought* (Working Paper 279). Pittsburgh, PA: University of Pittsburgh Graduate School of Business. Accessed from: https://bit.ly/3fGmmSv

Friedman, M. (1970). *The social responsibility of business is to increase its profits*. The New York Times Magazine. 13 September 1970, 6 pp. Accessed from: https://bit.ly/34V9L9K

George, G. (2006). Workplace ART programmes: Why do companies invest in them and are they working? *African Journal of AIDS Research, 5*(2), 179–188. https://doi.org/10.2989/16085900609490378. Accessed from: https://bit.ly/2ADN8fm.

Global Fund. (undated). *Global fund to fight AIDS, tuberculosis and malaria*. Accessed from: https://www.theglobalfund.org/en/overview/

Green Biz. (undated). *Corporate social responsibility articles*. Accessed from: https://bit.ly/38Brgjv

Hall, D., & Jones, S. C. (2008). Corporate social responsibility, condition branding and ethics in marketing. In D. Spanjaard, S. Denize, & N. Sharma (Eds.), *Proceedings of the Australian and New Zealand marketing academy conference* (pp. 81–88). Sydney, Australia: Australian and New Zealand Marketing Academy. Accessed from: https://bit.ly/2BgvL44.

Harris, F. (2011). Brands, corporate social responsibility and reputation management. In A. Voiculescu & H. Yanacopulos (Eds.), *The business of human rights -an evolving agenda for Corporate Responsibility* (pp. 29–54). Milton Keynes, UK: The Open University.

Hoen, E., Berger, J., Calmy, A., & Moon, S. (2011). Driving a decade of change: HIV/AIDS, patents and access to medicines for all. *Journal of the International AIDS Society, 14*(15) Accessed from: https://bit.ly/37vRaSW.

IFC. (2012). *Factsheet –IFC's updated sustainability framework*. Business and Human Rights, Supply Chains. [Online]. 2 pp. Retrieved March 15, 2015, from Available at www.ifc.org/sustainability

ISO. (2012). *Guidelines on social responsibility*. International Standard Organisation.

John Hopkins. (2020). *COVID-19 dashboard by the center for systems science and engineering (CSSE) at Johns Hopkins*. Accessed from: https://coronavirus.jhu.edu/

Kols, A. (2016). The social responsibility of international business: From ethics and the environment to CSR and sustainable development. *Journal of World Business, 51*(1), 23–34. Accessed from:. https://doi.org/10.1016/j.jwb.2015.08.010.

Kumar, T. (2017). The impact of CSR on pharmaceutical sector: Evidence from Incepta Pharmaceutical Company. *Journal of Business and Management, 2*, 26 August 2017. SSRN. Accessed at: https://bit.ly/3eAzwAN

Lampikoski, T., Westerlund, M., Rajala, R., & Moller, K. (2014). Green innovation games: Value-creation strategies for corporate sustainability. *California Management Review, 57*(1), 88–116.

Latapí Agudelo, M. A., Jóhannsdóttir, L., & Davídsdóttir, B. (2019). A literature review of the history and evolution of corporate social responsibility. *International Journal of Corporate Social Responsibility, 4*, 1. Accessed from: https://doi.org/10.1186/s40991-018-0039-y

Le Pors, C. (2020, April 2). *Why 176 institutional investors are calling out brands for failure to respect human rights*. Ethical Corporation. Available at: https://bit.ly/2wYDXnS

Lund-Thomsen, P., & Coe, N. M. (2015). Corporate social responsibility and labour agency: The case of Nike in Pakistan. *Journal of Economic Geography, 15*, 275–296.

Marks & Spencer. (2020). *Plan A 2025 Commitments*. Accessed from: https://bit.ly/2VPrATw

Mattan, D., & Moon, J. (2008). "Implicit" and "explicit" CSR: A conceptual framework for a comparative understanding of corporate social responsibility. *Academy of Management Review, 33*(2), 404–424. Accessed from: https://bit.ly/2Bab89S.

Meyer-Rath, G., Pienaar, J., Brink, B., van Zyl, A., Muirhead, D., Grant, A., Churchyard, G., Watts, C., & Vickerman, P. (2015). The impact of company-level ART provision to a mining workforce in South Africa: A cost–benefit analysis. *PLoS Medicine, 12*(9), e1001869. Accessed from: https://bit.ly/2Y36WS7.

Microsoft. (2019). *Microsoft 2019 CSR Report*. Accessed at: https://bit.ly/2ZgdvQN

Nestlé Foundation. (2019). *Nestlé Foundation: Report 2018*. Accessed at: https://bit.ly/2XZEWyO

Newell, A. (2012). *10 outstanding CSR reports*. Triple Pundit. Accessed at: https://bit.ly/30UsIcX

OHCHR. (2020). *Working Group on the issue of human rights and transnational corporations and other business enterprises*. Accessed from: https://bit.ly/2VMIemN

Overseas Development Institute. (2010). *Millennium development goals report card: Measuring progress across countries*. London: Author. Accessed from: https://bit.ly/3hWwng7.

Picciotto, S. (2003). Corporate social responsibility for international business. In: *The Development Dimension of FDI: Policy and Rule Making Perspectives*. Proceedings of Expert Meeting (Geneva, 6 to 8 Nov 2002). United Nations Conference on Trade and Development. UNCTAD/ITE/IIA/2003/4. ISBN 92-1-112596-0.

Porter, M. E., & Kramer, M. R. (2011). Creating shared value. *Harvard Business Review, 89*(1 and 2), 62–77.

Reuters Events. (undated). *Sustainable business*. Accessed from: https://www.reutersevents.com/sustainability/

Ruggie, J (2011). *Guiding principles for the implementation of the United Nations 'protect, respect and remedy' framework*. United Nations Human Rights Council, A/HRC/17/31 [Online]. Accessed from: http://tinyurl.com/ngsp3j4

SAB Miller. (2011). *Farming better futures*. Accessed from: https://bit.ly/2x2x4Cc

Sachs, J. D. (2012). From millennium development goals to sustainable development goals. Viewpoint. *Lancet, 379*, 2206–2211. Accessed from: https://bit.ly/2TWN9BC

Safaricom Foundation. (2019). *Partnering to transform lives*. Annual report for year ending 31 March 2019. Accessed at: https://bit.ly/2Y0ljXm

Schönherr N, Findler F, Martinuzzi F (2017) Exploring the interface of CSR and the sustainable development goals. *Transnational Corporations, 24*(3). Accessed from: https://bit.ly/3cFrQeJ

Schwartz, M. S., & Carroll, A. B. (2003). Corporate social responsibility: A three-dimensional approach. *Business Ethics Quarterly, 13*(4), 503–530.

Shell Foundation. (2018). *Report of the Trustees and Financial Statement 2018*. Annual Report 2018. Accessed at: https://bit.ly/3e1HemE

Thompson Reuters Foundation. (2019). *Thomson Reuters Foundation Annual Report and Accounts for the year ended 31 December 2019*. Accessed at: https://tmsnrt.rs/2NEcIE2

Transparency in Supply Chains Platform. (2019). Accessed from: https://tiscreport.org/

Triple Pundit. (undated). *The business of doing better*. Accessed from: https://bit.ly/2XSM8Nx

UK Crown. (2015). *Modern Slavery Act 2015, Chapter 30*. Accessed at: https://bit.ly/2xPVIGu

UK Government, Home Office. (2015). *Transparency in supply chains etc -a practical guide. Guidance issued under section 54(9) of the Modern Slavery Act 2015*. 46 pp. Accessed from: https://bit.ly/3ay5jii

UN. (2000). *The Millennium Development Goals*. Accessed from: www.un.org/millenniumgoals/

UN. (2015). *Transforming our world: the 2030 Agenda for Sustainable Development*. A/RES/70/1, 41 pp. Accessed from: https://bit.ly/2VLbeeQ

UN. (2017). *The Millennium Development Goals Report 2015*. 17 April 2017. Accessed from: https://bit.ly/34TEmVe

UN Global Compact. (2020). GC Active: Total number of GC active COPs received. April 2020. Accessed from: https://bit.ly/2VLW0WW

UNHCR. (1966). International Covenant on Economic, Social and Cultural Rights. Adopted and opened for signature, ratification and accession by General Assembly resolution 2200A (XXI) of 16 Dec 1966. Accessed from: http://bit.ly/J1E1V3

Unilever. (undated). Sustainable Living Plan. Unilever. Accessed from: https://bit.ly/2KlepV6

van Opijnen, M., & Oldenziel, J. (2011). In C. Anderton (Ed.), *Responsible supply chain management, potential success factors and challenges for addressing prevailing human rights and other CSR issues in supply chains of EU-based companies*. (p. 214). European Union.

Vodafone Group Plc. (2020). *Vodafone Group Plc UK's Annual Report 2020. We connect for a better a future*. Accessed from: https://bit.ly/3eQbr9t

Voiculescu, A. (2011). Human rights and the narrative ordering of global capitalism. In A. Voiculescu & H. Yanacopulos (Eds.), *The business of human rights -an evolving agenda for Corporate responsibility* (pp. 10–28). Milton Keynes, UK: The Open University.

WBCSD. (2006). *From challenge to opportunity: The role of business in tomorrow's society*. The Tomorrow's Leader's Group. World Business Council on Sustainable Development, 40 pp. Accessed from: https://bit.ly/2zzn7NO

WHO. (2020, March 11). *WHO Director-General's opening remarks at the media briefing on COVID-19*. World Health Organisation. Accessed from: https://bit.ly/3hRw59B

Witte, C., & Dilyard, J. (2017). Guest editors' introduction to the special issue: The contribution of multinational enterprises to the sustainable development goals. *Transnational Corporations, 24* (3), 1–8. Accessed from: https://bit.ly/2KsnGL0.

Zagelmeyer, S., & Sinkovics, R. R. (2019). *MNEs, human rights and the SDGs – the moderating role of business and human rights governance*. IHRMI Discussion Paper 2019/02. ISSN 2629-0189. Accessed from: https://bit.ly/2VrBPyw

Veronica Broomes LLM, PhD UK-based consultant with expertise in Sustainability, Corporate Social Responsibility, Local Content in Supply Chains, Sustainable Development and Impact Assessments.

A dynamic, resourceful and experienced professional who has worked at the interface of research, consulting and training/lecturing, Veronica served as a part-time lecturer to postgraduate students in the Human Resources Management master's degree programme at the University of Coventry in England. More recently, in late 2019, she was appointed as an associate lecturer in the School of Applied Social Studies at the Robert Gordon University in Aberdeen, Scotland. Her consultancy work has included research on sea-level rise and Commonwealth Island States, conduct of sustainability audits for businesses, environmental and social impact assessments for agricultural, infrastructure and forestry projects and evaluating grant funding applications for waste and food projects.

Having written widely on CSR, sustainability and associated topics for over 15 years, her publications include articles on CSR and the links with Governance, Human Resources and in leveraging inward investments. Her book 'Who Invests Wins: **Leveraging Corporate Social Responsibilityfor CEOs,Investorsand Policymakers to Create 'Triple Wins"** provides insights into how CSR can be used strategically to create wins for CEOs, investors and policymakers. In 2013, she launched the first 'State of CSR' survey. It aimed to find out both what organisations across various industries were doing in CSR as well as inviting views on the future of CSR.

She has made meaningful contributions, including conference presentations and scholarly articles, to discussions about environment, social and governance (ESGs) in the decision-making process of investors and CSR as a catalyst for achieving development objectives in Commonwealth countries. Her early career was in agricultural research in Guyana where she worked as a plant tissue culture specialist/biotechnologist and established the national facility for plant tissue culture. As a part-time lecturer at a specialist further education college, Dr Broomes contributed to updating the botany curriculum as her lectures covered topics of agriculture biotechnology, biodiversity and climate change, including sea-level rise and implications of then contemporary issues for agriculture.

She is an affiliate member of the International Association for Impact Assessment (IAIA) and former head of its Guyana Chapter. She supports the startup and growth of small- and medium-sized enterprises (SMEs) through her contributions as a Volunteer Speaker with the Corporation of London's City Business Library, Portobello Business Centre in London and the Business and IP Centre at the Sheffield Central Library. In Sheffield, she volunteers also with the Sheffield Sustainability Network where she is an active member in its working groups on: (a) Sustainability Audits and (b) Diversity and Outreach.

She has worked in the UK and internationally, holds a Masters of Law (LLM) degree from The Open University, PhD in Plant Science from University of Sheffield and postgraduate certificate in Environmental Impact Assessment from University of London.

Anchoring Big Shifts and Aspirations in the Day-to-Day: A Case for Deeper Decision Making and Lasting Implementation Through Connecting Change-Makers with Their Values

Christine Locher

Abstract Making the fundamental shifts required towards greater social responsibility and sustainability requires a fundamentally different outlook and a systemic approach—for organizations, for individuals who drive change on their behalf and for those making the shifts out there on a daily basis to make it take hold in practice.

For all these change-makers, keeping the line of sight between a big goal and what drives them personally is a challenge that can overwhelm and lead to inaction in the face of the current reality. Key elements that help making big, longer-term systemic shifts actionable in the day-to-day are values and decisions, paired with a "teal" mindset.

(1) Values can help keep the line of sight between the big goal and what drives people personally. How they evolve individually and collectively helps to understand the fundamental, systemic nature of the shifts we are looking to make both as organizations and as individuals. This helps to bridge the gap between the current reality and the aspirations by keeping this linked to what people truly care about along the way.

(2) A "teal" mindset with its way to see the whole system, to experiment and refine is the third key aspect to get this right—at least as an aspiration to work towards.

(3) Day-to-day decisions are where these big goals and aspirations get practical, one decision at a time, and values help with fine-tuning. Current frameworks for decision making don't identify values as a key driver and neglect the role values play for sustaining the change-makers and for creating a movement and community necessary to make that change. As people evolve, so do their values towards a more holistic, systemic way of being and doing, leading to better decisions and ultimately better outcomes.

C. Locher (✉)
Christine Locher Limited, London, UK
e-mail: info@christinelocher.me

1 Introduction

The world is in trouble—and, deep down, we know it. Climate change, social inequality, conflicts, displacement, the list of challenges is long, and business does not always play all that glorious a role; or keeps change at the "fig leaf" level which won't help address what is truly needed at a global level. While this can feel overwhelming, the range of challenges is so vast that everyone can find something they resonate with that will help shift things. For big organizations (comprised of the people working there), for entrepreneurs or those in spirit driving changes through CSR or building sustainable businesses, and for each individual as they go about their lives. We need everyone to become a change-maker.

The shifts required are going to have to go beyond just paying lip service, and they are going to have consequences of how we all live. This is possibly the biggest change initiative people could get involved with if we take it as seriously as it actually is. With a team size of 7+ billion with their drive and ingenuity, and all the resources of this planet at our disposal, there are no excuses not to get started and this article will point to some key aspects to drive and sustain this.

To make the enormity of challenges more palatable, this article will use the UN Sustainable Development Goals (UNDP (n.d.), United Nation (2015)) as a proxy for the sorts of things that need fixing. They touch upon sustainability, and a lot of CSR activities are linked to one or several of these goals, and they influence people's lives all around the globe. And yes, in making these changes there is business to be done as well by big organizations and entrepreneurs alike.

Bigger shifts like the ones on the horizon deserve a systemic lens in how to go about making them happen. The journey starts with the big goal, moves on to values (individual and collective) and how they evolve and how they influence what happens. We also look at key shifts currently under way. Decisions are then about how things become practical, how things happen in work and life, and this will again be tied to values.

The term "change-maker" will be used to denote anyone taking an active role, irrespective of job title, employment status or other demographics. Everybody makes decisions and influences the decisions of others, and everybody can drive change, wherever they are, at whatever size currently available to them. This is how change starts. While these change-makers might not be entrepreneurs who start their own business, they are, like entrepreneurs, often very connected to who they are and what sort of change they want to create. And yes, some of these efforts might result in new businesses or in big changes in how old businesses address new challenges in a different way. For good this time, hopefully. This has the potential to also create meaningful livelihoods, grow leaders for future transformations and build clusters/tribes of likeminded individuals across the globe. This is how change accelerates.

This article examines how to create and keep the line of sight between a big goal and the individual—the leader who wants to drive the change and also the individual who wants to make changes and shifts on the ground, to make that change actually stick. It will then explore how to make this practical and keep it grounded through everyday decisions. This draws on research in decision-making theory, social psychology, systems thinking and change, integral theory, leadership and personal development. It is also rooted in a decade of coaching practice with leaders, entrepreneurs, change- and decision-makers who never stop striving to make positive change.

2 Values: Our Unique Ways to Care

2.1 What Are Values?

Values are part of what makes a specific person that person. "Personal convictions about what is desirable and how people ought to behave" (Myers 2013, p. 10). "I am positing 'value' as the most fundamental component of any individual's cognitive structure" (Cummins 1973, p. 6), a "character structure" (Bateson 1972, p. 314), "the rules by which an individual 'construes' his experience" (George Kelly quoted in Bateson 1972, p. 315). Values have "an emotional component, the more fulfilled they are in somebody's life, the more fulfilled the person feels" (Armstrong 1979, p. 34).

Values are universal, "enduring evaluative beliefs about general aspects of life that go beyond specific objects and situations" (Sutton and Douglas 2013, p. 157). Values can either be seen as an "end state of existence" or "mode of conduct" (Rokeach 1973). Although the qualities might well be universal (e.g. "justice"), how they are felt and acted upon is deeply personal and specific to each individual. People use values to e*valu*ate (i.e. to place a value on something) and to then make decisions based upon that. Most of this is not conscious, though, so being clear about what one's values are helps with making better decisions that lead to the values being more present and fulfilled as a result (Locher 2019).

Values are typically abstract (like justice, freedom, success) and what they mean in practice and how they are lived out varies greatly from person to person. Values come at different aggregation levels. It is worth going deeper when values get mentioned to get to a set of core values. For example: In coaching leaders, on occasion, "money" comes up as a value. This can represent different things for different individuals: Security, adventure, progress, quality, success, love and care, contribution and innovation are some examples. Going deeper here is highly recommended for decision-makers to become more conscious of what qualities are truly behind this to better reconnect with what truly matters to them. This increased awareness then also increases flexibility in how to live and fulfill the values and what decisions would best serve the expression of their values (Locher 2019).

Differences in values can lead to conflict between people or to tensions within individuals as they act against one of their values (e.g. tradition versus disruption). A specific value like "success" can mean something quite different for different people (Bär-Sieber et al. 2007, p. 175f). Values are also shared by groups and are necessary for the survival of the social system and the people who live in it as a common anchor (Walton 1969, p. 24).

2.1.1 Values Drive Behavior and Decisions

Values influence attitude and therefore indirectly influence behavior (Sutton and Douglas 2013, p. 157). They drive our preferences because something becomes more or less important depending on how much it does or does not resonate with a person's values (Bär-Sieber et al. 2007, p. 175f). "To say a person 'has a value' is to say that he has an enduring belief that a specific mode of conduct or end state of existence is personally and socially preferable to alternative modes of conduct or end states of existence" (Rokeach 1973, p. 159).

Values help us make decisions and drive change: "A person's value system may thus be said to represent a learned organization of rules for making choices and resolving conflicts" (Rokeach 1973, p. 160). Most of this happens unconsciously. Using values to decide is "inescapable" (Walton 1969, p. IX), especially during situations of change or where things are new "by raising new questions which old formulations not always satisfactorily answer" (Walton 1969, p. IX). In these situations, values provide a stable ground to refer back to, to test options against. The more conscious people are of their values, the better they can make decisions that help these values flourish for themselves, for the group(s) they are a part of, and for the greater whole.

"The provision of a suitable value criterion is an essential component of the decision-making process: Indeed, without a criterion whereby the merits of the alternative actions can be judged the decision degrades to a lottery" (M'Phearson 1976, p. 1). This allows judgment of "the relative merit of a thing" as to "its rightness, goodness, desirability as a potential action" (M'Phearson 1976, p. 1).

2.2 *Values Evolve*

Values evolve over time, often along with major life changes and transitions. This is also true for values held by groups of people, e.g. organizations, countries etc. (Bär-Sieber 2007, p. 177). Like the possibly better-known stages of childhood development, there are various models of adult development that come with a number of stages (Wilber 2005, p. 8f; Beck and Cowan 1996; McIntosh 2007; Cook-Greuter 2002, 2013; Prinsloo 2012). People make a series of these shifts over time in their adult lives, but not everyone makes all the shifts. What is colloquially referred to as "midlife crisis" can be one of these changes where priorities shift drastically and the person's outlook changes. Some examples might be the corporate lawyer who starts a charity, the executive who retrains as a coach, the newspaper editor who becomes a teacher, or the sales person who starts making art. This means either weighting of a person's core values changes from level to level, or they can nominally stay the same but take on a different meaning. For example, "success" as a value could mean very different things at each stage (Brown 2006).

2.2.1 Stages of Consciousness or Stages of Development

Very simplified, as people move along in their growth journey, their gravitation point moves from egocentric to ethnocentric to world-centric (with various levels in-between. See Brown (2006) for an overview). This is significant as the bigger lens is what enables people to think and act bigger and operate more at the systems level and to better make sense of whole-systems goals. Other ways of framing it could be from selfish to care to universal care (Wilber 2006, p. 51). There are many ways to slice this into levels. Brown (2006) gives a good overview of the levels and the terminology, as does Wilber (2006, inset pp. 68/69). For simplicity's sake, Wilber's color terminology is briefly explained, and is used throughout.

These stages represent "the actual milestones of growth and development" (Wilber 2006, p. 5), as a person or a group of people passes through stages, greater capabilities become available and get converted into permanent traits (Wilber 2006, p. 5). Each stage "transcends and includes" the previous ones.

The stages named "Orange," "Green" and "Teal" are deemed as particularly salient for the scope of this chapter, and are therefore described here in more detail. This includes the shifts between the different stages as key inflection points for change that we are observing both in individuals, but also in societies and in how organizations evolve and how they start seeing their contribution (Table 1).

2.2.2 Key Shifts

Orange to Green

This is where a lot of organizations are (at least in mature markets), and CSR activities can often be first forays into a new way of being and contributing. For successful professionals, this can manifest as a "midlife crisis" that leads to (sometimes drastical) shifts in someone's personal and professional life.

Stepping Up to "Teal"

This is the most crucial shift (Beck and Cowan 1996, p. 274ff), and a key one for change makers. From a Teal vantage point, people gain a bigger insight into the whole system and the stages leading up to this. Other stages see themselves as the be-all and end-all of development where they are. Teal starts being able to flexibly work with all stages where they are to bring about holistic, systemic change. The truly integral view, Beck and Cowan (1996, p. 107) call this the "spiral wizard," Bär-Sieber (2007, p. 190) calls it the "globalist."

Table 1 Core stages of development (Selection, table based on overview by Prinsloo (2012) with input from Brown (2006), Cook-Greuter (2002), Beck and Cowan (1996), Laloux (2014))

Stage	"Orange" (StriveDrive)	"Green" (HumanBond)	"Teal" (FlexFlow)
Core values	Create value, strategic, meritocratic, flexible, self-interest, resilience, abundance, competition, opportunities, taking action, autonomy, logic, intelligence	Communitarian, relational, emotional, all views are equal, harmony, genuine care for others	Systemic, integrative, flexible, curious, freedom, independence, learning, change, integration
Shadow aspects	Narcissism, unrestrained capitalism, cold rationality	Slow process as everyone needs to be heard, dissenting voices are unheard—can lead to lack of diversity, entitlement	Difficult to understand for other stages
Likes	Playing the game	Caring for the community	Having the overview, connecting the dots
Dislikes	Setbacks, goals not realized, rules, obstacles	Distortions, rage, separation, lack of consideration	Stagnant, rigid, dull contexts, boredom and lack of stimulation
Leaders	Mentor, guide, set goals	Democratic, show emotions, facilitate process	Knowledge, understanding, competence and intuition supersedes rank, position status symbols and power, leave freedom HOW to achieve something
Circle of care	Self	Community	System
Decision-making	Analysis, sees possibilities	Consensus, compromise, collaboration	Systemic, integrative, macro-level, whole-earth view—tactics depend on what is needed and are picked from across all stages

These Teal change-makers have characteristics in common. Rather than feeling daunted by the task at hand, these individuals enjoy paradoxes and uncertainties. They think in whole systems, flows and rhythms. Few ideas are sacred and they are comfortable interacting with all of them. For Teal change-makers, all is open for examination with great curiosity. They have an awareness that this is merely one level in evolution with more to come, more to learn, more to grow, rather than the best of all stages in an absolute way. They are also pragmatic and don't waste time playing games, getting bogged down in the two biggest shift fights happening in the modern (Western) world, between "Orange" and "Green"—and sometimes both of these against more traditional forces.

Values are key here to ground and connect with everyone else regardless of where they are. "Integral consciousness thus recognizes the importance of values because it can see that its values themselves that are actually evolving within the realm of consciousness and culture" (McIntosh 2007, p. 79).

"Teal" views are more likely to enable the kind of systemic thinking that will make the kind of profound change possible that the current challenges need. These change-makers will enable work at the system level and attract the people passionate about figuring out how to make this happen, and to be able to hold and include a diverse set of views and bring this together to a common goal (Cook Greuter 2013). They have "a sense of knowing the great questions but needing coordinated action to implement answers" (Beck and Cowan 1996, p. 74). Different stages also call for different ways of structuring organizations, and for a broader view of what is inside and outside, and who the stakeholders are (Laloux 2014, see sections about "Teal organizations" and their various emergent practices). Big and more traditional organizations often need to find a "place" for their teal change-makers that fulfills the needs of both sides. CSR activities can be a starting point for that, sometimes this can then radiate out into other parts of the business.

2.2.3 Triggering and Encouraging Change

The different stages are built like a pyramid, for the next stage to emerge, the one below needs to be fully formed and reasonably stable. For transformation to move towards the next stage, the following conditions for change have to be met (Beck and Cowan 1996, p. 75ff):

(1) Potential: Humans tend to settle and be satiated if conditions are "good enough." How much wiggle room and interest are there for change? Where is at least a bit of openness?
(2) Solutions: Things need to stabilize at the stage where they are before they can start moving up. Fixing problems at the current level then frees up energy and bandwidth for possible change upwards. What are the needs people rally around? What is holding people back? How can they be solved, and solved in a way that also frees up energy and points to the next level?
(3) Dissonance: Things that rock the boat, that question the status quo. People might be sensing something is wrong, might get feedback that conflicts with how they previously assessed the situation, or they might experience key life events that force them to reconsider their priorities or there might be strong events happening on the outside that disrupt. All of these experiences point to gaps between the emerging problems and the current level ability to address them which is perceived as inadequate.
(4) Barriers: These can be external or internal. They must be identified and addressed, and ownership needs to be taken to remove them. Key here is a forward-looking, solution-focused approach to not attempt to retreat into the "good old days" which are not going to come back. When removing barriers, the key is to also stabilize so people don't put up different barriers in search of more stability.

(5) <u>Insight</u>: A systemic view from a "Teal" vantage point helps greatly here and provide insight into assumed causes, possible alternatives and solutions. The assumption is that what led into the situation is not likely what will lead out of it. It is key to invite creativity and involve the people concerned. This needs to connect to their daily realities, not remain abstract or just be focused at a removed elite.

(6) <u>Consolidation</u>: When making bigger changes, support during the process is important. The "New" is a bit fragile at first and needs support and protection to stabilize. This can be awkward at first and might need ongoing encouragement and success stories so people don't get tempted to rebuild old barriers.

When all conditions are met, transformation to the next stage can occur. For change makers or people aspiring to make change, it is crucial to address the change from the stage the people involved are at, not from their own center of gravity which might be quite different. Shared values here help greatly in building alignment and in sustaining and nourishing people though the change. Being tuned into the key stages present also helps find a language that will resonate with the values at each stage.

This is the way for change to get started, where people actually are and what they currently need. This sets things up for the long-term. These needs would be met in line with the Sustainable Development Goals, linked to core values. In successfully addressing these, they already become the next building blocks for change and start bringing people along.

2.3 Connecting Individual, Organizational and Bigger Global Lens

"As the level of complexity grows, it's impossible for any one person to hold that complexity and to manage it" (Kofman 2004). Each decision-maker is therefore tasked with going deep and connecting with their personal values, as well as keeping the bigger picture and overarching goal in mind. Values and decisions help keep a clear line of sight between the two.

Nobody needs to do this alone. Different people and organizations will play a part in this and subsystems will emerge, beyond the boundaries of current existing classical organizations. Organizations will look differently (Laloux 2014) and contribute in different ways. Communication, conflict resolution and coordination will be key (Kofman 2004). Shared values are an excellent foundation for this. Change-makers will use all of these resources and ways of working, and this will "resonate very powerfully with all the individuals around and (...) touch everybody wherever they are and then with that touch awaken in them the passion for creating something that transcends each one of them but involves the community" (Kofman 2004). "Work is more meaningful when it gives us a sense of connection to others as well as to the community at large" (Goffee and Jones 2015, p. 136). And, we may add, to bigger goals too such as the Sustainable Development Goals.

Community is key here, building alliances, tribes, ecosystems that will nourish and sustain these kinds of activities and ways of working in this change process. There are goals to reach, but part of this is also the journey, what gets built, what is learned as change-makers (literally) go about their business.

3 How Do We Start to Make This Real? Decisions: Making It Real, One Step, One Decision at a Time

3.1 What Do People Do When They Make a Decision?

Decisions are crucial to making the desired change real. It is how current reality moves closer to the goal, the vision and the purpose. It is the first key step in taking action. A decision is picking one option, which then rules out the other(s). In most cases in life, people are not in a position to split-test two options at the same time and then pick the one that works better, so decision-making has a degree of finality and brings commitment.

Decision making and what drives people are largely unconscious. It is "at times, such a dynamic process that it can be difficult to tell if a decision is being made or not" (Blackmore and Berardi 2006, p. 6). And often, people don't recognize they are making a decision, or that they should in fact actually be making a decision (Heath and Heath 2014, p. 28). There are four steps to a decision (Heath and Heath 2014, p. 18): Encountering a choice, evaluating options, making that choice and living with the results of that choice. Not making a decision is in that sense also a decision, as time and circumstances will then cast the dice.

3.1.1 Ratio, Biases, Emotions and the Subconscious

Decision making is not as rational as we would sometimes like it to be, and biases distort our view (Dunning 2012, p. 251ff). People often overemphasize short term over the long term, or don't think of the opportunity cost that comes with a specific option (Heath and Heath 2014, p. 42). Having the long-term focus of a big goal or vision can help counteract short-termism.

Active during decision making is the "adaptive unconscious" (Gladwell 2005, p. 11f), the unconscious part of our brain that is able to compute large data very rapidly and makes decisions based on this, before the conscious mind even notices. Depending on situational requirements, people flip flop back and forth between the rational decision-making attempts and the adaptive unconscious that makes snap judgments. Which, by the way, can be as good or better as the decisions based on lengthy deliberation.

Also, people underestimate the role emotions play in their decision making. Emotions help evaluate options. People whose brain damage makes them perfectly rational in their decision making are not actually "good" at making decisions. Good decisions come with an emotional valuation (Damasio 1994, p. 193). Values are a key compass for this, too.

Decisions are mostly based on rather crude heuristics and fraught with biases (Dunning 2012, p. 251ff). Decision-makers are not the rational actors they might think they are. Availability and representativeness matter, as does framing, compromise effects and anchoring. People don't realize information given might not be complete and they focus and compartmentalize and therefore miss out on key aspects. Confirmation bias is active. Certainty and fear of loss override other factors that hint to something contrarian.

During the decision-making process, reason is often argued to be, quite literally, an afterthought to "justify our own thoughts and actions and to produce arguments to convince" (Sperber and Mercier 2017, p. 7). Logic is merely used as a structuring device to help us sound more credible in communication. So, if humans are clearly not (exclusively) rational, what role do emotions play in our decision making? People seek pleasure and want to avoid pain. People in part base their decisions on how they think they will feel about this in the future, and emotions will influence how people see factors in the decision, even if they don't look like they have anything to do with the emotion. People don't understand the role emotions play in their decision making very well, and therefore can easily get hijacked (Dunning 2012, p. 263).

3.2 The Context for Decision Making

"Theories of decision-making underestimate the confusion and complexity surrounding actual decision making. Many things are happening at once; technologies are changing and poorly understood; alliances, preferences and perceptions are changing; problems, solutions, opportunities, ideas, people and outcomes are mixed together in a way that makes interpretation uncertain and their connections unclear" (March 1982, p. 168). Having a Teal-style, whole-systems view and way of operating, combined with a strong vision and goal grounded in personal values will help make this complexity less arbitrary and confusing—possibly even enjoyable.

The Heath Brothers (Heath and Heath 2014, p. 18) speak of the Four Villains of decision making, each tied to one of the steps in the process: If the frame is too narrow when a decision maker encounters a choice, they will miss options. If their analysis is riddled with biases (e.g. confirmation bias), they will look for information that confirms what they already think. If they are driven by short-term emotions, they will make choices that might not serve them in the long term. And, living with it, they might be overconfident about the future this will bring. A teal-style systemic view and reflective practice with iterations will help buffer some of this as well.

Values help link to the bigger picture goal to widen the frame and the range of options or steps available (Heath and Heath 2014, chapter 3). And for the subsequent implementation, values also help the change maker building a tribe, a group of people that resonate with the change, who might pursue similar goals. This then facilitates a trusting exchange of best practices.

"How is it that we are 'ethical' in regards to one group, but unethical in regard to another?" (Landheer et al. 1960, p. 8). This tension is one to be resolved in making the shift towards longer-term, ethical decisions that are grounded in one's own personal values but also keep the bigger picture and sustainable future in mind.

3.3 Incremental Systemic Change: One Decision at a Time

According to Gladwell (2005, p. 267f), when it is a relatively straightforward decision, analysis is best. When there are multiple variables involved in a more complex scenario, our instincts may well be superior. Decisions get even better when people with different fields of expertise are involved and bring their wealth of experience and insights to the table. Teal change makers take this into account.

This is where a systemic approach can offer valuable insight. As things are usually not that linear (certainly not with goals the size of the Sustainable Development Goals), there is no sure-fire way of knowing what is better. Often, change makers can merely get a sense of "does that lead me towards, or away from"—and possibly a sense of what might be closer relative to each other. This is one of the key ways of working systemically, working with bits of information that are "the difference that makes a difference" (Bateson 1972, p. 315) and focus on solutions rather than getting bogged down in "causes" (that are in the past and that we can't do anything about right now).

Decisions are a very practical way to make that incremental change and to link everything that is done to the big goal. Decision makers don't need a full set of information (that might well not be available). That is no excuse not to get going. People perceive the difference between where they want to go (the goal) and their current surroundings or status quo. And then they take decisions and take action, one step at a time, which will in turn start generating change also at the systems level (Bateson 1972, p. 315ff).

A system in change always carries the possibility of both states (old state vs. new desired state) within itself, and, like a thermostat, the change-maker compares status quo with where we want to go and then fine tunes it. What keeps things moving forward are the decisions, so they need to be aligned with a greater goal with a clear line of sight. The system we operate in is far larger than the system of our own self (Bateson 1972, p. 319). Change happens by taking decision after decision and then observing the outcome, and adjusting. Systemic change makers not only change their environment, but also themselves in the process (Bateson 1972, p. 446ff). Values are a key compass also for this aspect.

Often times, people expect big complex solutions to tackle a big complex issue. But, as Senge points out, big visions get real through a "small number of interventions that can affect fundamental patterns" (Senge et al. 1981, p. 86-j). That is how small decisions and their implementation drive the big picture. "Satisficing" is a key approach to keep in mind and to take some of the possible stress off the table: Making a decision that is "good enough for the moment" rather than trying to optimize everything in one strike. This term was coined by Herbert A. Simon (1957) and serves as a pragmatic way of getting things moving along in the face of complexity, knowing things can be refined based on feedback.

For this to work, intuition and reason need to be in balance: "Intuition is the vital resource that <u>directs</u> (sic) the gifted scientist's rational analysis (Senge et al. 1981, p. 86-n). Senge quotes Mintzberg's story about a CEO using intuition to deal with problems that are "too complex for rational analysis" (86-n). A balance between intuition and rational decision-making approaches is needed to deal with a world as complex as ours.

4 Conclusion and a Call to Action

We, the sorts of people likely to read this, are the lucky ones. We already have an education of sorts, a position in society or at least some sort of toehold that allows us to speak, listen, think, decide and act. We make things happen for a living, and on most days, some of it actually works. We have values, and increasingly start living them out loud.

"We don't know how to visualize mankind" (Landheer et al. 1960, p. 9), the global crisis is just too big and complex. We need to find a better way to link the big goals and aspirations reality, and to make the steps we need to take more practical. Take it one decision at a time. There are 7 billion of us (and counting), and that is a big enough team to get any job done, so none of us will have to do this alone.

People all too often compartmentalize. When asked about values and decisions, some leaders immediately ask "personal decisions or business decisions?" And yet, in deeper interviews[1] with leaders about values, they also mentioned "love" as a core value with astonishing frequency and spoke about it in a way that gives hope. Often, this is followed by a confession that nobody ever has these conversations, or asks these questions. Let's make these a part of how we work, how we do things, how we relate and let this drive the bigger changes.

It takes that courage, that trailblazing spirit and grit that true change-makers have. It takes the willingness to be uncomfortable, to be wrong, to pivot, to pick yourself up again by the strength of your passion and your conviction alone. Scary? It is a minor discomfort compared to waking up on the wrong side of history.

This is our planet. This is our home. And, it is not going well. The Sustainable Development Goals seek to alleviate some of the worst pain points, but they won't implement themselves. There is no shortage of differences that need making, and plenty of opportunities to contribute. Pick yours.

---

[1] Book forthcoming, the next statements refer particularly to the interviewees from the finance and entrepreneurial space.

References

Armstrong, A. E. (1979). *On values and decision making*. PhD Thesis submitted at the University of Bath.

Bär-Sieber, M., Krumm, R., Wiehle, H. (2007): *Unternehmen Verstehen, Gestalten, Verändern. Das Graves-Value-System in der Praxis*. Wiesbaden: Gabler.

Bateson, G. (1972). *Steps to an ecology of mind. Collected essays in anthropology, psychiatry, evolution, and epistemology*. London: Intertext Books.

Beck, D. E., & Cowan, C. C. (1996). *Spiral dynamics. Mastering values, leadership and change*. Malden, MA: Blackwell.

Blackmore, C., & Berardi, A. (2006). *Introduction to environmental decision making. Book 1*. Milton Keynes: Open University.

Brown, B. (2006). *An overview of developmental stages of consciousness*. Boulder, Integral Institute. Retrieved July 21, 2019, from https://integralwithoutborders.net/sites/default/files/resources/Overview%20of%20Developmental%20Levels.pdf

Cook-Greuter, S. (2002). *A detailed description of the development of nine action logics. Adapted from Ego Development Theory for the Leadership Development Framework*. Next Step Integral. Retrieved July 21, 2019, from http://nextstepintegral.org/wp-content/uploads/2011/04/The-development-of-action-logics-Cook-Greuter.pdf

Cook-Greuter, S. (2013). *Nine levels of increasing embrace in ego development: A full-spectrum theory of vertical growth and meaning making*. Retrieved February 11, 2018, from http://www.cook-greuter.com/Cook-Greuter%209%20levels%20paper%20new%201.1'14%2097p%5B1%5D.pdf

Cummins, H. W. (1973). *Mao, Hsiao, Churchill and Montgomery: Personal values and decision making*. Beverly Hills: Sage.

Damasio, A. (1994). *Descartes' Error. Emotion, reason and the human brain*. New York: Harper Collins.

Dunning, D. (2012). Judgment and decision making. In S. T. Fiske & C. N. Macrae (Eds.), *The SAGE handbook of social cognition*. Los Angeles, London, New Delhi, Singapore, Washington DC: Sage.

Gladwell, M. (2005). *Blink. The power of thinking without thinking*. New York, Boston, London: Back Bay Books.

Goffee, R., & Jones, G. (2015). *Why should anyone work here?* Boston, MA: Harvard Business School Publishing.

Heath, C., & Heath, D. (2014). *Decisive. How to make better decisions*. London: Random House.

Kofman, F. (2004). *A fresh perspective: A conversation with Fred Kofman*. Integral Leadership Review. Retrieved July 21, 2019, from http://integralleadershipreview.com/5762-a-fresh-perspective-a-conversation-with-fred-kofman/

Laloux, F. (2014). *Reinventing organizations*. Brussels: Nelson Parker.

Landheer, B., Van Der Molen G., Vlekke B. H. M., Thivy, J. A., Kwee S. L., Sprout H. et al. (1960). *Ethical values in international decision making. The Conference of June 16–20, 1958*. Institute of Social Studies. Publications on Social Change. The Hague: Van Keulen.

Locher, C. (2019): *Values-based: Career and Life Changes that make Sense*. Tunbridge Wells: Christine Locher Ltd.

M'Phearson, P. K. (1976). *Subjective values in decision making: The development of a value calculus*. University of Bath, Dissertation at the Department of Systems Science.

March, J. G. (1982). Theories of choice and making decisions. In R. Armson & R. Paton (Eds.), *(1994) Organizations: Cases, issues and concepts*. London: The Open University/Paul Chapman Press.

McIntosh, S. (2007). *Integral consciousness and the future of evolution. How the integral worldview is transforming politics, culture and spirituality*. St. Paul, MN: Paragon House.

Myers, D. G. (2013). *Social psychology*. New York: McGraw Hill.

Prinsloo, M. (2012). *Consciousness models in action. Comparisons.* Integral Leadership Review June 2012 (August–November 2012 issue). Retrieved July 21, 2019, from http://integralleadershipreview.com/7166-consciousness-models-in-action-comparisons/
Rokeach, M. (1973). *The nature of human values.* New York: Jossey-Bass.
Senge, P., Kiefer, C., Andersen, D. F., Morecroft J. D. W. (1981). Metanoic Organizations. In: *Proceedings of the System Dynamics Research Conference.*; System Dynamics Research Conference; Rensselaerville; NY, 1981; Oct, 1981, pp. 86–87
Simon, H. (1957). *Models of man. Social and rational mathematical essays on rational human behaviour in a social setting.* New York: Wiley.
Sperber, D., & Mercier, H. (2017). *The enigma of reason. A new theory of human understanding.* London: Allen Lane.
Sutton, R., & Douglas, K. (2013). *Social psychology.* London: Palgrave Macmillan.
UNDP. (n.d.). *The Sustainable Development Goals.* Retrieved July 21, 2019, from http://www.undp.org/content/undp/en/home/sustainable-development-goals.html
United Nations. (2015). *Resolution A/RES/70/1: Transforming our World: The 2030 Agenda for Sustainable Development.* Retrieved July 21, 2019, from http://www.un.org/en/ga/search/view_doc.asp?symbol=A/RES/70/1&Lang=E
Walton, C. C. (1969). *Ethos and the executive. values in managerial decision making.* Englewood Cliffs, NJ: Prentice Hall.
Wilber, K. (2005). *The Integral Operating System Version 1.0.* Boulder, CO: Sounds True.
Wilber, K. (2006). *Integral spirituality. A startling new role for religion in the modern and postmodern world.* Boston/London: Integral Books.

Christine Locher M.A. FRSA, FLPI is a consultant, coach and author, working with leaders looking to put their vision into practice based on their core values, following a first career in newspaper and radio journalism and a global career in learning and development. She has degrees in Communication, Intercultural Communication and Psychology (M.A. from Ludwig-Maximilian-University Munich, Germany with studies at Kyushu University Fukuoka, Japan) and postgrads in Systems Thinking in Practice (Open University Milton Keynes, UK) and Conflict Resolution (Fernuniversitaet Hagen, Germany) and a professional certificate in Solution Focused Business Practice (University of Wisconsin-Milwaukee, USA). Her book: 'Values-based: Career and Life Changes that make Sense' launched in May 2019, her article 'Stepping into Your Values (literally)' launched in November 2019 in the Book of Beautiful Business and her book on decision-making based on values will launch in 2021. She is a Fellow of the RSA (Royal Society for Arts, Manufactures and Commerce) and a Fellow of the Learning and Performance Institute (LPI). She is an active member of the Institute of Directors' Entrepreneurship network. She is a certified coach since 2007 (first ICF-accredited training in Germany). She is a trained yoga teacher studying yoga, mindfulness and various body-centred approaches with teachers in Germany, India and the USA. She is also licenced for psychotherapy in Germany (Heilpraktiker fuer Psychotherapie) trained in Client-Centred Therapy, Gestalt and Psychosynthesis/Transpersonal Psychotherapy. She coaches for WYSE (World Youth Service and Enterprise, a charity affiliated with the UN Department of Public information that serves emerging leaders all around the world) and serves on the board of the eLearning Network. She runs her own coaching and leadership company.

The Role of Corporate Social Responsibility in Business Sustainability

George Kofi Amoako

Abstract CSR is important not just because it helps businesses to help society and communities but because it can help firms to build trust and identify with the needs of communities. Establishing and keeping trust with customers, communities and regulators isn't simple and can be easily damaged or lost. To be successful in the long-term, firms need to think beyond what's affecting them today to what's going to happen tomorrow. This isn't just about addressing changes to technology or the needs of customers, but also taking into account alterations in social, environmental and governance issues A couple of studies in the past revealed that reputation and competitive advantage are the results of increased customer satisfaction after engaging in CSR. This study adds to knowledge by suggesting that Corporate strategy, Human resource strategy, Research and development strategy and Product strategy can affect CSR and hence the business sustainability of any given firm. The study proposes a conceptual model that can be tested empirically to confirm this assertion or otherwise.

1 Introduction

The concept and business awareness of CSR has evolved considerably since it first emerged in the 1950s (de Bakker et al. 2005). Over this time, the concept has developed from relatively uncoordinated and voluntary practices into more explicit commitments in response to stakeholder pressures and eventually into ongoing future commitments. According to the corporate watch report (2006), "The phrase Corporate Social Responsibility was coined in 1953 with the publication of Bowen's Social Responsibility of Businessmen". Bowen (1953) defined CSR as "the obligations of businesspersons to pursue their policies, to make their decisions or to follow their lines of action, which are desirable in terms of the objectives, and values of society". He argued that, businesspersons are responsible for the consequences of

G. K. Amoako (✉)
Faculty of Management, University of Professional Studies, Accra, Ghana

© Springer Nature Switzerland AG 2021
S. Vertigans, S. O. Idowu (eds.), *Global Challenges to CSR and Sustainable Development*, CSR, Sustainability, Ethics & Governance,
https://doi.org/10.1007/978-3-030-62501-6_11

their actions in a sphere somewhat wider than corporate financial performance, indicating the existence and importance of corporate social performance. Davis (1960) set forth his definition of CSR as "businessmen's decision and actions taken for reasons at least partially beyond the firm's direct economic or technical interest". By arguing that CSR was a blunt idea but discussed in a managerial context, he further suggested that, some socially responsible business decisions can be justified by the long-run economic gains of the firm, thus, paying back for its socially responsible behavior. Frederick (1960) however, saw CSR as a private contribution to society's economic and human resource and a willingness on the part of the business to see that those resources were, utilized for broad social ends.

From this multiplicity of definitions of CSR we choose the following definition of CSR: CSR is a stakeholder-oriented concept that extends beyond the boundaries of the organization, driven from an ethical understanding of the responsibility of the organization for the impact of its business activities, seeking in return the willingness of society to accept the legitimacy of the business (based on Gray et al. 1996). We decided for the definition because it is based upon the stakeholder concept and calls for a real integration of CSR into the organization's strategy. Also, the definition emphasizes that CSR should result in a win-win situation for the company and its stakeholders.

1.1 Why CSR Is Important?

'In the past, simply fulfilling economic performance alone was critical to ensuring the success of companies and their shareholders, but this traditional perspective is not compatible with society's current demands, as there is increasing social pressure on organizations to become more sustainable and reduce impacts on the environment, promoting sustainable results in their business processes' (de Souza Freitas et al. 2020). CSR is important because businesses are based on trust and foresight. Establishing and keeping trust with customers, communities and regulators isn't simple and can be easily damaged or lost. To be successful in the long-term, companies need to think beyond what's affecting them today to what's going to happen tomorrow. This isn't just about addressing changes to technology or the needs of customers, but also taking into account alterations in social, environmental and governance issues. Implementing CSR as part of company's strategy has been said by some management experts. Levine (2008) states that strategic CSR activities can increase competitive advantages range, such as brand reputation, awareness and employee morale and relationships with regulators and consumers. Graafland et al. (2003), Veríssimo and Lacerda (2010) and Kranz and Santalo (2010) found the companies implementing strategic CSR will improve performance and competitiveness. Recent study by Phillips et al. (2019) 'emphasized that empowerment of staff by their leaders has made a significant difference in all three areas of sustainable performance (social, environmental, economic) and that this can only be achieved where the approach is embedded in the core business by CSR leaders who can apply the concepts of value creation through a strong CSR culture'. This paper focuses on human resource management because most firms CSR activities are implemented through and in conjunction with the HR department. Stahl et al. (2020) argue that there is increasing pressure on modern firms to engage in corporate social responsibility (CSR) in order to address the current lack of confidence in business, align their activities with the needs and expectations of stakeholders. They argue that human resource management (HRM) has a potentially vital role to play in contributing to a firm's CSR activities.

2 Theoretical Framework

2.1 CSR Development and Implementation in Corporate Strategy

CSR strategy development and implementation could be considered as an organizational change process (i.e., moving from a present to a future state; cf. George and Jones 1996), or as a new way of organizing and working (Dawson 2003). The aim is to align the organization with the dynamic demands of the business and social environment through the identification and management of stakeholder expectations.

In order to embed CSR in corporate strategy formulation and implementation, companies must at minimum pursue the following course of action;

1. Senior leadership and management of the firm, including the board, must make an authentic, firm, and public commitment to CSR, and engage in it—

   For CSR to thrive, senior executives must be brought on board, commit it, and engage in it. A clear vision of CSR needs to be embedded within and reflect the core values of the firm, and to be linked to the mission, vision and values of the organization—realizing that it creates social or environmental value and also business value. CSR should be treated and managed as a core business strategy just like the strategies of marketing, R&D, capital expenditures and talent management.

2. Determine the top three business objectives and priorities of the company, and develop a CSR strategy that will contribute to the achievement of those business objectives—

   In developing a CSR strategy, the company's leaders must first determine what specific business objectives this strategy must support.

3. Align CSR strategy with core objectives and the firm's core competencies—

   To be effective, CSR goals must be aligned with two things:

   - Core business objectives
   - Core competencies of the firm

   The first step for a company is to align its CSR goals with its specific business objectives for a particular time period. Aligning CSR goals with business objectives is not as easy an exercise as it may appear. Although there may be many legitimate business objectives, to develop the most efficient and successful CSR goals, the company must prioritize its primary objectives for the time frame. Business objectives can and should change over time, and they should be routinely reexamined in relation to the CSR strategy. Aligning CSR strategy with the firm's core competencies is the next step for creating effective fit with CSR initiatives. This step requires focus and discipline. But regardless of the amount of effort involved, the time spent will be worth it. CSR initiatives originate from all parts of an organization and should be linked directly to what the firm actually knows, does, or is expert in. Creating a CSR strategy will unify your efforts, giving them much more power than if they remain a random mishmash of disparate initiatives.

4. Fully integrate CSR into the culture, governance, and strategy-development efforts of the company, and into existing management and performance systems—

   CSR can be both a risk-mitigation strategy and an opportunity-seeking strategy, and leaders should look for the "sweet spot" within their organizations—that is the intersection between business and social or environmental returns.

5. Develop clear performance metrics, or key performance indicators, to measure the impact of CSR strategies—

   CSR performance metrics should be internal such as reputation improvements, gains in market share, brand perception, increased sales, decreased operational expenditures, and employee satisfaction—as well as external, focused on society and the environment. With no performance metrics in place, there will be no way to prove that the strategy was effective, which means that the strategy will not be suitable over the long haul.

2.2 CSR Development and Implementation in Human Resource Management Strategy

Review of theoretical and empirical research point to four main areas in which human resource development can play an important role in promoting CSR in firms, this includes leadership development, education and training, culture change, and fostering critical reflection(Jang and Ardichvili 2020). Human resource managers are well positioned to play an influential role in assisting their organization attain its goals of becoming a socially and environmentally responsible firm—one which reduces its negative and increases its positive impacts on society and the environment. Additionally, human resource (HR) professionals in organizations that perceive successful corporate social responsibility (CSR) as a key driver of their financial performance can be vital in realizing on that objective. While there is considerable guidance to firms who wish to be the best place to work and for firms who seek to manage their employee relationships in a socially responsible way, there is a dearth of information for the HR manager who sees the importance of embedding their firm's CSR values throughout the organization, who wish to assist the executive team in integrating CSR into the company's DNA. Indeed, HR's mandate to communicate and implement ideas, policies, and cultural and behavioral change in organizations makes it central to fulfilling an organization's objectives to "integrate CSR in all that we do." That said, it is important to understand that employee engagement is not simply the mandate of HR. Indeed people leadership rests with all departmental managers. HR can facilitate the development of processes and systems; however, employee engagement is ultimately a shared responsibility. Perhaps the best way to look at the reality of HR strategy formulation is to remember Quinn, Mintzberg and James's (1988) statement that strategy formulation is about 'preferences, choices, and matches' rather than an exercise 'in applied logic'. It is also desirable to follow Mintzberg's analysis and treat HR strategy as a perspective rather than a rigorous procedure for mapping the future.

The more the HR practitioner can understand their leverage with respect to CSR, the greater their ability to pass these insights along to their business partners towards the organization's objectives in integrating CSR throughout their operations and business model. As human resources influence many of the key systems and business processes underpinning effective delivery, it is well positioned to foster a CSR ethic and achieve a high performance CSR culture. Human resource management can play a significant role so that CSR can become "the way we do things around here". HR can be the key organizational partner to ensure that what the organization is saying publicly aligns with how people are treated within the organization. HR is in the enviable position of being able to provide the tools and framework for the executive team and CEO to embed CSR ethic and culture into the brand and the strategic framework of the organization. It is the only function that influences across the entire enterprise for the entire 'lifecycle' of the employees who work there—thus it has considerable influence if handled correctly. HR is poised for this lead role as it is adept at working horizontally and vertically across and within the organization, so important for successful CSR delivery. Should such an organizational gap exist, the senior HR leader can champion, lead and help drive a CSR approach if necessary. In the coming years as CSR progressively becomes part of the business agenda and the fabric of responsible corporations, it will become a natural agenda for the HR practitioner.

2.3 Steps of Integrating CSR into HR Management

According to Strandberg (2009),Human resource professionals are highly tuned to considering CSR from both a values based and a business-case perspective. They work in a business function that readily identifies both the business benefits and the people benefits of fostering CSR alignment and integration. However, there is little guidance available to human resource leaders who wish to advance CSR within the firm. This section provides a starting point for managers mapping out their strategic approach. It can serve as a checklist for advanced managers who are well on the path, and it can provide a roadmap for the manager who is committed to make a difference in this way and is at the beginning of their journey. Ideally these steps would be followed more or less sequentially; however in practice this is often not possible and indeed, some managers may have already implemented certain components. It is therefore entirely feasible to start from the middle of this list and work in all directions towards the end goal: a CSR integrated company that is reaping the employee and business case benefits, while leveraging community sustainability. This guide has been developed recognizing the constrained economic environment of our times; the tools and tactics proposed in this roadmap are those which can readily be integrated into the HR practitioner's daily regimen. Indeed, as the foregoing business case analysis partly demonstrates, a CSR program can add significant business value. There are other business case benefits of CSR—for example, operational cost savings from reduced materials use—that can be significant, pointing to a financial rationale for the development of a strong CSR strategy and integration effort. The following lays out ten(10) steps HR practitioners can follow to support the integration of CSR throughout the business strategy and operations (Strandberg 2009).

- Step 1: Vision, mission, values and CSR strategy development
    Vision, Mission, Values
    Successful CSR requires a clearly articulated vision, mission and values. The HR practitioner could initiate or support the development, or upgrade, of a vision, mission and values foundation if CSR is not explicitly addressed.
- Step 2: Employee codes of conduct
    The HR function is typically responsible for drafting and implementing employee codes of conduct. As such, HR managers hold the pen on the principles contained in the employee codes. Since a number of recent high profile corporate frauds, boards of directors have become very concerned about the ethical culture within their organizations, looking for 100% sign-off on and compliance with codes of conduct which articulate their ethical values. This is an ideal home for the expression of an organization's commitment to socially and environmentally-based decision-making as it is one of the rare documents which all employees are bound by and come into contact with. As such it is a key tool for cultural integration of CSR norms.
- Step 3: Workforce planning and recruitment
    Workforce planning consists of analyzing present workforce competencies; identification of competencies needed in the future; comparison of the present workforce to future needs to identify competency gaps and surpluses; the preparation of plans for building the workforce needed in the future; and an evaluation process to assure that the workforce competency model remains valid and that objectives are being met. For a CSR oriented company, this consists of evaluating the need for skill sets and competencies central to the emergent sustainability economy—an economy of resource and energy scarcity, human and environmental security constraints, changing societal norms and government expectations.

- Step 4: Orientation, training and competency development

  During the orientation process employees should be given a thorough overview of the clear line of sight between the company's vision, mission and core CSR values and goals. To ensure maximum alignment and early employee 'buy-in' to the strategic CSR direction of the organization, this general orientation should be deemed mandatory for all levels of new employees.

- Step 5: Compensation and performance management

  Next to recruitment and competency development, compensation and performance management are central to the HR function. HR is involved in setting performance standards and expectations and monitoring results to performance objectives. CSR should be recognized in both the base job responsibilities as well as the annual performance objectives at the individual and team levels. Performance reviews could consider how the employee has advanced their personal and the organization's CSR goals over the period.

- Step 6: Change management and corporate culture

  Human resource practitioners are the keepers of the flame when it comes to corporate culture, team building and change management processes. Growing and adapting to the changing marketplace necessitates that firms pursue significant behavioral shifts from time to time. Keeping true to the CSR values compass is a critical guidepost to change management and team alignment. Additionally, the move to incorporate a CSR ethic throughout the firm necessitates a change management approach.

- Step 7: Employee involvement and participation

  As mentioned earlier, employees are among the key stakeholders for the development of any CSR strategy or program. Employee engagement has been acknowledged as a key driver of shareholder value in a firm and is becoming a key metric for monitoring corporate performance by Board and management. Research by Towers Perrin in 2007 revealed that an organization's reputation for social responsibility was one of the top 10 engagement drivers, along with senior management's interest in employee well-being, opportunities to improve skills and capabilities and input into decision-making. (Cited in European Alliance for CSR 2008, p. 11).

- Step 8: CSR Policy and Program Development

  HR is also in a position to drive policy development and program implementation in HR areas that directly support CSR values. Wellness, diversity, work-life balance and flextime policies are CSR programs directly within the HR manager's purview. In organizations committed to reducing their carbon footprint HR practitioners can develop programs enabling employees to use alternative transportation to get to work (e.g. providing showers, secure bike lock-ups, parking spots for van pools and co-op or hybrid cars, bus passes, etc.) and work remotely, including other forms of headquartering and "hoteling", teleworking, etc.

- Step 9: Employee Communications

  Every CSR strategy requires the development and implementation of an employee communication program to convey the corporate direction, objectives, innovation and performance on its CSR efforts. Intranets, websites, blogs, wikis,

social networking sites, podcasting, videos, forums, town hall meetings, regular team briefings, webcasts, voicemails, print and electronic newsletters and other forms of social media need to be deployed to bring the CSR message to the workforce—in ways that are attuned to the communication channels of the employee, which are changing rapidly in this age of web 2.0.
- Step 10: Measurement, Reporting—and celebrating successes along the way!

As what gets measured gets managed, it is vital that both CSR performance and employee CSR engagement be actively measured and reported to executive, the board of directors and publicly. Typically this is done in the form of an annual CSR report. Increasingly, many of these reports are disclosing employee engagement scores. A number of research report the positive relationship between CSR and HR. For example (Jamali et al. 2015) pointed out that CSR represents a relevant approach for the human resource management (HRM) field based on synergies that has be found to exist between CSR and HRM

2.4 CSR Development and Implementation in R&D and How It Affects Product Strategy

It has been proven that a high level of CSR is a strategy that firms can use to differentiate themselves (Hull and Rothenberg 2008; Mackey et al. 2007; Siegel and Vitaliano 2007) in order to obtain certain competitive advantage. Moore (2001) and Harrison and Freeman (1999) make an important point when they state that social performance and economic performance should not be separated, since in order to determine whether a firm is "good", it has to perform well on both counts. Research and development (R&D) is another way a firm can obtain competitive advantage (Hull and Rothenberg 2008), with the long-standing theoretical literature linking investment in R&D with improvements of the firm in the long run (McWilliams and Siegel 2000; Griliches 1979).

CSR can be viewed as a type of investment used as a mechanism for product differentiation, where CSR can be positioned in the context of 'resources', in which CSR policies would help to improve processes for developing products and services, and of 'outputs', where CSR policies and attributes would have a direct impact on a firm's product. For example, firms can maintain a level of CSR by having products with "CSR attributes (such as pesticide-free fruit) or by using CSR-related resources in their production processes (such as naturally occurring insect inhibitors and organic fertilizers)" (McWilliams and Siegel 2001). It has been found that the introduction of new and improved processes and products is positively related with R&D intensity (Hitt et al. 1996). Innovative strategies employed by firms have a substantial impact on processes; in order to create new products and services that have a competitive advantage, they must meet the four criteria described by the Resource Based Value theory, namely, they should be valuable, rare, inimitable, and the organization must be organized to deploy these resources effectively (Barney

1991). Using these criteria, resources that may lead to a competitive advantage include socially complex resources such as reputation, corporate culture, long-term relationships with suppliers and customers, and knowledge of assets (Barney 1986; Hillman and Keim 2001; Teece 1998).

At the same time, researchers contend that it is necessary for businesses to look beyond their narrow focus of social responsibility and take social concern into consideration in strategic management decisions, as this will ensure business interests in the long term by creating a close bond with their community (Carlson et al. 1993; Quazi and O'Brien 2000). Further research shows that consumers prefer products and invest in firms that care for the environment and maintain good citizenship behavior (Zaman et al. 1996; Gildia 1995; Quazi and O'Brien 2000), which helps the firm to build a good reputation and image as valuable resources that can create a competitive advantage for it. Schnietz and Epstein (2005) agree with McWilliams and Siegel (2001) and Lantos (2001) in that CSR creates a reputation that a firm is honest and reliable, giving financial value to the firm. In response to this reputation, consumers will typically assume that the products of these types of firms are of good quality, and they become difficult for other firms to imitate. In addition, firms in industries with skilled labor shortages have used CSR as a means to recruit and retain workers. Brammer and Pavelin (2006) state that depending on a firm's industry and environment, social responsibility actions must vary in order to fulfill general stakeholder expectations and build a good reputation.

At the same time, firms can profit through the use of R&D, since R&D intensive industries usually have 'entry barriers' where companies can achieve effects such as economies of scale and product differentiation (Porter 1979). These effects help firms to obtain a competitive advantage over other firms. How R&D investment affects firm productivity is a question that is of considerable interest to several researchers. There is the seminal work by Griliches (1981) and his hedonic model based on US firm-level data, which used market value as an indicator of the firm's productivity from investments in R&D. Several other researchers have used this same model to prove that there is a positive relationship between R&D investments and the market value of the firm (Cockburn and Griliches 1988; Hirschey 1982; Jaffe 1986). R&D is considered to be a form of investment in 'technical' capital that results in knowledge enhancement, which leads to product and process innovation. This innovative activity allows firms to enhance their productivity (McWilliams and Siegel 2000). Studies such as those by Ben-Zion (1984), Clark and Griliches(1984), Griliches (1998), Guerard et al. (1987), Hall (1999) and Lichtenberg and Siegel (1991) report similar results that confirm a positive correlation between R&D investment and firm growth. Investment in R&D involving innovation related with CSR processes and products is attractive to some consumers, such as recycled products or organic pest control. McWilliams and Siegel (2001) stated that using a differentiating strategy in order to obtain a competitive advantage through the use of CSR resources may also include investment in research and development (R&D). McWilliams and Siegel (2000) proved that CSR is positively correlated with R&D intensity "because both are associated with product and process innovation" . If CSR and R&D are highly correlated, an equation that includes CSR and does not include

R&D intensity as determinant of a firm's performance will turn out to be upwardly biased. Other researchers have also suggested that R&D should be included as a moderator in theoretical models that have received mixed or ambiguous empirical support (Han et al. 1998; Hull and Rothenberg 2008). Therefore, a longitudinal study of the interactions between R&D intensity and CSR variables is called for (Hull and Rothenberg 2008), as one that will provide insight and facilitate an understanding of the interaction that exists between these two variables. Based on the above arguments, we therefore suggest the following hypothesis: R&D intensity positively affects CSR.

Earlier research (Graves and Waddock 1994) has shown that there are clear differences between different industries in levels of investment in R&D (Waddock and Graves 1997). Furthermore, the characteristics of a firm's industry have been hypothesized to be a key influence on its social performance (McWilliams and Siegel 2000), since industries differ according to the stage of the product lifecycle they are in. The use of CSR as a differentiation strategy will be present depending on the industry's lifecycle, since little product differentiation is expected in the embryonic and growth stages because firms are focused on perfecting processes and satisfying growing demands (McWilliams and Siegel 2001). Some industries will be young and companies active in them will have a range of alternative investment projects, whereas mature industries offer fewer alternative investment opportunities to their companies (Brammer and Millington 2008). In addition, depending on the industry, companies may have a different view of CSR actions and the way they are implemented in their R&D processes. Quazi and O'Brien (2000) state that "the broader dimension of social responsibility, therefore, calls for innovation in production and marketing to reap the benefits of proactive social action". The authors give the example of pollution control and how some companies consider it to be an unnecessary expense, thereby perceiving it negatively in financial terms. Meanwhile, other firms may argue that pollution is a sign of inefficiency and flawed technology that also costs the firm money and affects the community. This second perspective is supported by Ahmed et al. (1998), who have found that environmentally friendly companies have better productivity and profitability than non-environmental firms. R&D intensity varies according to the industry, and is usually more intense in manufacturing industries than in non-manufacturing ones. For example, the automotive industry has initiated intensive R&D programs in order to develop a new kind of technology-based competition in response to current environmental changes, long-term increases in petrol prices and regulatory efforts to curb the threat of global climate change (Khaledabadi and Magnusson 2008). These types of changes and increasing stakeholder pressure on firms to tackle social issues are driving more and more companies to engage in CSR activities (Quazi 2003).

Moreover, R&D intensive industries such as pharmaceuticals may face particular incentives to engage in CSR activities that boost the long-term supply of highly skilled labor (Brammer and Millington 2008). Another reason that manufacturing industries might increase their CSR activities is that, according to Nicolleti and Scarpetta (2003), there are more industry-specific regulations are in manufacturing industries than in non-manufacturing ones. Williamson et al. (2006) state that manufacturing in processes have significant economic and environmental impacts, which have led firms to develop CSR practices that favour our environment. Many firms are adopting voluntary environmental management systems, signing international agreements such as the UN Global Compact, or have joined local projects to minimize waste. "These trends have largely been driven by an increasing demand for "transparency" from stakeholders, and perceived consumer demand for environmental quality" (Chapple et al. 2004). These national and industry forces create environments in which stakeholders and local competitors have different expectations of what the appropriate levels and types of corporate citizenship should be (Gardberg and Fombrun 2006), with more pressure being placed on manufacturing industries because they are believed to use up more resources, create more waste and have a higher intensity of R&D activities than their non-manufacturing counterparts simply because of the nature of their processes. Both, corporate social responsibility (CSR) and Sustainability development (SD), have been adopted widely by various organisations and governmental bodies to harness the opportunities to reduce carbon emissions. At the same time the adoption of these two concepts is a direct response to the increasing pressure on organization to be seen acting in a sustainable and ethical manner (Elmualim 2017).

3 Business Sustainability

Sustainability refers to an organization's activities, typically considered voluntary, that demonstrate the inclusion of social and environmental concerns in business operations and in interactions with stakeholders (van Marrewijk and Verre 2003).

According to Bocken et al. (2019) 'Sustainable business model innovation is about creating superior customer and firm value by addressing societal and environmental needs through the way business is done. Business models require intentional design if they are to deliver aspired sustainability impacts' the sustainability of business and other operations is at the heart of every top level management of institutions. Many businesses are still not enthralled to the idea of CSR even though the concept of CSR is not really a new subject; nevertheless, numerous corporations have employed the concept and have benefited from its merits such as business sustainability. If companies carefully integrate CSR into their strategy formulation and implementation (product, corporate, R&D, human resource management) as described in this literature can contribute significantly to improved company performance. Companies that effectively engage in corporate social responsibility experience many benefits. They enjoy a positive effect on staff retention, recruitment, and motivation. They experience increased customer satisfaction and loyalty, particularly when customers buy based on relationships, trust, reputation or brand. Some research link CSR, both in strategy and implementation level, with following consequences variables: organizational performance Prado-Lorenzo et al. (2008); Škerlavaj et al. (2007), profitability, Mittal et al. (2008), corporate sustainability,

Kampf (2007), a competitive advantage, reputation and corporate image, Arendt and Brettel (2010).

3.1 Conceptual Framework of CSR

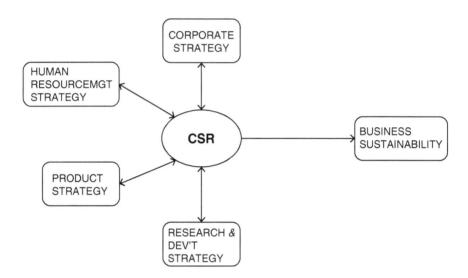

3.2 Explanation of Conceptual Framework

Author is proposing that embedding CSR into strategy formulation and implementation can lead to business sustainability. Thus when CSR is carefully integrated into product, corporate, R&D, human resource management strategy formulation and implementation, organizations experience many benefits.

3.3 Contribution to Theory

Implementation of Corporate Social Responsibility (CSR) is a concept in which companies pay attention to social and environmental integration in their business operations. Sustainability becomes a major a tenet when firms integrate CSR strategies and is one of the key issues for every industry. Clients are increasingly favoring agencies with the knowledge and know-how to help them achieve their sustainability goals. This paper supports kampf's study that integrating CSR into strategy formulation and implementation leads to corporate sustainability. Socially responsible companies, large and small, have better reputations, can better detect and respond to risk, and can anticipate the needs of their stakeholders with greater certainty. When things go wrong, they are more quickly forgiven for mistakes.

Embedding CSR into HR strategy formulation and implementation often provides cost-effective opportunities for training and innovation. Corporate volunteerism programs, a common initiative, are most commonly touted as a way to acquire new skills and develop teams, but employees can gain new skills through a number of CSR programs. And as the organization becomes better integrated with its customers and other stakeholders, and as employees become more engaged with the company's goals, there is often a significant increase in creativity and innovation.

Firms can profit from integrating CSR into R&D and product strategy formulation and implementation. For example, involving R&D related with CSR processes and products is attractive to some consumers, such as recycled products. This supports research showing that consumers prefer products and invests in firms that care for the environment and maintain good citizenship behavior by Zaman et al. (1996), Gildia (1995) and Quazi and O'Brien (2000). This paper further agrees with the resource based view theory which suggests that an organization can gain and sustain competitive advantage by developing valuable resources and capabilities that are relatively inelastic in supply.

3.4 Managerial Implications

In order to successfully embed CSR into strategy formulation and implementation at the corporate level, managers must first support the idea and further make a commitment to CSR and engage in it. They must ensure that CSR strategy is aligned with their firm's core objectives and core competencies and perhaps be treated and managed as a core business strategy just like the strategies of marketing& sales, R&D, talent management etc. CSR also needs to be a component across all senior manager performance plans, in order for it to flow to other levels of the organization.

As stated in the above literature, HR managers are well positioned in helping integrate CSR into their firms' DNA. They must communicate and implement concepts, policies, and cultural and behavioral change in the organization to make it central to fulfill the objective of embedding CSR into strategy formulation and implementation.

Managers need to employ innovative strategies in order to create new products and services to gain a competitive edge over their rivals, such as, investing in R&D innovations related with CSR processes and products. 'As globalization, ever-changing demographics and competition for the world's draining resources force transformational change, businesses would require leadership that is not only enlightened but sustainability-savvy as well to prosper. HR has a very critical role to play in aligning talent with these emerging realities' (Priyadarshini 2020).

Finally, managers must look beyond their narrow focus of social responsibility and take social concern into consideration in strategic management decisions.

4 Conclusion

Some earlier studies in different environmental management domains have predicted that customer satisfaction, reputation, and competitive advantage are three outcomes of CSR (e.g. Mulki and Jaramillo 2011; Salmones et al. 2009; Walsh and Beatty 2007). According to Saeidi et al. (2015) better reputation and competitive advantage are the results of increased customer satisfaction after engaging in CSR. This study adds to knowledge by suggesting that Corporate strategy, Human resource strategy, Research and development strategy and Product strategy affect the business CSR and hence the business sustainability of any given firm. This can be tested empirically to confirm this assertion or otherwise.

References

Arendt, S., & Brettel, M. (2010). Understanding the influence of corporate social responsibility on corporate identity, image, and firm performance. Management Decision, 48(10), 1469–1492.
Ahmed, N. U., Montagno, R. V., & Flenze, R. J. (1998). Organizational performance and environmental consciousness: An empirical study. *Management Decision, 36*, 57–62.
Barney, J. (1986). Organizational culture: Can it be a source of sustained competitive advantage? *Academy of Management Review, 11*, 656–665.
Barney, J. (1991). Firm resources and sustained competitive advantage. *Journal of Management, 17*, 99–120.
Ben-Zion, U. (1984). The R&D and investment decision and its relationship to the firm's market value: Some preliminary results. In Z. Griliches (Ed.), *R&D, patents, and productivity* (pp. 134–162). Chicago, IL: University of Chicago Press.
Bocken, N., Boons, F., & Baldassarre, B. (2019). Sustainable business model experimentation by understanding ecologies of business models. *Journal of Cleaner Production, 208*, 1498–1512. https://doi.org/10.1016/j.jclepro.2018.10.159.
Bowen, R. H. (1953). *Social responsibilities of the businessman* (1st ed.). New York: Harper.
Brammer, S., & Millington, A. (2008). Does it pay to be different? An analysis of the relationship between corporate social and financial performance. *Strategic Management Journal, 10*, published online.
Brammer, S., & Pavelin, S. (2006). Corporate reputation and social performance: The importance of fit. *Journal of Management Studies, 43*(3), 435–455.
Carlson, L., Grove, S. J., & Kangun, N. (1993). A content analysis of environmental advertising campaigns: A matrix method approach. *Journal of Advertising, 22*(3), 27–39.
Chapple, W., Morrison Paul, C. J., & Harris, R. (2004). Manufacturing and corporate environmental responsibility: Cost implications of voluntary waste minimization. *Structural Change and Economics Dynamics, 16*, 347–373.
Clark, K. B., & Griliches, Z. (1984). Productivity growth and R&D at the business level: Results from the PIMS database. In Z. Griliches (Ed.), *R&D, patents, and productivity* (pp. 393–416). Chicago, IL: University of Chicago Press.
Cockburn, I., & Griliches, Z. (1988). Industry effects and appropriability measures in the stock market's valuation of R&D and patents. *The American Economic Review, 78*(2)., Papers and Proceedings of the One-Hundredth Annual Meeting of the American Economic Association (May, 1988), pp. 419–423.
Davis, K. (1960). Can business afford to ignore social responsibilities? *California Management Review, 2*(3), 70–76.

Dawson, P. (2003). *Understanding organisational change: Contemporary experience of people at work*. London: Sage.

De Bakker, F. G. A., Groenewegen, P., & Den Hond, F. (2005). A bibliometric analysis of 30 years of research and theory on corporate social responsibility and corporate social performance. *Business & Society, 44*(3), 283–317.

de Souza Freitas, W. R., Caldeira-Oliveira, J. H., Teixeira, A. A., Stefanelli, N. O., & Teixeira, T. B. (2020). Green human resource management and corporate social responsibility. *Benchmarking: An International Journal*.

Elmualim, A. (2017). CSR and sustainability in FM: Evolving practices and an integrated index. *Procedia Engineering, 180*, 1577–1584.

Europe, C. S. R. (2008). *The European Alliance for CSR Progress Review 2007: Making Europe a pole of excellence on CSR*. European Commission, Brussels, 4.

Frederick, W. C. (1960). The growing concern over business responsibility. *California Management Review, 2*, 54–61.

Gardberg, N. A., & Fombrun, C. (2006). Corporate citizenship: Creating intangible assets across institutional environments. *Academy of Management Review, 31*, 329–346.

George, J. M., & Jones, G. R. (1996). *Understanding and managing organizational behavior*. Reading, MA: Addison-Wesley Publishing Co.

Gildia, R. L. (1995). Consumer survey confirms corporate social responsibility affects buying decisions. *Public Relations Quarterly, 39*, 20–21.

Graafland, J., Van De Ven, B., & Stoffele, N. (2003). Strategies and instruments for organizing CSR by small and large businesses in The Netherlands. *Journal of Bussiness Ethics, 47*, 45–60.

Graves, S. B., & Waddock, S. A. (1994). Institutional owners and corporate social performance. *Academy of Management Journal, 37*(4), 1035–1046.

Gray, R., Owen, D., & Adams, C. (1996). *Accounting and accountability: Changes and challenges in corporate social and environmental reporting*. Hemel Hempstead: Prentice Hall.

Griliches, Z. (1979). Issues in assessing the contribution of R&D to productivity growth. *Bell Journal of Economics, 10*(1), 92–116.

Griliches, Z. (1981). Market value, R&D, and patents. *Economics Letters, 7*, 183–187.

Griliches, Z. (1998). *R&D and productivity: The econometric evidence*. Chicago, IL: National Bureau of Economic Research for the University of Chicago Press, University of Chicago Press.

Guerard, J. B., Jr., Bean, A. S., & Andrews, S. (1987). R&D management and corporate financial policy. *Management Science, 33*, 1419–1427.

Hall, B. H. (1999). *Innovation and market value*. National Bureau of Economic Research Working Paper, 6984.

Han, J. K., Kim, N., & Srivastava, R. K. (1998). Market orientation and organizational performance: Is innovation a missing link? *Journal of Marketing, 62*(4), 30–45.

Harrison, J. S., & Freeman, R. E. (1999). Stakeholders, social responsibility, and performance: Empirical evidence and theoretical perspectives. *Academy of Management Journal, 42*(5), 479–485.

Hillman, A. J., & Keim, G. D. (2001). Shareholder value, stakeholder management, and social issues: What's the bottom line? *Strategic Management Journal, 22*, 125–139.

Hirschey, M. (1982). Intangible capital aspects of advertising and R & D expenditures. *The Journal of Industrial Economics, 30*(4), 375–390.

Hitt, M. A., Hoskisson, R. E., Johnson, R. A., & Moesel, D. D. (1996). The market for corporate control and firm innovation. *Academy of Management Journal, 39*(5), 1084–1119.

Hull, C. E., & Rothenberg, S. (2008). Firm performance: The interactions of corporate social performance with innovation and industry differentiation. *Strategic Management Journal, 29*, 781–789.

Jaffe, A. B. (1986). Technological opportunity and spillovers of R&D: Evidence from firms' patents, profits and market value. *American Economic Review, 76*, 984–1001.

Jamali, D. R., El Dirani, A. M., & Harwood, I. A. (2015). Exploring human resource management roles in corporate social responsibility: The CSR-HRM co-creation model. *Business Ethics: A European Review, 24*, 125–143.

Jang, S., & Ardichvili, A. (2020). Examining the link between corporate social responsibility and human resources: Implications for HRD research and practice. *Human Resource Development Review*, 1534484320912044.

Kampf, C. (2007). Corporate social responsibility. *Corporate Communications: An International Journal, 12*(1), 41–57.

Khaledabadi, H. J., & Magnusson, T. (2008). Corporate social responsibility and knowledge management implications in sustainable vehicle innovation and development. *Communications of the IBIMA, 6*, 8–14.

Kranz, D., & Santalo. (2010). When necessity becomes a virtue: The effect of product market competition on corporate social responsibility. *Journal of Economics & Management Strategy, 19*(2), 453–487.

Lantos, G. P. (2001). The boundaries of strategic corporate social responsibility. *Journal of Consumer Marketing, 18*(7), 595–630.

Levine, M. A. (2008). China's CSR expectations mature. *With PRC Stock Journal of Production Management, 10*(3), 32–45.

Lichtenberg, F., & Siegel, D. (1991). The impact of R&D investment on productivity: New evidence using linked R&D-LRD data. *Economic Inquiry, 29*, 203–228.

Mackey, A., Mackey, T. B., & Barney, J. B. (2007). Corporate social responsibility and firm performance: Investor preferences and corporate strategies. *The Academy of Management Review, 32*(3), 817–835.

McWilliams, A., & Siegel, D. S. (2000). Corporate social responsibility and firm financial performance. *Strategic Management Journal, 21*(5), 602–609.

McWilliams, A., & Siegel, D. S. (2001). Corporate social responsibility: A theory of the firm perspective. *The Academy of Management Review, 26*(1), 117–127.

Mittal, R. K., Sinha, N., & Singh, A. (2008). An analysis of linkage between economic value added and corporate social responsibility. *Management Decision*.

Moore, G. (2001). Corporate social and financial performance: An investigation in the U.K. Supermarket Industry. *Journal of Business Ethics, 34*, 299–315.

Mulki, J. P., & Jaramillo, F. (2011). Ethical reputation and value received: Customer perceptions. *International Journal of Bank Marketing, 29*(5), 358–372.

Nicoletti, G., & Scarpetta, S. (2003). Regulation, productivity and growth: OECD evidence. *Economic Policy, 18*(36), 9–72.

Phillips, S., Thai, V. V., & Halim, Z. (2019). Airline value chain capabilities and CSR performance: The connection between CSR leadership and CSR culture with CSR performance, customer satisfaction and financial performance. *The Asian Journal of Shipping and Logistics, 35*(1), 30–40.

Porter, M. E. (1979). The structure within industries and companies' performance. *The Review of Economics and Statistics, 61*(2), 214–227.

Prado-Lorenzo, J. M., Gallego-Álvarez, I., García-Sánchez, I. M., & Rodríguez-Domínguez, L. (2008). Social responsibility in Spain: Practices and motivations in firms. *Management Decision, 46*(8), 1247–1271.

Priyadarshini, S. (2020). HR professionals: Key drivers for implementing sustainable business practices. *Strategic HR Review*.

Quazi, A. (2003). Identifying the determinants of corporate managers' perceived social obligations. *Management Decision, 41*(9), 822–831.

Quazi, A., & O'Brien, D. (2000). An empirical test of a cross-national model of corporate social responsibility. *Journal of Business Ethics, 25*, 33–51.

Quinn, H. M., Mintzberg, H., & James, S. R. M. (1988). *The strategy process: Concepts, contexts and cases*. Englewood Cliffs: Prentice Hall.

Saeidi, S. P., Sofian, S., Saeidi, P., Saeidi, S. P., & Saaeidi, S. A. (2015). How does corporate social responsibility contribute to firm financial performance? The mediating role of competitive advantage, reputation, and customer satisfaction. *Journal of Business Research, 68*(2), 341–350. https://doi.org/10.1016/j.jbusres.2014.06.024.

Salmones, M. G., Perez, A., & Bosque, I. R. (2009). The social role of financial companies as a determinant of consumer behavior. *International Journal of Bank Marketing, 27*(6), 467–485.

Schnietz, K. E., & Epstein, M. J. (2005). Exploring the financial value of a reputation for corporate social responsibility during a crisis. *Corporate Reputation Review, 7*(4), 327–345.

Siegel, D. S., & Vitaliano, D. F. (2007). An empirical analysis of the strategic use of corporate social responsibility. *Journal of Economics & Management Strategy, 16*(3), 773–792.

Škerlavaj, M., Štemberger, M. I., & Dimovski, V. (2007). Organizational learning culture—The missing link between business process change and organizational performance. *International Journal of Production Economics, 106*(2), 346–367.

Stahl, G. K., Brewster, C. J., Collings, D. G., & Hajro, A. (2020). Enhancing the role of human resource management in corporate sustainability and social responsibility: A multi-stakeholder, multidimensional approach to HRM. *Human Resource Management Review, 30*(3), 100708.

Strandberg, C. (2009). The role of human resource management in corporate social responsibility: Issue brief and roadmap report.

Teece, D. (1998). Capturing value from knowledge assets: The new economy, markets for knowhow, and intangible assets. *California Management Review, 40*(3), 55–79.

Van Marrewijk, M., & Verre, M. (2003). Multiple levels of corporate sustainability. *Journal of Business Ethics, 44*(2/3), 107–110.

Veríssimo, J. M., & Lacerda, T. C. (2010). The new age of corporate social and ethical consciousness: Toward a new leadership mindset. In *Proceedings of the 12th Annual International Leadership Association Conference. Leadership* (Vol. 2).

Waddock, S., & Graves, S. (1997). The corporate social performance – Financial performance link. *Strategic Management Journal, 18*(4), 303–319.

Walsh, G., & Beatty, S. (2007). Customer-based corporate reputation of a service firm: Scale development and validation. *Journal of the Academy of Marketing Science, 35*(1), 127–143.

Williamson, D., Lynch-Wood, G., & Ramsay, J. (2006). Drivers of environmental behavior in manufacturing SMEs and the implications for CSR. *Journal of Business Ethics, 67*, 317–330.

Zaman, M., Yamin, S., & Wong, F. (1996). Environmental consumerism and buying preference for green products. In *Proceedings of the Australian Marketing Educators' Conference* (pp. 613–626).

George Kofi Amoako, Senior Lecturer at the Marketing Department at University of Professional Studies Accra, Ghana. He also lectures as adjunct lecturer at Lancaster University Ghana Campus and Webster University Ghana Campus. He has taught Market Research, Digital Marketing, and Introduction to Marketing, Marketing Management Essentials and Fundamentals of Marketing at Lancaster University. He is an academic and a practising Chartered Marketer (CIM-UK) with specialization in Branding, CSR and Strategic Marketing. He has many years of industry experience. He was the regional sales manager for Fan Milk Ghana Ltd and The head of Lucas Desk at Mechanical Lloyd Ltd in Accra, Ghana. He has consulted for Decathlon Ghana Limited, Ghana Insurance Commission, Ghana Civil Aviation Authority, Bank of Ghana and many others. He was educated in Kwame Nkrumah University of Science and Technology in Kumasi, Ghana and at the University of Ghana and the London School of Marketing (UK). He obtained his PhD from London Metropolitan University, UK. He also received Postgraduate Certificate in Academic Practice (International) from Lancaster University, UK in November 2018. He has considerable research, teaching, consulting and practice experience in the application of Marketing Theory and principles to everyday marketing challenges and management and organizational issues. He is a Chartered Marketer with The Chartered Institute of Marketing, UK. He was a member of the University Council of the University of Ghana in 2000–2001. He has consulted for public sector and private organizations both in Ghana and UK. He has a strong passion for marketing strategy, branding, service quality and CSR issues in the corporate world. He has published extensively in internationally peer-reviewed academic journals and presented many papers at international conferences in Africa, Europe, America and Australia. He has contributed to the latest marketing book 'Break Out Strategies for Emerging Marketing', edited by Professor J. Sheth and recommended by Philip Kotler. This book was launched in the USA in June 2016 and was published by Pearson. He spoke at TEDx Accra and British Council organized seminar on Customer Profitability on 19 August 2016. He is currently a quality assurance consultant for British Council Jobs for the Youth Program and also Educational Consultant for the British Council Connecting Classroom Program in Ghana. He is a twice recipient of Academy of Management Scholarship for faculty development in Africa for 2012 and 2013; a recipient of Academy of Marketing Science J. Sheth Scholarship Award in 2013 and AfricaCSRLeadership Award, Le Meridian (2015) in Mauritius.

Part V
Reporting

Institutional Pressures and CSR Reporting Pattern: Focus on Nigeria's Oil Industry

Uzoechi Nwagbara and Anthony Kalagbor

Abstract This chapter examines the nexus between institutional mechanisms and corporate social responsibility (CSR) reporting/disclosure pattern in Nigeria. It explores the role institutions play in framing and "legitimising" CSR reporting in Nigeria. The chapter focuses specifically on Nigeria's oil industry, which is habitually known for lack of accountability, legitimacy, responsibility and transparency in its operation. Institutional issues are considered as being formal and informal. It is often argued that the combination of formal and informal elements of institution, in any social setting, influences and frames the adoption or otherwise of CSR by organisations, through their adherence to acceptable governance regime. Like any jurisdiction, CSR reporting in Nigeria, is heavily influenced by politico-cultural, regulatory, governance and institutional realities, which constrain and/or enable the adoption of a specific type of CSR reporting: good or bad. Consequently, organisations report CSR activities in response to regulatory regimes; cognitive pressures that facilitate an understanding and interpreting CSR practice correctly; and, socio-cultural values enforcing and regulating the same practice. However, in corrupt regime like Nigeria, CSR reporting tends to foreground social performance and philanthropy at the expense of accountability and business responsibility, which disadvantages wider stakeholders' interests. Through prior literature reviewed, this paper highlighted and found that multinational corporations (MNCs) in Nigeria use CSR reporting to advance shareholder interest, legitimise their behaviour and advance profit maximisation, given a lack of poor institutional regulations

---

U. Nwagbara (✉)
University of the West of Scotland, Paisley, Scotland

Cardiff Metropolitan University, Cardiff, UK

Coal City University, Enugu, Nigeria
e-mail: uzoechi.nwagbara@sunderland.ac.uk

A. Kalagbor
University of Cumbria in London, Cumbria, UK
e-mail: Anthony.kalagbor2@cumbria.ac.uk

influencing CSR reporting. This CSR practice results in a lack of accountability, legitimacy and responsibility in CSR (reporting) in Nigeria.

1 Introduction

Reporting information about social and environmental performance of organisations (hereafter called CSR reporting), is gradually becoming the norm and increasingly considered as *"the de factor law for business"* (Tilt 2016; KPMG 2017). CSR reporting is a process in which organisations report on their activities in society to wider stakeholders (Nwagbara and Belal 2020). Although businesses are not mandated and/or required by law to report on their activities; they are currently expected to do so (Hofman et al. 2017). The Global Reporting Initiative (GRI) as well as other international regulatory bodies including AccountAbility (AA1000), Social Accountability International (SA8000) and the UN Global Compact provides benchmarks and guidelines, which have become commonly accepted rubrics for businesses to report on their activities in terms of content, materiality and format (Belal 2008). Studies on corporate disclosure is undertaken in several labels such as sustainability disclosure (Jensen and Berg 2012), CSR reporting (Vurro and Perrini 2012), CSR disclosure (Cho et al. 2015), and corporate social reporting (Belal 2008). Information disclosed in CSR reports is stand-alone and voluntary.

Several studies have examined CSR reporting from a plethora of approaches. For example, writers have focused on CSR reporting (pattern) in different countries (see Bhatia and Makkar 2019; Visser and Tolhurst 2010); across industries (Young and Marais 2012; Campbell et al. 2006); and across countries (Visser and Tolhurst 2010; Chapple and Moon 2005). Other research emphasises that CSR reporting is used by organisations for reputational effect (Hooghiemstra 2000), financial performance (Shabana et al. 2016), legitimacy management (Campbell 2007), social licence (Denedo et al. 2017), repairing of image (Nwagbara and Belal 2020), and mitigating risk (Bebbington et al. 2008). Taken together, there is a consensus that CSR (reporting) is an instrument in the hands of organisations to further the interest of shareholders (Carroll 1991; Freeman 1984). This contention has precipitated stakeholder criticism of corporate practice and activities including CSR reporting (Frynas 2005, 2009; Idemudia and Ite 2006). Other writers have focused on organisational size, industry classification, and recently institutional factors as determinants of CSR reporting (Shabana et al. 2016).

Institutional oriented writers have leveraged on institutional theory to demonstrate how organisations can adopt new strategies and forms of reporting in a manner that signifies more symbolism than substance (Shabana et al. 2016) due to external and/or internal pressures. This includes formal and informal institutions. The former includes regulations, contracts, constitutions, and forms of governing bodies; while the latter refers to customs, traditions, norms, values and moral beliefs and behaviour that have passed the test of time (North 1990). Central to how CSR is reported is the role played by institutions (Khan et al. 2018).

Accordingly, Matten and Moon (2008) have noted that CSR is located in broader responsibility systems in which legal, social, governmental, and business actors operate in line with some measure of mutual interdependency, responsiveness, capacity and choice all shaped by institutional works operative in a specific business environment (DiMaggio and Powell 1983). Thus, business success depends not only on control and coordination of productive activities, but also on the capacity of organisations to be isomorphic with the institutional milieu (Scott 1995; DiMaggio and Powell 1983). As noted by North (1990) institutional theory both constrains and enables organisational actions and behaviour. It constrains through guidelines, negative sanctions or punishments; and enables via the use of rewards, incentives and/or positive mechanisms (Belal 2008; Campbell 2007). These institutions shape the way CSR is practised and reported in Nigeria, which has been described as unethical, managerial, unaccountable and irresponsible (Frynas 2005).

Research (O'Dwyer 2003; Owen et al. 2000) has indicated that the nature of CSR reports by oil multinational corporations (MNCs) in Nigeria, specifically Chevron, Shell, ExxonMobil, TotalFinaElf, and Agip (the big five) regarding their CSR for environmental, ethical and social performance is undermined by the methods they are done (Emeseh and Songi 2014). In Nigeria, issues of accountability, transparency and legitimacy about CSR disclosure are well documented (Ogiri et al. 2012) as stakeholders have constantly accused MNCs of shirking the facts in issues disclosed (Emeseh and Songi 2014). Some scholars have referred to this issue as "managerial capture" of CSR reporting (O'Dwyer 2003) signalling a situation in which organisations dominate the entire process of reporting and making public only the information that advances their strategic objectives (Baker 2010). It is to this end that Emeseh and Songi (2014) have argued that MNCs in Nigeria "... routinely publish glowing reports on their activities ... It is unfortunate ... [it is not] easy to check their veracity of claims owing to lack of reliable data or information from regulatory agencies" (p. 137) as well as weak, corrupt institutions that permit unaccountable CSR reporting/practice in Nigeria (Amaeshi et al. 2016). Several important reports (see Amnesty 2015; UNEP 2011) have documented the adverse social and environmental consequences of oil exploration activities in Nigeria, and specifically, in 2013 Shell, Nigeria's highest MNC, was legally held liable for social and environmental devastation of the Niger delta, the region where oil is explored in Nigeria (Denedo et al. 2017).

Therefore, a firm's adoption of CSR practices is predicated on pressures from institutional environment. In the contemporary world of business venturing, it has been recognised that there exists growing external (and internal) pressures from different stakeholder groups (including the government, the media, employees, communities, trade unions and others) on organisations to fulfil wider environmental and social responsibilities and/or goals through embracing CSR practices, specifically to dedicate resources to CSR (McWilliams and Siegel 2001). As a result, organisations are now including CSR as an essential aspect of their strategic goals due to its significance to organisational success and sustainability (Porter and Kramer 2006). An essential facet of institutional theory is its relationship with legitimacy (Khan et al. 2018; Donaldson and Preston 1995). Organisations can use

CSR as reputational capital, to obtain legitimacy, and to be isomorphic. Thus, one of the significant facets of institutional theory is isomorphism, which refers to homogenisation or similarities implying that any institutionalised practice makes it unethical and/or unacceptable for organisations to do things differently in a specific context, which might result to loss of legitimacy (Scott 1995). Therefore, unaccountable CSR reporting pattern in Nigeria is framed by wider, institutional problems and weak governance mechanisms in Nigeria, which permit such institutionalised form of reporting (Amaeshi et al. 2016).

Institutional imperatives both create and constrain incentives for organisations to behave through imposing demand and expectations on companies, granting them with varying degrees of legitimacy and offering disparate levels of returns on CSR action (McWilliams and Siegel 2001). However, CSR practice and action vary across countries, depending upon the institutions that frame firms operation (Khan et al. 2018). As a result, many scholars (Campbell 2007) concur that more attention should be paid to institutional mechanisms, which may influence an organisational practice including CSR. The type of institutional pressure explored in this paper is external pressure. Although prior research has provided insights into the incentives for CSR practice, the CSR concept is oftentimes grounded in the voluntary actions and/or behaviours of firms and is rarely expanded into the broader institutional context (Shabana et al. 2016). For instance, Jamali and Mirshak (2007) argue that CSR is an initiative by an organisation that goes beyond its narrow legal requirements and interests. Research from institutional lens can add to nuanced understanding of the broader institutional contexts in which CSR is embedded (Yin and Zhang 2012) and might broaden insights on which institutional issues affect MNCs' CSR engagement and practice in Nigeria.

While CSR (reporting) is proliferating in the west, it is rather ignored in developing counties (Belal 2008). One of the main reasons for a dearth of research on institutional incentives for CSR practice in a developing country like Nigeria is that most inquiry into this concept is undertaken in the west, where the institutional frameworks are relatively stable and effective (Scherer and Palazzo 2011). Thus, broadening insights into the roles played by institutions may be significant in understanding MNCs' strategic positioning in Nigeria, where institutional frameworks are less stable, corrupt and weak (Amaeshi et al. 2016). Relatively, little is known about the roles institutions can play in advancing knowledge about CSR reporting and initiatives. In filling this void, this paper is conceived. Therefore, the research question that is intended to be answered here is: what are the institutional mechanisms that frame CSR reporting pattern in Nigeria? In responding to this question, this paper leverages on institutional theory to explore institutional mechanisms that frame CSR reporting pattern.

As a result, the overall aim of this chapter is to explore how institutional pressures in Nigeria frame its CSR reporting pattern with a focus on its oil industry. Literature on CSR, CSR reporting and institutional theory as well as related concept will be used to actualise this intention. This paper hypothesises that CSR reporting is shaped by institutional frameworks operative in the country, which make it challenging for business responsibility as these institutional frameworks are steeped in corruption and corporate capture, undermining stakeholder engagement, accountability and legitimacy (Belal 2008). The paper is limited to the big five oil MNCs in Nigeria as stated above. Using these MNCs, who constitute over 90% of Nigeria's GDP (Idemudia and Ite 2006), this paper will shed light on the activities of oil companies in Nigeria in the era of heightened demand from societies for organisations to behave in a socially responsible and transparent manner (Campbell 2007).

This paper contributes to the CSR reporting literature in several ways. First, while extant literature confirms the effect of institutional players on CSR (see Campbell 2007), there is paucity of research on how different types of institutional players that impact CSR reporting. This chapter adds to the repertoire of the CSR literature by providing institutional incentives for CSR reporting. Second, this paper will benefit from an understanding of CSR beyond mere voluntary and/or philanthropic behaviour (Matten and Moon 2008). Third, the institutional approach will facilitate a contextual understanding of the CSR phenomenon (Idemudia and Ite 2006) by essentially focusing on the institutional antecedents that are less proclaimed in developing countries as well as cross-cultural insights (Maignan and Ralston 2002). This paper therefore responds to wider call (Yin and Zhang 2012; Matten and Moon 2008; Campbell 2007) and fills this "blind spot" (Campbell 2007, p. 948) by examining the roles of institutions in CSR reporting. In this vein, we contribute to the sparse CSR reporting literature from the developing countries' perspective (Vertigans et al. 2016; Belal 2008).

This paper proceeds as thus. First, the literature on CSR reporting, CSR reporting in Nigeria and the institutional approach to organisational practice is presented. Next, the context of Nigeria is highlighted followed by the rationale for the nature of CSR reporting in Nigeria. Finally, conclusion and further research direction and limitations are presented.

2 CSR Reporting and Motivation

Recent global regulatory reforms such as Sarbanes-Oxley Acts (SOA), Regulation Fair Disclosure and GRI amongst others have acknowledged the importance of CSR reporting (KPMG 2017). CSR reporting generally means a form of corporate disclosure regarding organisational CSR activities, which are not regulated by law (Belal 2008; Idowu and Papasolomou 2007). It is also a timely, deliberate, and formal release of voluntary (or required) information including press releases, annual report, and sustainability report, outlining several social and environmental activities undertaken by a firm (Gray et al. 1995). Given that CSR reports are voluntary (Shell 2017; Gray et al. 1995) they can serve as an instrument enabling firms to respond to wider stakeholders who ceaselessly claim accountability and transparency from them in order to determine if they are trustworthy, accountable, responsible and legitimate in what they report (Nwagbara and Belal 2020).

2.1 Accountability and Responsibility Issue

CSR reporting broadens the accountability of a firm to wider stakeholders beyond its traditional preparation of financial accounts to shareholders such that the interests of other users of information contained in CSR reports, including communities in the Niger delta, the government and regulatory bodies, are put into consideration. CSR reporting can be proactive, which is done when firms anticipate concerns/expectations of wider stakeholders and report on CSR related issues about such expectations. There is a couple of widely held justification for CSR reporting (Gray et al. 1995). Accordingly, Holland and Gibbon (2001, p. 279) noted that "in order to be accountable to their constituents – their stakeholders, or those agencies that fall within the company's sphere of influence" firms leverage on CSR reporting. As argued by Siltaoja and Onkila (2013) the voluntary nature of CSR reporting makes it susceptible to problem of legitimacy and unaccountability as well as raises doubts in the minds of wider stakeholders (Emeseh and Songi 2014).

2.2 Organisational Legitimacy Issue

Organisational legitimacy can be defined as societal approval of corporate behaviour and actions (Suchman 1995). Dowling and Pfeffer (1975) note that firms exist within the environment of a super-ordinate system, within which they can enjoy legitimacy as long as their behaviour and actions are congruent with wider set of rules, regulations, laws, values and social expectations embodied in the "super-ordinate system" (Campbell 2007; Parsons 1960). Legitimacy is crucially needed for organisational success and sustainability (Suchman 1995). The centrality of legitimacy to organisational survival has precipitated effort by firms to achieve legitimacy at all cost for their selfish interests (Campbell 2007).

Firms can achieve legitimacy via CSR reporting. For example Carroll and Buchholtz (2000) contend that ethical and legitimacy responsibilities involve what is generally expected by a firm over and above legal and economic expectations (Freeman 1984). In achieving legitimacy through CSR reporting, firms have run the risk of rather achieving contrary effects of their actions, which resonates with what Ashford and Gibbs (1990, p. 188) identified as "self-promoter's paradox" . This concept highlights that organisations over-emphasise their CSR initiatives and commitment in a manner that often leads to contradictory expectations (Ashford and Gibbs 1990). As firms engage in CSR reporting, corporate managers become "self-seduced" and "self-absorbed" in their bid to report on CSR for legitimacy, which is exclusive of stakeholders' interests and engagement.

2.3 Stakeholder Engagement Issue

CSR reporting is one of the effective methods to understand how social and environmental issues are disclosed to wider stakeholder groups (Morsing and Schultz 2006). This suggests that each stakeholder group has a "stake" or interest in the firms as well as in the process of their wealth creation and benefit (Freeman 1984). Scholars have identified that the strategic and shareholder-centric nature of CSR reporting makes it difficult and almost impossible for firms to accurately and honestly engage diverse stakeholder groups because of interest contestation (Belal 2008). Solomon and Lewis (2002) observe that there is a propensity for firms not to report negative or hypothetical harmful information, and instead disclose only good practices making them good corporate citizens (Idemudia and Ite 2006). In the Nigerian context, stakeholders maintain that CSR reporting is not engaging and participatory. It is rather used by MNCs to further their selfish interest and it is designed to marginalise stakeholder voice and inclusion. Hence, it is controlled by management and as such the veracity and legitimacy of its contents and materiality are in doubt as well as methods of stakeholder engagement. In corroborating this contention, Thomson and Bebbington (2005) concluded that there is little or no relationship between wider society's interests and the number of stakeholders appointed by UK firms, which makes CSR reporting process underrepresented and flawed.

2.4 Organisational Performance, Power and Strategy Issue

CSR (reporting) can be used to highlight the nexus between CSR reporting and a firm's financial performance, strategic position and stakeholder power (Ullman 1985). Specifically, stakeholder power ensures that MNCs in Nigeria will give greater salience to issues that relate to stakeholders (for example MNCs) with more powers than less economically powerful stakeholder, such as the communities in the Niger delta (Mitchell et al. 1997). Stakeholder salience is the level of importance attached to stakeholders' concerns by an organisation. As indicated by Ullman (1985) CSR reporting can also be used by organisations for strategic posture given MNCs' relative strategic position in contrast with wider stakeholders. Consequently MNCs, who have more active strategic position, can influence wider stakeholders by making CSR reports available for strategic gains. Regarding performance, Roberts' (1992) empirical work illustrates that there is a positive correlation between socially performing firms and high level of CSR reporting. This contention is validated by several studies (Aras and Crowther 2008; Ullman 1985) that have stressed the positive relationship between CSR reporting, strategic gain, power and financial performance; hence, why organisations report on their CSR activities.

2.5 Risk Reduction and Reputation Building Issue

Several scholars have argued that stakeholders need CSR reports to actually understand the risks, to which firms are exposed via environmental and social performance, and how proactive firms are in regards to socio-environmental and community issues (Idemudia and Ite 2006). CSR reporting is heavily influenced by risk reduction and reputation building by organisations (Bebbington et al. 2008). MNCs operating in Nigeria use CSR reports to address these issues their sullied image i as seen in popular press owing to their culpability in the destruction of Nigeria's environment as well as involvement in deepening poverty (Idemudia and Ite 2006). Also, they use CSR reports to reduce risk associated with harmful practices that undermine corporate responsibility and accountability (Frynas 2005).

Thus, insurers, investors, and underwriters use CSR reports to source relevant information pertaining environmentally and socially responsible business initiative as well as potential risks including fines for noncompliance. Amongst other reasons, another rationale for CSR reporting deals with repairing/building reputation (Hooghiemstra 2000). As noted by Hooghiemstra (2000) and Nwagbara and Belal (2020), the most important motives for CSR reporting are corporate image and brand management, which allow firms to reduce operating costs, increase capacity to attract investors and advance profit maximisation for shareholders. For example, leveraging on non-managerial approach to CSR reporting, Belal and Roberts' (2010) study found that "current disclosure practice" in Bangladesh has failed "to meet the expectations of stakeholders" (p. 321), which can lead to corporate impression management and risk reduction (Owen et al. 2001).

2.6 The Institutional Approach to Organisational Practice

As stated earlier, this paper focuses on the external pressures that shape CSR reporting pattern in Nigeria. Accordingly, this paper is based on the macro perspective to CSR reporting institutionalisation. First, institutions can be considered as a network of cultural roles shaping and guiding organisational behaviour (Scott 1995) including CSR practice and reporting (Yin and Zhang 2012). As observed by North (1990) institutions are formal and informal apparatuses, which permit efficient exchanges and interactions between social actors such as MNCs and the communities in Nigeria (Amaeshi et al. 2016). Institutions are thus a set of mechanisms, values, practices, myths, instruments, relationships and belief systems that help in maintaining relatively stable forms of practices including institutionalised CSR reporting (Kostova and Roth 2002). As a result, institutions comprise societal "higher order" factors and phenomena beyond an individual organisation, constituting or restraining the interests and political participation of various actors without requiring recurring collective mobilisation or authoritative interference to achieve certain regularities (DiMaggio and Powell 1983). Consequently, institutional rationality relies not on collection of patterned or individual action between groups such as MNCs and other stakeholders, but on institutions that frame behaviours and actions (Scott 1995).

Doh and Guay (2006) have noted that attention needs to be paid to the institutional mechanisms that result in the framing and implementation of CSR (reporting). For example, leveraging on Matten and Moon's (2008) framework and new institutionalism approach, Hofman et al. (2017) explored CSR (reporting) under China's authoritarian capitalism. They found two types of CSR: CSR as an instrument for local reputation; and CSR as a reflection of global and national social expectation— all based on CSR responding to institutional pressures. Similarly, Shabana et al. (2016) have used "two-stage model" of isomorphic mechanism to shed light on "the institutionalisation" (p. 1) of CSR reporting. Institutional theory emphasises myths, legitimacy and isomorphism that shape organisational actions (Berger and Luckmann 1966). Also, isomorphism can be divided into *competitive* and *institutional* isomorphism. The former is concerned with how competitive forces drive firms towards adopting efficient, managerial and least-cost systems and practices. On the other hand, according to DiMaggio and Powell (1983) institutional factors are further divided into coercive, mimetic and normative sub-categories. Scott (1995) went further to present a version of institutional theory that emphasises regulative and/or cultural-cognitive (Scott 1995).

Organisational response to institutional pressures is becoming more complex. In responding from these institutional pressures, organisations use CSR reporting to address issues regarding social, environmental, ethical and community issues. However, doubt persists about MNCs' sincerity of purpose, commitment to CSR and stakeholder engagement (Emeseh and Songi 2014; Owen et al. 2000). As argued by Campbell (2007, p. 947) research has indicated that "the way corporations treat their stakeholders depends on the institutions within which they operate". Therefore, inasmuch as "corporations influence the political system or operate in failed states, like Nigeria (Dobers and Halme 2009), "without any democratic mandate or control, we need to consider how we can close the democracy gap and make corporate decisions more accountable" (Scherer and Palazzo 2011, p. 921).

3 Contextualising Nigeria: CSR Reporting, Corruption and Weak Institutions

Nigeria is Africa's largest economy, amounting mainly from the country's huge earnings from oil exports, yet more than 70% of its population lives below the poverty line owing to the incidence of oil exploration (Frynas 2005). Apart from ailing economy, poverty and inept political governance, corruption and weak institutional arrangements have redoubled problems in Nigeria (Adegbite and Nakajima 2012). Scholars have however noted that effective CSR (reporting) can significantly contribute to the lives of Nigerians, majority of whom have lost confidence in the nation's political leadership and governance regarding the provision of their basic necessities, but look up to businesses, particularly MNCs, to help deal with providing social amenities (Scherer and Palazzo 2011). Nevertheless, corporate corruption, incessant incidents of accounts manipulation, non-transparent disclosure, auditors' compromise, poor corporate governance and other fraudulent behaviours continue to thrive in Nigeria owing to ineffective institutions and "patronage-based Nigerian society" (Bakre et al. 2017, p. 1288), which fuel corporate malfeasance by MNCs (Egbon et al. 2018). Broadly, the political economy of Nigeria is implicated in CSR practice and reporting. This scenario is characteristic of developing country (Amaeshi et al. 2016). It is based on this context, that this paper explores the institutional pressures that frame CSR reporting pattern focusing on Nigeria's oil industry, which has a history of corruption, corporate-stakeholder conflict and inept CSR practice.

There is a concerted effort to institutionalise and regulate CSR reporting in Nigeria. This includes efforts by Nigeria Stock Exchange (NSE), Companies and Allied Matters Act (CAMA, 2004), Securities and Exchange Commission (SEC) and the Nigerian Accounting Standards Act (2003) resulting in the establishment of Financial Reporting Council of Nigeria in 2011 (formerly Nigerian Accounting Standards Board 1982). The creation of Petroleum Industry Bill (2012), the Nigeria Extractive Industry Transparency Initiative bill (NEITI 2007) and the Fiscal Responsibility Act (2007), among other measures were aimed at making corporate disclosure more transparent and accountable (Egbon et al. 2018; Nwagbara and Ugwoji 2015). However, performance and effectiveness of these regulatory institutions is in question given institutional weaknesses for enforcement (Bakre et al. 2017).

In the literature, Wallace (1988) found that "the environment of corporate reporting in Nigeria" is complex and challenging on the heels of Nigeria's peculiar political economy of oil extraction and production. He further stated that the country's inept institutional governance frameworks and regulatory regime provide weak "control mechanisms" (p. 352) to police corporate activities. In a comparable work, Disu and Gray (1998) explored "social reporting and the MNCs in Nigeria" (p. 13) outlining that it is steeped in advancement of corporate interests. Similarly, Adelopo's (2011, p. 338) research on listed companies in Nigeria using annual report noted that there is a substantial positive association between voluntary disclosure and firm size. Additionally, Hassan and Kouhy's (2013) work investigated carbon emission and reporting in Nigeria regarding gas flaring. It found that corporate interest dominates information disclosure about gas flaring, which can explain the lingering corporate-stakeholder conundrum in Nigeria resonating with incessant conflict, war, violence and social and environmental devastation of Nigeria's Niger delta region (UNEP 2011).

Based on the above, Agbiboa (2013) has described corruption in Nigeria as *semper et ubique*. In a corrupt country such as Nigeria, driving accountability and transparency has proved very challenging. Hence as argued by Emeseh and Songi (2014) in corrupt regimes there is palpable diversion of resources and political goods from the public (stakeholders) to private consumption. Various regulatory measures have been taken by the Federal Government of Nigeria to ensure business responsibility, transparency and accountability are praiseworthy; yet, there is a lack of effective implementation of these measures (Adegbite and Nakajima 2012). According to the Global Corruption Report 2003, which is produced by Transparency International, Nigeria is often ranked very high on corrupt countries globally.

Some writers (Adeleye et al. 2020; Mair and Marti 2009; Khanna and Palepu 1997) suggest that a lack of effective regulation is premised on institutional voids, where institutional frameworks that support market, business responsibility and accountability are weak and/or absent or fail to realise the role expected of them (Mair and Marti 2009). For example, institutional voids represent contexts that lack the capacity to police firms' activities including CSR reporting activities. As noted by Khanna and Palepu (1997) institutional void does not denote absence of institutions; it rather stresses their inability to be functional and effective (Mair and Marti 2009). In corrupt countries like Nigeria, organisations try to create legitimacy and moral behaviour by showcasing their social and environmental activities to different stakeholder groups through for example philanthropy and charity giving that takes attention away from their unaccountable CSR activities. For instance, Idemudia and Ite (2006) painted a gloomy picture of CSR (and reporting) in Nigeria, where philanthropy and related green-washing strategies conceal irresponsible CSR practice and reporting.

4 Understanding *raison d'etre* of CSR Reporting Pattern in Nigeria: Ineffective Institutions and CSR Reporting

Institutional theory has been adopted and tested in several, prior research, which is too broad to sufficiently unpack here, given this paper's remit (see Suchman 1995 and Scott 1995 for detail). For example, Kostova (1999) leveraged on "three-dimensional country institutional profile" to account for how a country's governmental guidelines and rules impact domestic business activities. Likewise, Rathert's (2016) research found that MNCs utilise CSR in their business internationalisation to seek legitimacy in their host countries. Even though writers on institutionalism differ in relation to emphasis, the main theoretical claim is that firms operating in analogous business setting are likely to seek legitimacy (and recognition) by adopting and practising dominant, structures, processes and practices in their environment (Scott 1995). These institutions (formal and informal) are closely linked with each another, producing, fostering and transmitting unaccountable CSR reporting pattern, business irresponsibility, lack of transparency and stakeholder exclusion mechanism making CSR institutionally corrupt (Adeleye et al. 2020; Amaeshi et al. 2016).

Institutionally writers have identified that institutional approach can be instrumental in closing legitimacy, accountability and stakeholder engagement "gap" in Nigeria. This is part of the preoccupation of the institutional approach to CSR reporting adopted in this paper. As noted by Ogiri et al. (2012) regrettably in Nigeria "CSR regulatory body has created a huge implementation problem of CSR policies"

(p. 267). Specifically, "little attention has been paid to utilising social reporting as a means of promoting the responsibility of corporations in the country" (Amao 2008, p. 100). This is as a result of institutionalised corruption and poor regulatory frameworks through which CSR (reporting) is regulated. The below section will present the major reasons for the nature of CSR reporting in Nigeria.

4.1 Weak Legal Institutions

One of the reasons for the pattern of CSR reporting in Nigeria is the nature of the country's legal institutions (Okike 2007), which are responsible for regulating corporate behaviour for corporate governance. There are various laws and regulatory instruments, which provide basis for business responsibility and accountability (and financial reporting) in Nigeria. One of the main (legal) institutionalised frameworks is Corporate Affairs Commission, which was created by Companies and Allied Matters Acts (CAMA) of 1990. CAMA has some provisions including requirements for disclosures, preparation and publication of financial and social transactions and auditing. Also, CAMA mandates Registrar of companies at the Corporate Affairs Commission to monitor compliance with various requirements and specifies penalties in cases of non-compliance by firms.

As noted by Okike (2007) there are adequate laws regulating corporate activities including CSR practice, what Nigeria lacks is effective implementation and legal enforcement of CSR, which leaves the MNCs with immense power to operate irresponsibly (Frynas 2005). As noted by Amao (2008), before 1968 there was no provision for mandatory disclosure provisions in Nigerian company law. The Companies Act of 1968 introduced the idea of mandatory disclosure, which was modelled on British Companies Act 1948. However, the scope and applicability of these disclosure/reporting provisions have been amended under CAMA (Sections 331). Since Obasanjo administration in 1999, several provisions and laws have been made to regulate CSR and related phenomena with little positive effect (Emeseh and Songi 2014). One of these is Nigeria Extractive Industry Transparency Initiative (NEITI), which has been described as shambolic. Similarly Petroleum Industry Bill (PIB) is a charade to further deepen and marginalise minorities in the Niger delta (Agbiboa 2013). In sum, there are unworkable laws and regulations in Nigeria, which makes business responsibility problematic.

4.2 Managerial and Corporate Capture

Corruption, bribery, auditors' compromise, sharp practices, accounts manipulation, and non-transparent disclosure impede legitimacy and accountability, which are crucial for CSR (Frynas 2009). This practice can be considered as a form of managerial capture; which is a situation in which organisations are in control of reporting given their relationship with government and their position powers (O'Dwyer 2003). In this instance, corporate corruption and unethical practices as a consequence facilitates capturing of (CSR) reporting that is antithetical to the tenets of business responsibility and accountability. Additionally, it has been alleged that external and/or third party verifiers in Nigeria are routinely bribed by firms to validate their CSR reporting, thereby legitimising their CSR (reporting) (Bakre et al. 2017). Okike (2007) warns that in a corrupt society the quality of reporting and accounting system including the materiality of issues reported may be incredulous.

Furthermore, this type of capture is evident in opaque and skewed language used in preparing CSR report, directors' statements, annual reports, and supplementary annual reports as well as related reports. For example, criticism of Shell's activities in the Niger delta since it started oil exploration in Nembe Oloibiri, the present Bayelsa State in 1933, including CSR reporting by UNEP and influential world bodies like Friends of the Earth corroborate the company's complicity in environmental damage of the region. More importantly the falsified nature of Shell's corporate reports, which alleges that the militants are responsible for oil spillage in the Niger delta, demonstrates managerial capture and business irresponsibility. Accordingly, Bakre (2007) unambiguously asserted that "[A]s a consequence of the many cases of fraud, falsification and deliberate overstatement of companies' accounts and other professional misconducts (such as poor institutional mechanisms for ensuring accountability in CSR disclosure), many ... public and private companies have ..." (p. 278) been criticised as not being transparent and responsible in their reporting. In addition, given that stakeholders' voices are marooned at best or excluded at worst in CSR reporting, this situation chimes with "complete absence of any mechanism by which stakeholder views can feed directly into the reporting process" (Belal and Owen 2007, p. 474).

4.3 Lack of Checks and Balances

Constant checks and balances are required for effective CSR implementation (Amao 2008). This process serves as a way of promoting accountability and transparency and also a way of gaining public trust in CSR (Frynas 2009). This resonates with what has been described as "false dawns in past decades" (Silberhorn and Warren 2007, p. 352) in relation to corporate irresponsibility, precipitated by apparent lack of checks and balances. Lack of check and balances has also materialised in 2008 global financial crisis and the grand global business scandals such as climate change, Enron, Worldcom, Parmalat, and Niger delta conundrum, intensifying stakeholder clamour for accountability in CSR reporting and practice (Belal 2008). Lack of check and balances negatively impacts enforceability, stakeholder empowerment and fair play. Evidence has shown that when businesses and government are left unchecked they can become irresponsible and oppressive. For example, the World Bank has recommended improved government oversight as well as checks and balances to effectively control gas flaring as part of its Global Gas Flaring Reduction (GGFR) public-private partnership that started in 2002. Promoting regulation through check and balances has become the norm in some countries of the world including Canada and Norway. Nevertheless, Nigeria lags behind in ensuring checks and balances translate into responsible business and CSR reporting.

4.4 Poor Political Leadership and Corporate Governance

Leadership is fundamental to shaping national politics and organisational behaviour (Burns 1978). Leadership, which is the ability of a leader to galvanise and mobilise support and influence which can bring change and/or new way of doing things (Kotter 1990) is fundamental to ethical practice (Rotberg 2012). Rotberg (2012) has noted that leaders and their leadership style exert strong influence on corporate sector as well as its overall working. Thus, effective and enabling political leadership is a precondition for responsible business operations, especially in fragile and/or weak institutions like Nigeria. Business managers and leaders follow the nature and mode of leadership in a country, which shapes their practice, for legitimacy and isomorphism (Amaeshi et al. 2016; Gill 2006). The inseparable link between political leadership and organisational culture and practice including CSR reporting cannot be over-emphasised. In this context, in weak institutions business managers, who act in the interests of shareholders, rely on the brand of political leadership to further the interest of the firm. Therefore, there no doubt that political leadership hugely influences corporate governance.

Inept political leadership behaviour as seen in Nigeria unavoidably moderates corporate governance and practice triggering poor governance and unacceptable CSR practice (Rotberg 2012). Effective political leadership is reflected in positive institutional and corporate governance, guiding the vision and strategies of MNCs on the path of business responsibility and accountability. However, in context where there is diminished credibility and/or inept leadership corporate actions can be unchecked thereby promoting unethical and irresponsible business dealings. This situation can also create problem of compliance and enforceability of regulations and laws (Adegbite and Nakajima 2012). In buttressing this, Amao (2008, p. 102) has argued that whereas other countries of the world introduced effective codes of corporate governance in the past decades, shaped by leadership, that regulate business behaviour, the Nigerian codes of corporate governance are an outstanding exception, retaining the usual shareholder-centric approach to corporate governance and stakeholder engagement. This context is clearly a sign of institutional void. As noted by Amaeshi et al. (2016) institutional voids and poor leadership can precipitate the emergence and strengthening of ineffective institutions that can lead to poor CSR reporting and practice.

4.5 Incapacitated Social Movement/Activism

All over the world, social movements can be instrumental in engendering change and consideration of stakeholder interest in government and corporate dealings including CSR activities (Utting 2005). One main reason CSR (reporting) has attracted so much attention is that it is being driven by powerful sets of actors/ movements: non-governmental organisations (NGOs) and trade unions concerned with sustainable business and development as well as human rights agitation and disapproval of corrupt practices. Social movements can serve as opposition to managerialism, fierce capitalism and irresponsible CSR reporting regime (Georgallis 2017). Social movements have been described as forces from below that force government and institutions to rethink the interest of wider stakeholders and entire citizens for accountability and social justice. However, in situation in which social movements are incapacitated, there is a tendency for businesses and governance to be unethical and unaccountable (Kolk and Lenfant 2015).

Through the coalition of NGOs, corporations and governments the world over have been brought to justice. For example, protest following Arab Spring, international campaigns and collaborative works by social movements have precipitated change in governance mobilising and advocating that voices of the marginalised and repressed as seen in the Niger delta are heard for greater corporate-stakeholder engagement and capacity building (Idemudia 2017; Abramov 2010). Nevertheless, it has been noted that in regimes that decry social movement/activism, like the case of Nigeria, bringing genuine stakeholder engagement and lasting change is problematic (Georgallis 2017). In the wake of the Millennium Declaration Goals (MNGs) and subsequently Sustainable Development Goals (SDGs) as well as the World Summit on Sustainable Development (WSSD), focus has been drawn to community and international development, stakeholder engagement in CSR reporting and poverty alleviation through CSR (Idemudia and Ite 2006), which social movements facilitate. The involvement of social moments has also materialised in asking MNCs to rethink their practice for more sustainable business. As contended by Abramov (2010) and Kolk and Lenfant (2015) the activities of social movements have been described as essentially "counter-hegemonic" and/or emancipatory social initiatives to drive sustainable, responsible business.

5 Failed/Fragile State Syndrome

Failed and/or fragile states are countries with weak institutional frameworks that do not enable effective regulation of activities including CSR activities by oil MNCs (Kolk and Lenfant 2015). In fragile states, collaborative efforts and public-private partnerships, which leverage on inclusive engagement strategies and accountability, are deactivated triggering a lack of accountability and transparency (Abramov 2010; Kolk and Lenfant 2010). As advised by Kolk and Lenfant (2015) owing to the limits of nation-state governance, particularly in frail/fragile states, there is need to reconceptualise private social actors, like civil society and NGOs, to champion open deliberation processes aimed at enhancing fairer and more just business environment that will benefit wider stakeholders, not just the greedy few in developing countries. This process will be instrumental in defining priorities, engendering responsible business and promoting democratic culture that will positively impact CSR (Idemudia and Osayande 2018; Kolk and Lenfant 2015; Scherer and Palazzo 2011). However, in failed state system it is challenging to have a form of CSR, which is empowering, inclusive, accountable, and sustainable.

6 Conclusion

The focus of this chapter is to explore the relationship between institutions (formal and informal) and CSR reporting pattern in Nigeria's oil industry, which has a history of corporate-stakeholder conflict, poor CSR and problematic corporate practice. Indeed, as demonstrated in the preceding sections institutions influence and shape the adoption or otherwise of CSR practice and reporting, through firms' adherence to acceptable and/or unacceptable governance regime prevalent in a social space. The roles played by institutions in shaping corporate practice, for example, CSR practice/reporting cannot be over-emphasised. Extant, relevant literature reviewed and analysed here substantiates the centrality of institutions in de/legitimising and institutionalising CSR reporting pattern as they exert cognitive pressures that facilitate correct understanding and interpreting CSR practice and socio-cultural norms and values enforcing and regulating the same practice. As has been demonstrated, Nigeria like any social setting is unavoidably influenced by its political, historical, cultural, regulatory and governance realities, which constrain and/or enable the adoption of CSR reporting either acceptable or unacceptable.

Given that there is institutionalised corruption, weak intuitions and poor regulatory regime in Nigeria CSR reporting tends to foreground social performance and philanthropy at the expense of business accountability, transparency, legitimacy and responsibility, which put the interest of MNCs' shareholders against wider stakeholders, like the communities in the Niger delta. Furthermore, this paper has revealed that various motivation for business engagement in CSR reporting. It was found that amongst other reason, MNCs in Nigeria use it to legitimise their practice; they also use CSR reporting to comment on their philanthropic endeavours, increase profitability, promote accountability, advanced their power/strategy as well as reduce risk and build their reputation. Furthermore, a couple of reasons were identified as responsible for the dynamics of CSR reporting pattern in Nigeria including poor political leadership and corporate governance, lack of checks and balances, weak legal institutions and managerial and corporate capture as well as incapacitated social movement/activism and failed/fragile state syndrome.

References

Abramov, I. (2010). Building peace in fragile states–Building trust is essential for effective public–private partnerships. *Journal of Business Ethics, 89*, 481–494.

Adegbite, E., & Nakajima, C. (2012). Institutions and institutional maintenance: Implications for understanding and theorising corporate governance in developing economies. *International Studies of Management and Organisation, 42*(3), 69–88.

Adeleye, I., Luiz, J., Muthuri, J., & Amaeshi, K. (2020). Business ethics in Africa: The role of institutional context, social relevance, and development challenges. *Journal of Business Ethics, 161*, 717–729.

Adelopo, I. (2011). Voluntary disclosure practices amongst listed companies in Nigeria. *Advances in Accounting, 27*(2), 338–345.

Agbiboa, D. E. (2013). Between corruption and development: The political economy of state robbery in Nigeria. *Journal of Business Ethics, 108*(3), 325–345.

Amaeshi, K., Adegbite, E., & Rajwani, T. (2016). Corporate social responsibility in challenging and non-enabling institutional contexts: Do institutional voids matter? *Journal of Business Ethics, 134*(1), 135–153.

Amao, O. O. (2008). Corporate social responsibility, multinationals corporations and the law in Nigeria: Controlling multinationals in host countries. *Journal of African Law, 52*(1), 89–113.

Amnesty. (2015). *Niger Delta: Shell's manifestly false claims about oil pollution exposed, again.* London: Amnesty International.

Aras, G., & Crowther, D. (2008). Developing sustainable reporting standards. *Journal of Applied Accounting Research, 9*(1), 4–16.

Ashford, B. E., & Gibbs, B. W. (1990). The double-edge of organisational legitimation. *Organisation Science, 1*(2), 177–194.

Baker, M. (2010). Re-conceiving managerial capture. *Accounting, Auditing & Accountability Journal, 23*(7), 847–867.

Bakre, O., Lauwo, S. G., & McCartney, S. (2017). Western accounting reforms and accountability in wealth redistribution in patronage-based Nigerian society. *Accounting, Auditing & Accountability Journal, 30*(6), 1288–1308.

Bakre, O. M. (2007). The unethical practices of accountants and auditors and the compromising stance of professional bodies in the corporate world: Evidence from corporate Nigeria. *Accounting Forum, 31*, 277–303.

Bebbington, J., Larrinaga, C., & Moneva, J. M. (2008). Corporate social reporting and reputation risk management. *Accounting, Auditing & Accountability Journal, 21*(3), 337–361.

Belal, A., & Roberts, R. (2010). Stakeholders' perceptions of corporate social reporting in Bangladesh. *Journal of Business Ethics, 97*(2), 311–324.

Belal, A. R. (2008). *Corporate social responsibility reporting in developing countries: The case of Bangladesh.* Aldershot: Ashgate.

Belal, A. R., & Owen, D. (2007). The views of corporate managers on the current state of, and future prospects for, social reporting in Bangladesh: An engagement based study. *Accounting, Auditing and Accountability Journal, 20*(3), 472–494.

Berger, P. L., & Luckmann, T. (1966). *The social construction of reality.* New York: Doubleday.

Bhatia, A., & Makkar, B. (2019). Extent and drivers of CSR disclosure: Evidence from Russia. *Transnational Corporations Review, 11*(3), 190–207.

Burns, J. M. (1978). *Leadership.* New York: Harper & Row.

Campbell, D. J., Moore, G., & Shrives, P. J. (2006). Cross-sectional effects in community disclosure. *Accounting, Auditing & Accountability Journal, 19*(1), 96–114.

Campbell, J. L. (2007). Why would corporations behave in socially responsible ways? An institutional theory of corporate social responsibility. *Academy of Management Review, 32*(3), 946–967.

Carroll, A. B. (1991). The pyramid of corporate social responsibility: Toward the moral management of organisational stakeholders. *Business Horizons*, (July–August), 39–48.

Carroll, A. B., & Buchholtz, A. K. (2000). *Business and society: Ethics and stakeholder management*. Ohio: South-Western.

Chapple, W., & Moon, J. (2005). Corporate Social Responsibility (CSR) in Asia: A seven-country study of CSR web site reporting. *Business & Society, 44*(4), 415–441.

Cho, C. H., Michelon, G., Patten, D., & Roberts, R. B. (2015). CSR disclosure: The more things change...? *Accounting Auditing & Accountability, 28*(1), 14–35.

Denedo, M., Thomson, I., & Yonekura, A. (2017). International advocacy NGOs, counter accounting, accountability and engagement. *Accounting, Auditing & Accountability Journal, 30*(6), 1309–1343.

DiMaggio, P. J., & Powell, W. W. (1983). The iron cage revisited: Institutional isomorphism and collective rationality in organisational fields. *American Sociological Review, 48*(2), 147–160.

Disu, A., & Gray, B. (1998). An exploration of social reporting and MNCs in Nigeria. *Social and Environmental Accounting Journal, 18*(2), 13–15.

Dobers, P., & Halme, M. (2009). Corporate social responsibility and developing countries. *Corporate Social Responsibility and Environmental Management, 16*(5), 237–249.

Doh, J. P., & Guay, T. R. (2006). Corporate social responsibility, public policy, and NGO activism in Europe and the United States: An institutional-stakeholder perspective. *Journal of Management Studies, 43*(1), 47–73.

Donaldson, T., & Preston, L. E. (1995). The stakeholder theory of the corporation: Concepts, evidence and implication. *Academy of Management Review, 20*(1), 65–91.

Dowling, J. B., & Pfeffer, J. (1975). Organisational legitimacy: Social values and organisational behaviour. *Pacific Sociological Review, 18*(1), 122–136.

Egbon, O., Idemudia, U., & Amaeshi, K. (2018). Shell Nigeria's Global Memorandum of Understanding and corporate-community accountability relations: A critical appraisal. *Accounting, Auditing & Accountability Journal, 31*(1), 51–74.

Emeseh, J., & Songi, O. (2014). CSR, human rights abuse and sustainability report accountability. *International Journal of Law and Management, 56*(2), 136–151.

Freeman, R. E. (1984). *Strategic management: A stakeholder approach*. Boston: Pitman.

Frynas, J. G. (2005). The false developmental promise of corporate social responsibility: Evidence from multinational oil companies. *International Affairs, 81*(3), 581–598.

Frynas, J. G. (2009). *Beyond corporate social responsibility: Oil multinationals and social challenges*. Oxford: Oxford University Press.

Georgallis, P. (2017). The link between social movements and corporate social initiatives: Toward a multi-level theory. *Journal of Business Ethics, 142*, 735–751.

Gill, R. (2006). *Theory and practice of leadership*. London: Sage Publications.

Gray, R., Kouhy, R., & Lavers, S. (1995). Corporate social and environmental reporting: A review of the literature and a longitudinal study of UK disclosure. *Accounting, Auditing & Accountability Journal, 8*(2), 47–77.

Hassan, A., & Kouhy, R. (2013). Gas flaring in Nigeria: Analysis of changes in its consequent carbon emission and reporting. *Accounting Forum, 37*(2), 124–134.

Hofman, P. S., Moon, J., & Wu, B. (2017). Corporate social responsibility under authoritarian capitalism: Dynamics and prospects of state-led and society-driven CSR. *Business & Society, 56*(5), 651–671.

Holland, L., & Gibbon, J. (2001). Processes in social and ethical accountability: External reporting mechanisms. In J. Andriof et al. (Eds.), *Perspectives on corporate citizenship: Context, content and processes*. Strongsville: Greenleaf Publishing.

Hooghiemstra, R. (2000). Corporate communication and impression management – New perspectives why companies engage in corporate social reporting. *Journal of Business Ethics, 27*(1–2), 55–68.

Idemudia, U. (2017). Environmental business-NGO partnerships in Nigeria: Issues and prospects. *Business Strategy and the Environment, 26*(2), 265–276.

Idemudia, U., & Ite, U. E. (2006). Corporate community relations in Nigeria's oil industry: Challenges and imperatives. *Corporate Social Responsibility and Environmental Management, 13*, 194–206.

Idemudia, U., & Osayande, N. (2018). Assessing the effect of corporate social responsibility on community development in the Niger Delta: A corporate perspective. *Community Development Journal, 53*(1), 155–172.

Idowu, S. O., & Papasolomou, I. (2007). Are the corporate social responsibility matters based on good intentions or false pretences? An empirical study of the motivations behind the issuing of CSR reports by UK companies. *Corporate Governance, 7*(2), 136–147.

Jamali, D., & Mirshak, R. (2007). Corporate social responsibility (CSR): Theory and practice in a developing country context. *Journal of Business Ethics, 72*(3), 243–262.

Jensen, J. C., & Berg, N. B. (2012). Determinants of traditional sustainability reporting versus integrated reporting: An institutionalist approach. *Business Strategy and the Environment, 21*(5), 299–316.

Khan, M., Lockhart, J., & Bathurst, R. (2018). Institutional impacts on corporate social responsibility disclosures: A comparative analysis of New Zealand and Pakistan. *International Journal of Corporate Social Responsibility, 3*(1), 1–13.

Khanna, T., & Palepu, K. (1997). Why focused strategies may be wrong for emerging markets. *Harvard Business Review, 75*(4), 41–51.

Kolk, A., & Lenfant, F. (2010). MNC reporting on CSR and conflict in Central Africa. *Journal of Business Ethics, 93*(2), 241–255.

Kolk, A., & Lenfant, F. (2015). Partnerships for peace and development in fragile states: Identifying missing links. *Academy of Management Perspectives, 29*(4), 422–437.

Kostova, T. (1999). Transnational transfer of strategic organizational practices: A contextual perspective. *The Academy of Management Review, 24*(2), 308–324.

Kostova, T., & Roth, K. (2002). Adoption of an organisational practice by subsidiaries of multinational corporations: Institutional and relational effects. *Academy of Management Journal, 45*(1), 215–233.

Kotter, J. (1990). *Leading change.* New York: Free Press.

KPMG. (2017). *KPMG international survey of corporate responsibility reporting 2017.* Amsterdam: KPMG Global Sustainability Services.

Maignan, I., & Ralston, D. A. (2002). Corporate social responsibility in Europe and the US: Insights from businesses' self-presentations. *Journal of International Business Studies, 33*(3), 497–514.

Mair, J., & Marti, I. (2009). Entrepreneurship in and around institutional voids: A case study from Bangladesh. *Journal of Business Venturing, 24*(5), 419–435.

Matten, D., & Moon, J. (2008). "Implicit" and "explicit" CSR: A conceptual framework for a comparative understanding of corporate social responsibility. *Academy of Management Review, 33*(2), 404–424.

McWilliams, A., & Siegel, D. S. (2001). Corporate social responsibility: A theory of firm perspective. *Academy of Management Review, 26*(1), 117–127.

Mitchell, R. K., Agle, B. R., & Wood, D. J. (1997). Towards a theory of stakeholder identification and salience. *Academy of Management Review, 22*(4), 853–886.

Morsing, M., & Schultz, M. (2006). Corporate social responsibility communication: Stakeholder information, response and involvement strategies. *Business Ethics: A European Review, 15*(4), 323–338.

North, D. C. (1990). *Institutions, institutional change and economic performance.* Cambridge: Cambridge University Press.

Nwagbara, U., & Belal, A. (2020). Persuasive language of responsible organisation? A critical discourse analysis of corporate social responsibility (CSR) reports of Nigerian oil companies. *Accounting, Auditing & Accountability Journal, 32*(8), 2395–2420.

Nwagbara, U., & Ugwoji, A. C. (2015). CSR reporting and accountability: The case of Nigeria. *Economic Insights – Trends and Challenges, IV, LXVII*(1), 77–84.

O'Dwyer, B. (2003). Conceptions of corporate social responsibility: The nature of managerial capture. *Accounting, Auditing and Accountability Journal, 16*(4), 523–557.

Ogiri, I. H., Samy, M., & Bampton, B. (2012). Motivations of legitimacy theory for CSR reporting in the Niger Delta region of Nigeria: A theoretical framework. *African Journal of Economic and Sustainable Development, 1*(3), 265–283.

Okike, E. N. M. (2007). Corporate governance in Nigeria: The status quo. *Corporate Governance, 15*(2), 173–193.

Owen, D. L., Swift, T. A., Humphrey, C., & Bowerman, M. (2000). The new social audits: Accountability, managerial capture or the agenda of social champions? *European Accounting Review, 9*(1), 81–90.

Owen, D. L., Swift, T., & Hunt, K. (2001). Questioning the role of stakeholder engagement in social and ethical accounting, auditing and reporting. *Accounting Forum, 25*(3), 264–282.

Parson, T. (1960). *Structure and process in modern society*. New York: Free Press.

Porter, M. E., & Kramer, M. R. (2006). Strategy and society, the link between competitive advantage and corporate social responsibility. *Harvard Business Review, 84*(12), 78–92.

Rathert, N. (2016). Strategies of legitimation: MNEs and the adoption of CSR in response to host-country institutions. *International Business Studies, 47*(7), 858–879.

Roberts, R. W. (1992). Determinants of corporate social responsibility disclosure: An application of stakeholder theory. *Accounting, Organisation & Society, 17*(6), 595–612.

Rotberg, R. I. (2012). *Transformative political leadership: Making a difference in the developing world*. Chicago: The University of Chicago Press.

Scherer, A. G., & Palazzo, G. (2011). The new political role of business in a globalised world: A review of a new perspective on CSR and its implications for the firm, governance, and democracy. *Journal of Management Studies, 48*(4), 899–931.

Scott, W. R. (1995). *Institutions and organisations*. Thousand Oaks, CA: Sage.

Shabana, K. M., Buchholtz, A. K., & Carroll, A. B. (2016). The institutionalisation of corporate social responsibility reporting. *Business & Society, 56*(8), 1107–1135.

Shell. (2017). Our activities in Nigeria, Sustainability Report. Http.www.reports.shell.com › managing-operations › our-activitie... (Accessed 17/08/2020).

Silberhorn, D., & Warren, R. C. (2007). Defining corporate social responsibility: A view from big companies in Germany and the UK. *European Journal Business Review, 19*(5), 352–372.

Siltaoja, M. E., & Onkila, T. J. (2013). Business in society or business and society: The construction of business–society relations in responsibility reports from a critical discursive perspective. *Business Ethics: A European Review, 22*(4), 357–373.

Solomon, A., & Lewis, L. (2002). Incentives and disincentives for corporate environmental disclosure. *Business Strategy and the Environment, 11*(I3), 154–169.

Suchman, M. C. (1995). Managing legitimacy: Strategic and institutional approaches. *Academy of Management Review, 20*(3), 571–610.

Thomson, I., & Bebbington, J. (2005). Social and environmental reporting in the UK: A pedagogic evaluation. *Critical Perspectives on Accounting, 16*(5), 507–533.

Tilt, C. A. (2016). Corporate social responsibility research: The importance of context. *International Journal of Corporate Social Responsibility, 1*(1), 1–10.

Ullman, A. A. (1985). Data in search of a theory: A critical examination of the relationships among social performance, social disclosure, and economic performance of U.S. firms. *Academy of Management Review, 10*(3), 540–577.

UNEP. (2011). *Environmental assessment of Ogoniland*. Nairobi: United Nations Environment Programme (UNEP).

Utting, P. (2005). Corporate responsibility and the movement of business. *Development in Practice, 15*(3 & 4), 375–388.

Vertigans, S., Idowu, S. O., & René, S. (Eds.). (2016). *Corporate social responsibility in Sub-Saharan Africa: Sustainable development in its embryonic form*. Heidelberg: Springer Publishing.

Visser, W., & Tolhurst, N. (Eds.). (2010). *A country-by-country analysis of corporate sustainability and responsibility*. London: Greenleaf Publishing.
Vurro, C., & Perrini, F. (2012). Making the most of corporate social responsibility reporting: Disclosure structure and its impact on performance. *Corporate Governance: International Journal of Business in Society, 11*(4), 459–474.
Wallace, R. S. O. (1988). Corporate financial reporting in Nigeria. *Accounting and Business Research, 18*(72), 352–362.
Yin, J., & Zhang, Y. (2012). Institutional dynamics and corporate social responsibility (CSR) in an emerging country context: Evidence from China. *Journal of Business Ethics, 111*, 301–316.
Young, S., & Marais, M. (2012). A multi-level perspective of CSR reporting: The implications of national institutions and industry risk characteristics. *Corporate Governance: An International Review, 20*(5), 432–450.

Uzoechi Nwagbara holds both BA and MA in English as well as MSc and PhD in human resource management and management, respectively, from University of Wales, UK. He has been teaching in higher education for upwards of 19 years. He is a published academic, management consultant and researcher having published 6 books and over 100 publications in peer-reviewed international journals including ABS rated journals (1, 2 and 3 star journals). He is also on the editorial boards of some international, peer-reviewed journals including *Journal of Sustainable Development in Africa* and *Economic Insights: Trends and Challenges*. He is ad hoc reviewer for some ABS rated journals including *Accounting, Auditing and Accountability Journal (AAAJ)*, *International Journal of Human Resource Management* and *Employee Relations,* among others. He is a visiting professor at Coal City University, Nigeria and Director of Studies (DoS) at Cardiff Metropolitan University and University of the West of Scotland as well as teaches at University of Sunderland. Her research interests are eclectic involving corporate social responsibility, corporate social responsibility reporting, sustainable development, human resource management, leadership, management, corporate governance and postcolonial studies.

Anthony Kalagbor holds BSc and MBA, respectively, from Plymouth University and University of Sunderland in marketing and international human resource management, respectively. He is currently lecturing at Cumbria University London Campus. He also lectured at Greenwich School of Management (GSM) London and was a tutorial lecturer at Rivers State University of Science and Technology, Nigeria as well as worked at Mercantile Bank of Nigeria plc. for upwards of six years before moving to Europe for further studies. While in Europe, he worked as sales manager in Drettmann Yachts, a reputable company in Europe (Germany). He also has been awarded Chartered Management Institute (CMI) Level 7 Diploma in Strategic Management and Leadership. He is a prospective doctoral candidate and interested in CSR research and consults in the areas of business responsibility and sustainability.

Corporate Social Indices: Refining the Global Reporting Initiative

Claus Strue Frederiksen, David Budtz Pedersen, Morten Ebbe Juul Nielsen, and Samuel O. Idowu

Abstract The objective of this paper is twofold. First, we demonstrate that the Global Reporting Initiative (GRI) Standards do not enable stakeholders to compare the relevant social performance of different organizations. We argue that the GRI Standards are based on a misguided claim about informed decision-making, which does not fully respect the conceptual difference between transparency and comparability. Secondly, we demonstrate that it is possible to improve the GRI Standards by incorporating a set of social indices which can be compared cross-organizationally. The inspiration for this framework is found in the UN's Human Development Index and Sen's capability approach.

1 Introduction

The Global Reporting Initiative (GRI) reporting recommendations was created in order to assist organizations (regardless of size, sector, and location) in preparing sustainability reports that matter, i.e. reports that actually disclose an organization's

---

C. S. Frederiksen (✉)
Gubra Aps, Copenhagen, Denmark
e-mail: csf@gubra.dk

D. B. Pedersen
Department of Communication and Psychology, Aalborg University, Copenhagen, Denmark
e-mail: davidp@aau.dk

M. E. J. Nielsen
Philosophy and Institute for Food and Resource Economics, Copenhagen University, Copenhagen, Denmark
e-mail: mejn@hum.ku.dk

S. O. Idowu
Guildhall School of Business and Law, London Metropolitan University, London, UK
e-mail: s.idowu@londonmet.ac.uk

sustainability status and enable stakeholders to make informed decisions.[1] In line with the famous Brundtland Report GRI defines sustainability as "development that meets the needs of the present without compromising the ability of future generations to meet their own needs" (GRI Standards Glossary 2016, p. 17). In this regard GRI notes that sustainable development includes the economic, social and environmental dimensions. The first full version of the GRI guidelines was released in 2000, and since then the GRI framework have become the most widely used set of international guidelines for sustainability and CSR reporting (Lozano and Huisingh 2011; Marimon et al. 2012; Moneva et al. 2006; Roca and Searcy 2012; Skouloudis et al. 2009).

The objective of this paper is twofold. First, we demonstrate that the GRI Standards do not sufficiently or adequately enable stakeholders to make informed decisions. We argue that the main problem with the GRI Standards is that they do not enable stakeholders to compare the relevant social performance of different organizations.[2] More specifically, two aspects of the GRI Standards are scrutinized namely (a) the informed decision claim and (b) the stakeholder approach. In this context we argue that the GRI Standards are based on a misguided claim about informed decision-making, which does not fully respect the conceptual difference between transparency and comparability. In relation to the stakeholder approach we argue that the lack of cross-organizational comparability cannot be justified by reference to a stakeholder-based argument.

Second, we suggest a strategy for improving the GRI Standards in order to approximate sufficient comparability. Importantly, the indices we suggest are meant to supplement, not substitute the GRI Standards. The indices should be seen as an additional tool in the toolbox focusing on improving the comparability between organizations social performance. The conceptual source of inspiration for improving *cross-organizational* comparability is the UN Human Development Index (HDI), which measures human development at the level of nation-states. We argue that the HDI can be used as an inspiration for making *comparisons* of corporate social performance. The most important lesson to be learned from the HDI is the use of comparable social data. Like the creators of the UN Human Development Index, we use Amartya Sen's capability approach as the conceptual foundation in order to select the relevant data. We conclude the paper by presenting specific suggestions of how to create social indices that better enable comparability and thus improve stakeholders' ability to make informed decisions. As far as we are aware, no scholarly contribution has previously specified plausible ways of constructing comparable corporate social indices. Even Elkington (1997, 2004), the father of the Triple Bottom Line (TBL) presents no concrete means by which

---

[1] Over the years GRI has developed and presented a number of reporting guidelines, GRI Standards from 2016 is the latest version.

[2] In this paper, we focus exclusively on social reporting, which means that, strictly speaking, our arguments can only be used to conclude that the GRI Standards do not secure transparency on social matters. That said, we believe that most of our arguments also present a great deal of insight into environmental reporting.

corporations can compare social data. In this paper, we attempt to sketch out how such a comparison of corporate social data may look like by constructing corporate social indices.

Before we begin, however, we would like to point out that we realize that social and environmental reporting in its various shapes and forms for many years have been the subject of substantial critique. Especially Gray et al. (e.g. 1995), Gray (2002, 2010) has discussed and criticized the practice of social accounting and sustainability reporting. In a recent paper Milne and Gray (2013) argue that the current approach to sustainability reporting, dominated by the GRI framework, is probably doing more harm than good, because it tend to reinforce business-as-usual, which is highly unsustainable. The main problem, according to Milne and Gray, is that GRI is confusing sustainability and sustainable development with any reference to triple bottom line rhetoric, i.e. economic, social and environmental accounting. Gray and Milne also argue that the GRI guidelines are both partial and incoherent. They are partial because they do not include all the relevant social and environmental actions and inactions of the companies. They are incoherent because they are not based on an over-arching theory, which guides the selection of indicators and explains the relationships they have to each other.

We agree with Milne and Gray that the current use of the term 'sustainability' by GRI is problematic. However, since we are focusing exclusively on the social aspect of corporate reporting, we have chosen to not to enter the sustainability-debate. Also, when it comes to the social aspect of the GRI Standards, we believe that it is possible to improve the guidelines in a manner, which makes them substantially different from reporting-as-usual. Including our suggestions into the GRI Standards will enable stakeholders (at least to some degree) to compare companies' social performance. The selected social indicators do not provide a full picture; however, we believe that the indicators we have chosen are the most salient ones. And unlike the GRI, our social indicators are (as will be demonstrated later) selected by an over-arching theory, namely Sen's capability approach.

2 The GRI Standards

The GRI Standards are very comprehensive and consists of 37 separate documents. 6 of these include reporting recommendations on economic reporting, e.g. one document on how to report in regards to anti-corruption. 8 documents concern environmental reporting, including documents on how to report on issues like emission and water withdrawal. 19 documents concerns social reporting, including documents on how to report in regards to child labour and safety issues.[3] Each of these 33 documents hence contains a number of concrete reporting

---

[3]The last 4 documents concerns foundations for GRI, general reporting guidelines, management and glossary.

recommendations on a specific economic, environmental or social issue. In total, the GRI Standards contains of more than 500 concrete reporting recommendations. The design of these recommendations varies. Some guidelines require specific numbers (e.g. on tax payments, water usage, etc.), others ask for specific information (e.g. about suppliers, legal frameworks, etc.), and others still ask for statements (e.g. on values, actions, etc.). In the context of the present paper, two aspects of the GRI Standards are especially worthy of note, namely (a) the informed decision claim and (b) the stakeholder approach. We will present and discuss both of these elements.

2.1 The Informed Decision Claim

The main purpose of sustainability reporting, according to GRI, is to enable organizations to monitor (and if necessary improve) their sustainability performance and enable stakeholders to make informed decisions about the sustainability profile of the company (GRI 101, 2016, p. 3). In this context there is similarity with financial reporting which is defined by the American Accounting Association (1957, p. 536) as concerning "the process of identifying, measuring and communicating economic information to permit informed judgments and decision by users of the information." As with financial reporting, the GRI (101, 2016, p. 3) Standards "allows internal and external stakeholders to form opinions and to make informed decisions about an organization's contribution to the goal of sustainable development."[4] GRI (101, 2016, p.3) also claims that "The Standards are designed to enhance the global comparability and quality of information on these impacts, thereby enabling greater transparency and accountability of organizations." Importantly, according to GRI, transparency implies *comparability*:

> Comparability is necessary for evaluating performance. It is important that stakeholders are able to compare information on the organization's current economic, environmental, and social performance against the organization's past performance, its objectives, and, to the degree possible, against the performance of other organizations. (GRI 101, 2016, p. 14)

The GRI framework is based on the belief that stakeholders can evaluate the social and environmental performance of organizations within the GRI reporting initiative, at least in relation to an (1) an organization's financial, social, and environmental objectives; (2) past performance; and (3) to an unspecified degree relative to other organizations. It remains unclear, however, what GRI requires in order to enable stakeholders to make informed decisions and on what grounds we should believe that its guidelines fulfill these conditions. In the next section, we will analyze and discuss the informed decision claim.

---

[4]We will not discuss to which degree the GRI guidelines are based on some notion of decision usefulness and the resulting problems that stems from this. For a thorough discussion of decision usefulness see Williams and Ravenscroft (2015).

2.2 Discussing the Informed Decision Claim

In this section, we argue that (1) informed decision-making relies on the ability to distinguish clearly between transparency and comparability (instead of seeking to tie them together, as GRI does); (2) that GRI Standards do not, in a sufficient manner, enable stakeholders to compare the social performance of organizations; and (3) that *informed* decision-making in relation to organizations' social performance implies the ability to compare the social performance of different organizations. (Informed decisions implies information about the relevant alternatives; if we cannot compare, our decision is flawed from the perspective of "informed decision". The result is that the GRI Standards do not adequately enable stakeholders to make informed decisions.

When it comes to the ability to make informed decisions, GRI operates with two concepts, namely transparency and comparability. As mentioned above, GRI claims that transparency involves at least some level of comparability, including the ability to compare organizations' present and past performances. However, when it comes to comparing the performance of different organizations, it is unclear to what degree GRI finds comparability necessary in order for stakeholders to make informed decisions. However, instead of seeking to tie transparency and comparability together, we argue that GRI should recognize the conceptual difference between the two. Transparency, conceptually speaking, is a relatively straightforward notion: transparency is secured through full and unbiased disclosure of relevant facts about an issue. Problems with transparency are not conceptual; instead, they are practical (how to secure transparency) and to some extent definitional or extensional (what, for instance, are 'the relevant facts'). In contrast, much ink has been spilled in various philosophical discussions over the notion of comparability and its various companion notions such as incommensurability or incomparability (see e.g. the anthologies edited by Chang (1997) as well as by Crisp and Hooker (2000); for a less abstract discussion, see also the interesting paper on various aspects of making different gas emissions commensurable by MacKenzie (2009)). However, our purpose in this paper is not to engage in a highly abstract debate over comparability. Instead, we wish to suggest a broader but more practical understanding of comparability: Adequate *piecemeal* comparability between two organizations on a particular social issue is achieved if we can make at least a minimally rationally defensible ordinal ranking of their performance on a particular issue, for example, how well they respect workers' rights. Adequate, *complete* comparability is achieved if we can make an at least minimally rationally defensible ordinal ranking of their performance on *all* social issues. However, complete comparability is hopelessly difficult to achieve, and piecemeal comparability (although relevant in special cases) naturally says little about the overall social profile of a given organization. Accordingly, we shall suggest a more workable conception of comparability of social issues, which steers between these extremes. Our way of doing this is to highlight social indices as being of particular salience.

Accordingly, we will not discuss whether or not GRI secures transparency. Instead we argue that GRI does not (at least not sufficiently) ensure comparability. GRI does not operate with any social (or environmental) indices but instead with various data on issues such as workers' safety, gender equality, etc. There are two main reasons why this approach (in its current form) fails to ensure comparability. First, GRI offers no baselines when it comes to how much is enough (or too much) with respect to organizations' financial, social, and environmental impact. This makes it very difficult (if not impossible) for the reader to compare data from different organizations, especially with regards to cross-sectorial comparisons. For example, we cannot conclude that a company that produces computer hardware (let us call it 'CH') is more socially responsible than an oil company (let us call it 'OC') just because CH had fewer work related accidents last year than did OC. The problem is that it seems unfair to say that OC is not taking the issue of workers' health and safety as seriously as CH just because OC's employees are more likely to get hurt at work than CH's employees since it is hard to make oil rigs as safe as a hardware assembly line. Trying to compare data regarding work-related accidents thus seems a lot like comparing apples and oranges unless we have some clear baseline stating when a given amount of work-related accidents is too high. In other words: One may have full transparency in a certain sense; e.g., one might have full access to all relevant data on two different organizations' performance vis-à-vis worker's rights, philanthropic efforts and so on and so forth; but if we have no robust way to measure their relative effort, we do not have comparability. This is why we think it is fruitful to keep transparency and comparability analytical distinct (but naturally, comparability is impossible to achieve without some level of transparency). The problem is, it seems impossible to establish such a universal baseline since the different characteristics of organizations are in some respects highly relevant for evaluating their relative efforts.

In relation to some social data, it does not make sense to disregard an organization's specific line of business and the social (and environmental) challenges produced by that business. Note that we are not claiming that incomparable cross-sectional data should be excluded from social reports. The inclusion in social reports of data concerning, for instance, the number of work-related accidents should keep organizations on their toes and thereby result in fewer accidents. What we are claiming is that social reports (also) need to include cross-comparable data to enable stakeholders to make informed decision. In addition, GRI offers no guidance on how to compare the different types of data, e.g. how are we to compare data on workers' safety with data describing the level of workforce gender equality? The ability to compare different social data is necessary in order to be able a comparison of the overall social performance of different organizations. What is needed is a common social currency (as we shall see later, we seek to construct social indices by using Amartya Sen's capability approach as a social currency) that enables stakeholders to compare and rank different social (and environmental) performance measures. Without such a social currency or guidelines specifying how to compare different data, stakeholders are unable to compare different organizations' social performance.

Yet, comparability is necessary in order for stakeholders to make informed decisions with regards to organizations' social performance. Note that when it comes to financial reporting, comparability seems highly important because investors wish (at least to some degree) to be able to compare the financial performance and status of different organizations in order to decide in which organizations they should invest their money. Imagine a market in which financial reporting does not in any way enable investors and other stakeholders to compare different organizations' financial performance and status. Such a market would be very difficult to operate in because the decision of whether to invest in Company A, B, C, or D would be a game of pure chance.

Two things are worth noticing. First, we are not claiming that financial reporting enables investors (and other stakeholders) to make *completely* informed decisions. We acknowledge that choosing in which organizations to invest on the basis of their financial reports is not a risk-free business. That said, we still believe that organizations' financial reports to some degree enable investors and stakeholders to compare organizations' financial performance and status and to make informed decisions on that basis. Second, we are not claiming that GRI Standards are leaving stakeholders totally in the dark when it comes to comparing the social performance of different organizations since the guidelines (if followed correctly) might enable stakeholders to distinguish the really good organizations from the bandits. For instance, if Company A is paying bribes, uses forced labour, and is not paying any taxes whereas Company B is competing in a fair and transparent manner, paying reasonable wages, and is paying taxes in the countries in which it operates, it does not seem unreasonable to conclude that Company B's social performance is better than Company A's. However, since most cases are less clear cut, we need more than the GRI Standards to ensure a higher level of comparability with regards to organizations' social performance. This echoes the conclusion in a recent report published by the UN Global Compact. The report is based on interviews with prominent business leaders and CSR scholars, many of whom express a need for better metrics to measure organizations social and environmental impact. In the report Marilyn Carlson Nelson, Co-CEO Carlson Holdings, suggests a need for a corporate social and environmental ranking system and adds that such a system would be a strong motivation for action, since "nobody wants to be at the bottom of those lists" (UN Global Compact 2015, p. 26).

When reviewing the literature it becomes clear that only a few scholars believe that the GRI framework (even if followed) largely enable stakeholders to evaluate and compare organizations' sustainability performance. The problem is, however, that none of these scholars present any substantive argument in defense of this conclusion. For instance, Willis (2003) and Marimon et al. (2012) simply seem to *assume* that GRI actually meets the objective of securing transparency and comparability, yet such an unfounded assumption is unacceptable. It goes (almost) without saying that the existence of an objective does not necessarily mean that this objective will be met. In this regard, it does not help much when Willis (2003, p. 237) adds that, instead of being designed in "some distant ivory tower insulated from real world needs and practices or driven by a narrow set of interests and disciplines," the quality of the guidelines is ensured by the stakeholder process. Instead of presenting any kind of argument, Willis just assumes that GRI actually delivers as promised. Prado-Lorenzo et al. (2009) and Sutantoputra (2009) emphasize that the GRI framework is widely accepted (by organizations), thereby indirectly arguing that the framework ensures transparency and comparability (since so many people cannot be wrong). However, we should not accept popularity as a proof: Popular products do not necessarily deliver what they claim. Prado-Lorenzo et al. (2009, p. 99) try to strengthen the argument by adding that the GRI reporting recommendations guidelines include "an obligation for certain information to be expressed numerically and monetarily so as to facilitate its comparison." But as any accountant would know the inclusion of some numerically or monetary data is insufficient for securing transparency. Note, that many supporters of the GRI guidelines seek to support their position by referring to the work of other scholars, who conclude that the GRI guidelines secure transparency and comparability. The problem is that the works referred to in this manner cannot be used as a basis for concluding that the GRI guidelines meet their own objectives. For instance, Sutantoputra (2009) refers to Willis and Clarkson et al., Marimon et al. refers to Prado-Lorenzo and Clarkson et al., and Prado-Lorenzo et al. refer to Clarkson et al. However, as demonstrated above, neither Willis nor Prado-Lorenzo et al. present any substantial arguments in defense of GRI, meaning that it all (at least when it comes down to the academic contributions discussed above) depends on Clarkson et al. (2008), which does not consider social issues but only environmental issues. This implies that even if the work did demonstrate that GRI secures transparency and comparability, it would only be with regards to environmental issues.

2.3 The Stakeholder Approach

The GRI guidelines have been developed through an international stakeholder process, and according to GRI (Due Process for the GRI Reporting Framework 2007, p. 2), the "GRI Working Groups [main stakeholders] are the primary means for developing and revising the text of GRI Framework documents." With respect to the development and revision of the GRI guidelines, it seems reasonable to interpret the stakeholder element in the traditional way, as presented and defended by Freeman et al. (2011). GRI seems to have adopted (or at least aligned with) two of the key elements of Freeman's traditional stakeholder theory. First, stakeholders are those who seem to have something at stake vis-à-vis the organization (traditionally those who can affect or be affected by the organization). In the GRI Standards (101, 2016, p. 8)stakeholders are considered to be "entities or individuals that can reasonably be expected to be significantly affected by the reporting organization's activities, products, or services; or whose actions can reasonably be expected to affect the ability of the organization to implement its strategies or achieve its objectives." Second, the policy of the organization should reflect the interests of its stakeholders.

One way of defending the stakeholder approach and the associated guidelines is by claiming that although they might not be perfect, they are the least-bad option. This conclusion can be reached via a three-step argument. The first step is to claim that full transparency with regard to social impact (and environmental impact) is impossible: Despite all of the TBL rhetoric, we cannot construct a social (or environmental) bottom line similar to the financial bottom line. Norman and MacDonald (2004, p. 252) state that "In the language of moral philosophers, the various values involved in evaluations of corporate behaviour are 'incommensurable'; and reasonable and informed people, even reasonable and informed moral philosophers, will weigh them and trade them off in different ways." In addition, Norman and MacDonald (2004, p. 253) argue that not only are the values incommensurable; some of them are also impossible to aggregate since they concern rights and obligations, which agents should fulfil but not seek to increase or maximize. For instance, keeping a promise is a good thing, but *ceteris paribus*, you do not become more ethical by making and keeping two promises than by making and keeping one. Focus on maximization disregards important aspects of ethics (Norman and MacDonald 2004, p. 253). For the time being, let us assume that it is correct that social and environmental evaluations of organizations involve incommensurable (and non-aggregative) values about which reasonable agents disagree. This brings us to the second step in the argument, namely that the impossibility of creating fully comparable social and environmental bottom lines leaves us with two options. First, we can abandon the idea of social and environmental reporting altogether and go back to traditional financial reporting. Second, we can accept that social and environmental reporting will never secure transparency and comparability at the same level of consistency and precision as financial reporting can secure financial transparency and comparability. We can nevertheless conclude that some kind of social and environmental reporting is better than nothing (because it inspire organizations to focus on more than just the financial bottom line). The question, then, is: What is the best procedure for designing social and environmental reporting guidelines? This leads us to the third and final step in the argument. If social and environmental reporting is a subject upon which reasonable people disagree, then it seems reasonable to allow everybody (or at least the relevant stakeholders) to have a say in the process. On this basis, we can conclude that the best way of designing social and environmental reporting guidelines is by taking a stakeholder approach.

At first glance, this argument might appear convincing, but we believe it should be rejected. We will focus here only on the first step regarding incommensurable values and reasonable disagreement and on the arguments presented by Norman and MacDonald in their influential critique of TBL rhetoric and the impossibility of constructing a social (and environmental) bottom line.[5] Before we begin to elaborate

---

[5]Note that even though Norman and MacDonald critique is directed at TBL rhetoric it also poses a challenge for social indices, since social indices like social bottom lines are based on selectively aggregated social data.

upon their argument, it is important to note that Norman and MacDonald (2004, p. 251) claim that:

> There are fundamental philosophical grounds for thinking that it is impossible to develop a sound methodology for arriving at a meaningful social bottom line for a firm. There is a strong and a weak version of the argument: the strong version says that it is in principle impossible to find a common scale to weigh all of the social "positives" and "negatives" caused by the firm; and the weak version says, from a practical point of view, that we will never be able to get broad agreement (analogous, say, to the level of agreement about accounting standards) for any such proposed common scale.

Norman and MacDonald (2004, p. 251) later state that they only seek to demonstrate the weak, yet we note that they continue the discussion as if seeking to defend the strong claim. Four points are worthy of note. First, the issue of whether we will ever reach broad agreement on how to create social and environmental bottom lines is an empirical question, and Norman and MacDonald present no data in support of the weak claim. This indicates that they believe the plausibility of the weak claim to rely upon the plausibility of the strong claim. Second, broad agreement has actually been reached on similar areas: The Human Development Index and its related indices are based on seemingly incommensurable social values. Third, their key argument (Norman and MacDonald 2004, pp. 251–252) concerns legitimate disagreement about value judgments:

> There are many relevant and objective facts that can be reported and audited, any attempt to "weigh" them, or tot them up, will necessarily involve subjective value judgments, about which reasonable people can and will legitimately disagree (Norman and MacDonald 2004, p. 252).

This seems to be an argument in favour of the strong claim since disagreement would not be reasonable if it were possible to weigh the values against each other. To take a simple example, reasonable people can disagree about personal preferences and taste, e.g. whether Mozart is better than Beethoven, but they cannot reasonably disagree about whether a company that uses slave labor is, all else being equal, doing more harm than a company with reasonable working conditions.

Finally, it seems unreasonable to reject a standard just because it cannot achieve broad consensus: actual social reporting that would judge some corporations as socially irresponsible would probably be unpopular in some business circles and would therefore not achieve broad acceptance. Based on these reasons, our focus will be on the strong rather than the weak claim.

Even though it might be the case that some of the social data recommended by the GRI Standards (and included in current corporate reports) is incommensurable or based on non-aggregative ethical values, this does not make it impossible to create a social reporting system with some level of cross-organizational comparability. We are not claiming that our solution (which will be presented in a moment) results in social reporting on par with financial reporting, only that it results in social reporting somewhat similar to financial reporting, especially with regard to the system's ability to enable stakeholders (at least to some degree) to make informed decisions. Several points are worth observing in this respect. First, as Pava (2007) correctly notice, and as MacDonald and Norman (2007) also accept, the financial statement does not consist of a single number that represents a company's financial status. On the contrary, the financial statement includes several relevant financial figures. Constructing a social report somewhat similar to the financial report is not about aggregating all social values in order to present a single social bottom line. Second, reasonable disagreement about the weight and trade of data is not exclusively connected with social and environmental data but also presents a challenge for financial reporting.[6] However, this has not prevented the construction of financial bottom lines that generally enables stakeholders to make informed decisions. That said, the level of disagreement and incommensurability seems deeper when it comes to social and environmental data than in the field of finance, which might be due to the lack of social and environmental currencies. Third, accepting that some ethical values are not aggregative (e.g. because they involve rights and obligations) does not mean that they cannot be included in an adequate social report (which is similar to the financial report); it just means that they cannot be transformed into social currency. We suggest that, in relation to non-aggregative values (for instance, in cases of discrimination and the use of forced labour), organizations should report whether they support any *international standards* (e.g. the UN Global Compact) and whether they have had any violations of these standards. Fourth, in order to create social indices that are somewhat similar to the financial bottom lines, we need *social data* that is comparable across organizations. Again, this does not mean that the social data recommend by GRI and included in many current reports (e.g. number of work-related accidents) should not be included in social reports; it only means that such data cannot be included in the social indices, which are necessary to ensure some level of comparability (and hence the ability to make informed decisions). In the next section, we will present a way for creating social indices. In this respect, our point of departure will be the HDI, which is among the most influential indices when it comes to measuring national development.

3 Creating Social Indices

In order to effectively overcome the limitations of the GRI framework, we wish to sketch a new route for comparing social data among companies. This route is inspired by the Human Development Index (HDI), which is a composite statistic of life expectancy, education, and income indicators used to rank countries against each other with regard to key aspects of human development. The framework was created by the economists Amartya Sen and Mahbub ul Haq in 1990 and has since been adopted as a standard by the United Nations Development Program. The HDI measures human development on a national level and was originally introduced as an alternative to Gross Domestic Product (GDP) and other purely economic

---

[6]Again, for a thorough discussion of decision usefulness see Williams and Ravenscroft (2015).

measures of societal development. It is based on three dimensions: health (life expectancy per capita), education (expected years of schooling per capita), and income per capita. Mahbub ul Haq (2010, p. 87) created the index in order to establish a simple but substantial framework for measuring "the actual level of human development" based on Amartya Sen's capability approach.

The HDI has led to the creation of additional indices, including the Inequality-adjusted Human Development Index (IHDI), which adjusts for distributive inequalities in the three aforementioned dimensions, and the Gender Inequality Index (GII), which adjusts for inequalities between sexes. Note that the HDI and its related indices are neither designed nor intended to capture the complete extent of a country's development. The objective is more modest, namely to emphasize (by including non-economic data) that people and their capabilities, and not just economic growth, should matter when evaluating a country's level of development (UNDP 2014). In line with the architects of HDI, we believe that it is better to construct a simple social index based on a few key categories. Shifting perspective from the national level to the company level, the HDI-inspired indices for corporate social reporting might not provide the complete picture but certainly will provide more comparable and relevant information than previous reporting methods.

At first glance it might seem like a good idea to create a corporate human development index that is identical to the HDI or one of its related indices. The problem is that the HDI and its related indices cannot be used to directly measure corporate social performance. The HDI indices evaluate states, which are characterized by controlling social institutions within demarcated territories and by having (sometimes) legitimate power and responsibility over their citizens. Corporations lack both of these characteristics, which imply that national data regarding life expectancy, years of schooling, and indices of income are irrelevant in a corporate context. Unlike government policies, individual corporations do generally not have any (substantial) influence on life expectancy, years of schooling, and income per capita in a given country. Or so the standard story goes. While we acknowledge these limitations, it is still possible, as will become clear in a moment, to focus on the same tiers of human development in a corporate context as the ones chosen by HDI (i.e. health, education, and living standards).

There are several reasons as to why we have chosen HDI as point of departure for creating corporate social indices. First, HDI demonstrates that is possible to make comparable indices based on social data. Second, despite initial scepticism, HDI seems to have passed the test of time—various organizations and agencies has used it for 25 years, and since its introduction it has become the most widely used and accepted measure of national development (Fukuda-Parr 2003). Third, HDI has proven to be very flexible—it response to criticism by making adjustment e.g. by designing additional indices that captures specific elements which, according to the critics, are missing in the original index. Such flexibility might also be needed in regards to corporate social indices. Fourth, it is based on a solid philosophical foundation, namely Sen's capability approach (which will be discussed further below).

3.1 Social Data: Two Conditions

In order to translate the HDI framework from national evaluations to a corporate context, we suggest that the following two conditions must be met before the provision of social data (information) can be included in organizations' social indices:

1. The data concerns protecting or promoting peoples' freedom to do things they have reason to value (this condition is heavily inspired by Sen's capability approach, see below).
2. The data must be cross-organizational comparable, i.e. it can be compared with data from other organizations.

In the next section, we will explain why these two conditions are necessary in order to construct social indices and what kind of social data/information can fulfil these conditions.

3.2 First Condition

In order to fulfil the first condition and hence qualify as social data that can be included in organizations' social indices, the data must concern preserving or promoting people's freedom to do things they have reason to value. As mentioned above, this condition is heavily inspired by Sen's capability approach, which is also the normative foundation of HDI and its related indices.

The capability approach was pioneered by Sen in his theory of freedom and was presented in his famous Tanner Lecture in 1979 entitled 'Equality of What'.[7] Since Sen's original formulation, numerous scholars, most notably the American philosopher Martha Nussbaum, have supported and further developed Sen's conceptualization. In this paper, we focus exclusively on Sen's version of the capability approach. In Sen's view we should, when evaluating our lives, consider not only 'being' and 'doing' (e.g. being healthy and doing philosophy) but also our effective freedom to choose between different ways of living. In short, the capability approach is a measure for evaluating individual advantages and making interpersonal comparisons on overall advantages, focusing on the substantive (and not just formal) freedom that agents must enjoy in order to value the things they have reason to value. Several points are worth considering. First, the capability approach is not a theory of justice, e.g. it does not define fair distribution (Sen 2009). Second, the capability approach is first and foremost an alternative to the Rawlsian focus on primary goods, though as Robeyns (2003) notes, this does not mean that the capability approach denies that resources can have major influence on people's well-being. Third, the

---

[7] For a thorough and more elaborate discussion of the capability approach, see the Brighouse and Robeyns anthology *Measuring Justice, Primary Goods and Capabilities* from 2009.

capability approach is underspecified and does not imply which capabilities should be taken into account.

In relation to this last point, Casper (2007) argues that the vague definition increases the risk that "anything goes", especially when it comes to putting the approach into practice. To illustrate his point, Casper argues that the income category in HDI undermines the capability approach's original rationale since income is not a reliable proxy for wellbeing (Casper 2007). Four points are worthy of notice. First, many proponents of the capability approach, e.g. Robeyns (2006), acknowledge that the capability approach is radically underspecified but argue that this is not a problem as long as users of the approach explicitly justify their choice of capabilities and their measurements of them. In relation to HDI, the human development community followed, according to Fukuda-Parr (2003), two criteria in order to determine which capabilities should be considered most important. First, the capabilities must be universally valued by people regardless of culture, nationality, etc. Second, the capabilities must be basic, meaning that without them, many other capabilities are foreclosed. On this basis, the HDI's focus on health, education, and decent living standards seems reasonable since these areas are both basic and universally valued by reasonable people. Additionally, as Robeyns (2006, p. 355) stresses, data availability tends to play a crucial role when moving from ideal theory to practical application. This means that we cannot always rely on the best data (seen from an ideal perspective) but must rely on the best available data, and in this regard, income seems to represent the best available data as far as people's living standards are concerned. In other words, even though the data does not *directly* measure people's capabilities, it still functions as a reliable proxy for people's access to health, education, and resources.[8]

3.3 Second Condition

The second condition, that data must be cross-organizationally comparable, is necessary in order to ensure comparability, which is essential for all kinds of accounting, including financial accounting. One of the implications of this condition is the exclusion of data that might seem socially relevant but that cannot be transformed into comparable data. The product portfolio is perhaps the most telling example.

A company's product portfolio can have a huge influence on the company's social impact. GRI assigns a whole document under the title 'Customer Health and Safety' to product responsibility (GRI 416, 2016). The focus of GRI's reporting

---

[8]Using the capability approach is not a necessary but only a sufficient condition. One could also choose other plausible evaluation tools, e.g. the promotion of equality of opportunity. However, since we are using the HDI as inspiration for our construction of corporate social bottom lines, we have chosen to also use its theoretical foundations.

recommendations is mainly on the company's compliance with health and safety standards, marketing regulations, and customer satisfaction. These are, of course, relevant issues; however, none of them capture the tremendous social impact (positive or negative) that a product can have. When it comes to promoting capabilities, organizations' products can be highly important. To take a simple example, a pharmaceutical company producing and selling life-saving medicine (at affordable prices) seems, all else being equal, to have a much larger positive social impact than a company that produces and sells paper clips. Social reporting guidelines that do not take this into account seem unable capture the full picture of organizations' social performances. The problem, however, is that there seems to be no cross-organizationally comparable data when it comes to organizations' product portfolios. Even if we use health as a proxy for promoting capabilities, it is still very difficult to determine the social status of a product portfolio. First, there are many cases in which it is hard to determine the health-promoting status of a product. Is ice cream unhealthy? Yes, in large doses, but so are meat and cheese. Second, many organizations produce and sell both healthy (understood as nutritious) and unhealthy (understood as non-nutritious) products, and determining the social impact of such organizations proves very complicated. Third, and most importantly, many organizations do not produce or sell products that normally have a direct impact on consumers' health status, e.g. clothing manufacturers, construction organizations, bank and finance organizations, hotel and other service providers. This last point means that even though some organizations' product portfolios seem to have a substantial direct social impact (e.g. on the health status of the consumer), the vast majority of organizations' portfolios have little such direct influence. Excluding the product portfolio from the social indices is imperfect, but the lack of comparable data makes this exclusion necessary. Note that the suggested indices are not meant to exclude other kinds of organizational data from social (and environmental) reports. On the contrary, the indices are meant to supplement, not substitute, the GRI Standards. Even though a cigarette company might score high on the social indices, this score would not hide the fact, that the company's product is extremely unhealthy. And since we are not suggesting that the social indices are the only thing stakeholders should take into account, the inclusion of the indices in social reporting does not prevent the stakeholders from making informed decisions. On the contrary, the indices increase the possibility for informed decision making.

4 Three Corporate Social Indices

In the following, we will present three social indices inspired by and modelled upon the indices of the HDI framework (health, education, and income). Remember that data availability and reliability is crucial for all three dimensions when moving from ideal theory to practical application at the company level.

Health This social index contains the following three proxy indicators for health promotion and protection: (i) The percentage of revenue[9] donated to healthcare institutions and donations to non-profit organizations that focus mainly on improving people's health conditions. (ii) The percentage of revenue from the company's tax payments that is spent on healthcare by its national government. No company has this data in-house, of course, but we imagine a simple measure based on the country's own state budget. Imagine that Company A pays 25% of its revenue to Country B and that Country B spends 15% of its national budget on healthcare. In this case, it is reasonable to conclude that 3.75% of Company A's revenue is spent on healthcare. Again, this indicator is based on the correlation of data from the country's official budget and tax reports from the company. Finally, health promotion and protection can be measured by (iii) the percentage of revenue spent on healthcare initiatives directed at employees, including health insurance and safe working environment (including psycho-social health). Adding (i), (ii), and (iii) together results in the *promoting health index*.

Education This social index consists of three proxy indicators for promoting education: (i) The percentage of revenue donated to educational institutions and donations to non-profit organizations that focus mainly on improving peoples' educational level and proficiency. (ii) The percentage of revenue from the company's tax payments that is spent on education by its national government. Imagine that Company A pays 25% of its revenue to Country B and that Country B spends 15% of its national budget on education. In this case, it is reasonable to conclude that 3.75% of Company A's revenue is spent on education. (iii) The percentage of revenue used on educational initiatives (including initiatives aimed at improving work-related competencies) directed at employees. Depending on data availability, this bottom line can be expanded to include research and educational activities more broadly. Most advanced companies in the Western and non-Western world have substantial research and development departments and budgets, which spill over into education and building of critical mass in the country's or region's workforce. Adding (i), (ii), and (iii) together thus results in the *educational index*.

Living Standards This social index consists of three proxy indicators for promoting people's living standards: (i) The percentage of revenue donated to institutions or non-profit organizations that focus mainly on securing or improving people's living standard. (ii) The percentage of revenue from the company's tax payments[10] that is spent on securing or promoting people's living standard by its national government. Imagine that Company A pays 25% of its revenue to Country B and that Country B spends 15% of its national budget on securing or improving people's living standard. In this case, it is reasonable to conclude that 3.75% of Company A's revenue is spent on improving or securing people's living standard. (iii) The percentage of revenue

---

[9] We focus on percentage of revenue instead of percentage of surplus because focusing on percentage of surplus would punish effective organizations and reward ineffective ones.

[10] Or other kinds of similar transfer returns e.g. community dividend.

used on paying employee salaries (gross pay, before tax). Although such data is excluded from many existing sustainability and CSR reports, it seems perfectly legitimate to consider, for instance, salary levels as part of a corporation's social impact. Adding (i), (ii) and (iii) together results in the *living standards index*.

Notice the selected proxy indicators are just suggestions and should not be considered non-revisable. On the contrary, we realize that the selected indicators might need to be revised in order to more fully capture e.g. the health impact companies have on society. We just wanted to demonstrate that it is possible to refine the GRI Standards by creating social indices that improve stakeholders' ability to compare companies' social performance.

Also, the effect of this exercise is the establishment of a number of (simple) numerical indicators designating the level of Human Development by company (based on a combination of company and country data). Data availability is, of course, a crucial element for the success of the present proposal. The benefit of using the HDI framework is that country data is already centrally available and certified through the UN system and is part of the existing world map of the Human Development Index by country (the latest version is based on 2013 data; published July 24, 2014).[11] With regards to company-level data, data availability is less straightforward in the sense that data can have (or can be thought to have) implications for competitive advantages and, ultimately, market capture. We do not believe that this obstacle is generally impossible to overcome. To the contrary, companies that already today and within the existing CSR framework are working to optimize their social profiles and footprints have an explicit interest in publishing the aforementioned indices in a coherent and comparable fashion. Note that comparable data of the sort we are proposing here would allow prospective investors and customers to gain a more systematic and transparent picture of the company's social profile before investing in or purchasing products from the company. This is crucial in two instances: First, if the potential customers are nation-states or national authorities. Public buyers are likely to focus on the entire spectrum of externalities and benefits rather than on just getting the cheapest product. To get a low price from a company that places its revenue in tax havens might not, all things considered, be a good deal. Second, our proposal might have positive impacts for investors operating under specific ethical codes, including some pension funds. Such financial entities not only focus on expected returns on investment but also on wider social issues that might benefit or otherwise be in the interest of their investors.

In addition, many companies today compete against each other in a non-transparent and rigged manner: If Company A is submitting a bid for delivering e.g. a water supply system in a developing country and is met by competitors B and C, which are submitting what seems to be more financially attractive bids, then

---

[11] Only a few countries are not included (for various reasons), mainly because of the unavailability of certain crucial data. The following United Nations Member States were not included in the 2014 (inequality-adjusted) HDI report: North Korea, Marshall Islands, Monaco, Nauru, San Marino, Somalia, South Sudan, and Tuvalu (UN 2014).

Company A has an evident self-interest in including social data as part of the selection and decision-making process on behalf of the customer (perhaps a state or regional authority or another company). Availability and usage of social data will enhance the prospect of fair competition, giving investors and customers a more coherent and thus complete picture of the positive externalities a given investment or purchase may or may not have on human development. Consider again the delivery of a water supply system (or other large infrastructure in energy, healthcare, urban planning, etc.). The issue of whether tax will be paid in the country in which the enterprise is to be carried out has substantial effects on the overall evaluation of the investment. Likewise, whether or not a pharmaceutical company that is delivering medicine in a specific country or region is actively participating in and financially encouraging the education of health personal and improvement of health infrastructures may substantially affect the social and financial evaluation of the investment. In other words, companies that generally promote a socially responsible business profile will benefit from including HDI-based data in their CSR reports and will thus have an interest in making the data available.[12] Finally, it is worth considering the added value of a HDI-style corporate reporting system from the point of view of business management and strategy. As we have argued in this paper, and as is commonplace in the literature, there is a lack of comparability of data within the GRI framework. Overcoming this deficit will make an HDI-inspired approach more likely to be integrated into business planning and executive deliberations. Business leaders will be able to navigate which countries generally value social impact profiles and will thus be able to map on which regions of the world (or at a smaller scale, which regions or sectors) the company should concentrate. In addition, business leaders will be provided with a strong tool for strategic communication when dealing with potential investors precisely because the corporations will be able to present comparable data on positive (and negative) social impacts.[13]

5 Conclusion

We have argued that social reporting that matters must enable stakeholders to make informed decisions. In our view, this implies that stakeholders should (at least to some degree) be able to compare companies' social performance. In this regard, we have argued that the GRI Standards do not enable stakeholders to compare the relevant social performance of different organizations. Even though we recognize

---

[12]Technical details regarding accountancy of data addressing the three specified dimensions as well as more specific concerns regarding data management and disclosure are not discussed in this paper.

[13]Notice, we are not assuming a "buy in" from organisations. Instead, governments might take charge and make it obligatory by law to include the social indices into the CSR report. Also, even though some organizations might lobby against including social indices into their CSR report, such indices are clearly in the interest of some actors, including pension funds (with a social profile), public buyers and all those companies that would score relatively high on these indices.

that complete comparability of social issues is technically and methodologically difficult to achieve, we believe it possible to improve upon the GRI Standards by incorporating a number of cross-organizationally comparable social indices. The source of inspiration for these social indices is HDI. We realize that these indices might need to be developed further. For instance, since the indices will have to rely at least partially on reported financial information, problems about the trustworthiness and comparability of this information will emerge. This and many other problems are beyond the scope of this paper. However, we have demonstrated the possibility of creating social indices that substantially improve stakeholders' ability to make informed decisions with regard to social issues. Our hope is that we have provided inspiration for future developments in social reporting that can take us one step closer to comparability.

References

American Accounting Association. (1957). Accounting and reporting standards for corporate financial statements, 1957 revision. *Accounting Review, 32*(4), 536–546.

Casper, D. (2007). What is the capability approach? Its core, rationale, partners and dangers. *The Journal of Socio-Economics, 36*, 335–359.

Chang, R. (Ed.). (1997). *Incommensurability, incomparability and practical reason.* Cambridge, MA: Harvard University Press.

Clarkson, P. M., Yue, L., Richardson, G. D., & Vasvari, P. F. (2008). Revisiting the relation between environmental performance and environmental disclosure: An empirical analysis. *Accounting, Organizations and Society, 33*(4-5), 303–327.

Crisp, R., & Hooker, B. (Eds.). (2000). *Well-being and morality: Essays in honour of James Griffin.* Oxford: Oxford University Press.

Elkington, J. (1997). *Cannibals with forks: The triple bottom line of 21st century business.* Oxford, UK: Capstone Publishing Ltd.

Elkington, J. (2004). Enter the triple bottom line. In A. Henriques & J. Richardson (Eds.), *The triple bottom line: Does it all add up?* (pp. 1–17). London: Earthscan.

Freeman, E. R., Harrison, J. S., Wick, A. C., Parmar, B. L., & De Colle, S. (2011). *Stakeholder theory: The state of the art.* Cambridge University Press, Cambridge, UK.

Fukuda-Parr, S. (2003). The human development paradigm: Operationalizing sen's idea on capabilities. *Feminist Economics, 9*(2-3), 301–317.

Global Reporting Initiative. (2007). *Due process for the GRI Reporting Framework.* www.globalreporting.org

Global Reporting Initiative. (2016). *GRI Standards.* www.globalreporting.org

Gray, R. (2002). The social accounting project and accounting organizations and society privileging engagement, imaginings, new accountings and pragmatism over critique? *Accounting, Organizations and Society, 27*(7), 687–708.

Gray, R. (2010). Is accounting for sustainability actually accounting for sustainability...and how would we know? An exploration of narratives of organisations and the planet. *Accounting, Organizations and Society, 35*(1), 47–62.

Gray, R., Kouhy, R., & Lavers, S. (1995). Corporate social and environmental reporting: A review of the literature and a longitudinal study of UK disclosure. *Accounting, Auditing & Accountability Journal, 8*(2), 47–77.

Lozano, R., & Huisingh, D. (2011). Inter-linking issues and dimensions in sustainability reporting. *Journal of Cleaner Production, 19*, 99–107.

MacDonald, C., & Nornam, W. (2007). Rescuing the baby from the triple-bottom-line bathwater: A reply to Pava. *Business Ethics Quarterly, 17*(1), 111–114.

MacKenzie, D. (2009). Making things the same: Gases, emission rights and the politics of carbon markets. *Accounting, Organizations and Society, 34*, 440–455.

Marimon, F., Alonso-Almeida, M. M., Rodriguez, M. P., & Alejandro, K. A. C. (2012). The worldwide diffusion of the global reporting initiative: What is the point? *Journal of Cleaner Production, 33*, 132–144.

Milne, M., & Gray, R. (2013). W(h)ither ecology? The triple bottom line, the global reporting initiative, and corporate sustainability reporting. *Journal of Business Ethics, 118*, 13–29.

Moneva, J. M., Archel, P., & Correa, C. (2006). GRI and the camouflaging of corporate unsustainability. *Accounting Forum, 30*, 121–137.

Norman, W., & MacDonald, C. (2004). Getting to the bottom of "triple bottom line". *Business Ethics Quarterly, 14*(2), 242–263.

Pava, M. (2007). A response to 'getting to the bottom of "triple bottom line"'. *Business Ethics Quarterly, 17*(1), 105–110.

Prado-Lorenzo, J. M., Gallego-Alvarez, I., & Garcia-Sanchez, I. M. (2009). Stakeholder engagement and corporate social responsibility reporting: The ownership structure effect. *Corporate Social Responsibility and Environmental Management, 16*, 94–107.

Robeyns, I. (2003). Sen's capability approach and gender inequality: selecting relevant capabilities. *Feminist Economics, 9*(2–3), 61–92.

Robeyns, I. (2006). The capability approach in practice. *The Journal of Political Philosophy, 14*(3), 351–376.

Roca, L. C., & Searcy, C. (2012). An analysis of indicators disclosed in corporate sustainability reports. *Journal of Cleaner Production, 20*, 103–118.

Sen, A. (2009). *The idea of justice*. Cambridge, MA: Harvard University Press.

Skouloudis, A., Evangelinos, K., & Kourmousis, F. (2009). Development of an evaluation methodology for triple bottom line reports using international standards on reporting. *Environmental Management, 44*, 298–311.

Sutantoputra, A. W. (2009). Social disclosure rating system for assessing firms' CSR reports. *Corporate Communications: An International Journal, 14*(1), 34–48.

ul Haq, M. (2010). *The real wealth of nations: Pathways to human development*. New York: Human Development Report. United Nations

United Nations. (2014). *Human Development Report 2014 – "Sustaining Human Progress: Reducing Vulnerabilities and Building Resilience"*. HDRO (Human Development Report Office) United Nations Development Programme. Retrieved 25 July 2014

Williams, F. P., & Ravenscroft, S. P. (2015). Rethinking decision usefulness. *Contemporary Accounting Research, 32*(2), 763–788.

Willis, A. C. A. (2003). The role of the global reporting initiative's sustainability reporting guidelines in the social screening of investments. *Journal of Business Ethics, 43*, 233–237.

Claus Strue Frederiksen Ph.D., is a trained philosopher and has been working with CSR for more than a decade. He has published in a variety of scientific journals and books, including in *Journal of Business Ethics, International Journal of Applied Philosophy* and in Springer's *Encyclopedia of Corporate Social Responsibility*. Currently, he is working as a sustainability consultant at the Danish biotech company Gubra, where he has developed and implemented ethical approach to CSR.

David Butdz Pedersen (b. 1980) is professor of science communication and impact studies and director of the Humanomics Research Centre at Aalborg University, Denmark. His research focuses on responsible research and innovation, research impact and science and technology policy. He has published and edited about 150 scientific papers, monographs, chapters, policy reports and op-eds.

Morten Ebbe Juul Nielsen Ph.D., associate professor in philosophy, Institute for communication, Copenhagen University. He works in practical philosophy, with an emphasis on political and applied philosophy, including business ethics and CSR. He has published extensively internationally.

Samuel O. Idowu is a senior lecturer in accounting and corporate social responsibility at London Guildhall School of Business and Law, London Metropolitan University, UK. He researches in the fields of Corporate Social Responsibility (CSR), Corporate Governance, Business Ethics and Accounting and has published in both professional and academic journals since 1989. He is a freeman of the City of London and a Liveryman of the Worshipful Company of Chartered Secretaries and Administrators. He is the Deputy CEO and First Vice President of the Global Corporate Governance Institute. He has led several edited books in CSR, he is the editor-in-chief of three Springer's reference books—the Encyclopedia of Corporate Social Responsibility, the Dictionary of Corporate Social Responsibility and the Encyclopaedia of Sustainable Management (ESM). He is an editor-in-chief of the International Journal of Corporate Social Responsibility (IJCSR), the editor-in-chief of the American Journal of Economics and Business Administration (AJEBA) and an associate editor of the International Journal of Responsible Management in Emerging Economies (IJRMEE). He is also a series editor for Springer's books on CSR, Sustainability, Ethics and Governance. One of his edited books won the most Outstanding Business Reference Book Award of the American Library Association (ALA) in 2016 and another was ranked 18th in the 2010 Top 40 Sustainability Books by, *Cambridge University, Sustainability Leadership Programme.* He is a member of the Committee of the Corporate Governance Special Interest Group of the British Academy of Management (BAM). He is on the editorial boards of the International Journal of Business Administration, Canada and Amfiteatru Economic Journal, Romania. He has delivered a number of keynote speeches at national and international conferences and workshops on CSR and has on two occasions 2008 and 2014 won Emerald's Highly Commended Literati Network Awards for Excellence. To date, he has edited several books in the field of CSR, Sustainability and Governance and has written seven forewords to CSR books. Samuel has served as an external examiner to the following UK universities—Sunderland, Ulster, Anglia Ruskin, Plymouth, Robert Gordon, Aberdeen, Teesside, Sheffield Hallam, Leicester De Montfort, Canterbury Christ Church and Brighton. He has examined PhD theses for universities in the UK, South Africa, Australia, The Netherlands and New Zealand.

Environmental, Social and Governance (ESG) and Integrated Reporting

Selina Neri

Abstract Corporate Social Responsibility (CSR) and sustainable development address the relationship between business and society. In the 1980s, the Brundtland Report by the World Commission on Environment and Development (WCED, Our common future (Burdtland Report). Retrieved from https://sustainabledevelopment.un.org/content/documents/5987our-common-future.pdf, 1987) coined the term sustainable development and its principle of sustainability for economic prosperity. These were reinforced by the United Nations (UN) 2030 Agenda for Sustainable Development, the blueprint for a more sustainable future. In parallel, CSR has received increased attention in practice and in theory, and has been defined by the European Commission as "the responsibility of enterprises for their impacts on society" (A renewed European Union Strategy 2011–2014 for Corporate Social Responsibility, 2011, p. 6). Calls for CSR have been framed through a mix of voluntary and law mandated corporate actions which have led to conflicting results. In a twenty-first century bedevilled by challenges and macro systems' disruptions (e.g. social institutions, natural resources and technologies), environmental, social and governance (ESG) factors represent risks and opportunities that are strategically relevant and increasingly inform how businesses are run and investment decisions made. To reach their full potential, ESG need to be measured, included in managerial and investment decision-making and accounted for in integrated reporting.

1 Introduction

Demands on corporations (and their investors) to address societal expectations are on the rise and represent a key aspect of the contemporary business landscape. Global initiatives such as the UN Global Compact (UN 2020) and multiple OECD guidelines (OECD 2011a, 2011b, 2015, 2018) aim to encourage and guide corporate

S. Neri (✉)
HULT International Business School, Dubai Internet City, Dubai, United Arab Emirates
e-mail: selina.neri@faculty.hult.edu

and investment behaviour. Although it is beyond the scope of this chapter to elaborate on the heterogeneity of investor types and their short or long term investment horizons, it is critical to note that if corporations are to become socially responsible and create sustainable value, they need, among others, the support of their investors, their capital providers. Since the late 1980s (Avetisyan and Hockerts 2017), ESG encapsulate environmental (contribution to the climate transition or circular design of products and services), social (community impact, inclusion or combating modern slavery) and governance (executive compensation, disclosure, short and long term financial health of a company) dimensions of business strategies. ESG remain difficult to measure (Brest et al. 2018), yet represent an opportunity for standardization in the field of sustainable value creation (Avetisyan and Hockerts 2017).

Studies have investigated a range of issues surrounding CSR and sustainable development: these include managerial perceptions of social responsibility (Bansal et al. 2014; Singhapakdi et al. 1996), the importance of social responsibility in managerial decisions (Vitell and Paolillo 2004), executive perceptions of CSR (Skouloudis et al. 2015), external factors affecting socially responsible decision-making (Vashchenko 2017), the effect of managerial perceptions of ethics and morals on sustainable value creation (Wang et al. 2018), and managerial motivation for social and environmental disclosures (Jackson et al. 2020; Shafer and Lucianetti 2018). One of the key insights across these studies is that companies face increasing societal pressures. CSR and sustainable development remain difficult to measure and to incorporate in business decisions. In parallel, while earlier studies questioned the relevance of ESG (Campbell and Slack 2011; Deegan and Rankin 1997), more recently these factors have been shown to relate to significant economic and financial effects (Dhaliwal et al. 2011; Cheng et al. 2014; Grewal et al. 2019; Khan et al. 2016). ESG consideration varies from merely superficial interest to ESG forming a cornerstone of investment and business decisions (Kiernan 2007; Neri 2019; Vasuveda et al. 2018).

The path to socially responsible corporate behaviour remains bedevilled by tensions. Fundamentally, it is legitimate and reasonable both for society to expect corporations to contribute to sustainable development, and for investors to maximise the value of the assets entrusted in their care, and corporations need, among other things, to be financially healthy in order to remain in business. However, how these expectations and needs can be reconciled is a matter filled with tensions.

Many companies and investors "remain ambiguous about what it means to be socially responsible" (Neri et al. 2019, p. 442). ESG factors can act as indicators of risks and opportunities that may affect the financial bottom line (Van Duuren et al. 2016) and the ability to create value (Neri 2020b), thus they encapsulate a potential contribution to sustainable development. For instance, they can point to upcoming regulation (as was the case in the 2016 Facebook Cambridge Analytica scandal which accelerated the European General Data Protection Regulation) or they can be instrumental in shifting consumer preferences (e.g. towards healthier food and lifestyles), or affect a corporation's ability "to produce people" (Hollensbe et al. 2014, p. 1229), in other words committed and dedicated employees.

Against this backdrop, this chapter focuses on the lack of globally accepted ESG standards of measurement and the resulting lack of integrated reporting. This chapter will proceed as follows: first, it elaborates why it is critical for companies and investors to be able to measure the potential and real impact of ESG risks and opportunities; second, it presents the challenges that a lack of integrated reporting poses to responsible corporate behaviour.

2 ESG Risks and Opportunities

Crises have a tendency to accelerate trends. The turmoils of 2020, from the climate emergency and the global pandemic to the racial justice movement have propelled ESG into the mainstream conversation on CSR and sustainable development (Mooney 2020). ESG have emerged as strategically relevant and important because they represent many of the externalities around us (Neri 2020a; Powell 2020), encapsulating material risks and opportunities for companies and their investors (Espahbodi et al. 2019; Van Duuren et al. 2016). ESG represent "a tragedy (...) and a golden opportunity for positive system change" (Kiernan 2007, p. 478), as they boil down to what can make or break the very existence of a business.

"Society has reached its planetary limits for growth" (Bansal 2019). The world seems to be waking up and the ESG tide appears to be shifting (Goyer and Jung 2011; Semenova and Hassel 2019). The "bandwagon is rolling" (Cornell 2020) and ESG have been called the acid test of responsible capitalism (Powell 2020) and the new corporate *Zeitgeist* (Nauman 2020). Business is under extreme pressure to act upon its social responsibility and contribute solutions to the challenges of our time. Responsible business is no longer a distant chimera: it is a real and pressing *business* matter. As Kathleen McLaughlin, Walmart's Chief of Sustainability stated: "You can't separate environmental, social and economic success" (Financial Times 2020). Given the importance for companies to understand, account and report on ESG, it is both surprising and worrying that to date globally accepted ESG standards of measurement are missing and that the so-called "integrated reporting" is, in fact, not integrated at all. Furthermore, ESG factors continue to be referred to as non-financial and, when at all, accounted for in non-financial statements, most often separately from financial reports.

ESG isn't a new concept. The acronym dates back to the 1980s (Avetisyan and Hockerts 2017; Neri 2020b) but the underlying ideas are as old as humanity. Throughout history, nature has been disruptive, yet the climate crisis is accelerating the urgency of tackling environmental problems. Social unrest has also always existed, whether in ancient Greece due to wealth and land inequality (Fuks 1984), in the city of Norwich in Tudor England following disastrous harvests (Hoskins 1964), in Somalia, Rwanda and the former Yugoslavia following poverty and unemployment in "the lost decade of the 1980s" (Chossudovsky 1997, p. 1786), or in Hong Kong in the summer of 2019 (Korner 2019). Corporate governance issues (and scandals) are also not new, as the failings of the Italian Medici Bank in 1494 indicate (Dinesen 2020), with regulation aimed at improving governance effectiveness and strengthening control. What is different now is the immediacy, the scale and the global effects of the consequences of such events, giving added urgency and importance to responding and anticipating them, hence the increasing centrality of ESG factors in business and investment decisions. What is also different now is that companies' main sources of value have increasingly shifted from tangible to intangible assets (Hanson 2013; Haskel and Westlake 2018), however current financial standards and accounting practices were created at a time when tangible assets dominated value creation. Consequently, there is a need for urgent adjustment.

Some management scholars and practitioners argue that there is a risk that companies and their investors pay lip service to ESG adoption (Armstrong 2020; World Economic Forum 2019), "claiming to be socially responsible" (Reghunandan and Rajgopal 2020, p. 1). In other words, there is a risk of facing "incidences where an organization's 'talk' does not match its 'walk' " (Glozer and Morsing 2020, p. 363). However, there is also a promise that a mainstream adoption of ESG standards might enable financial reporting that accounts for the risks and opportunities intrinsic to all aspects of contemporary business, thus helping corporations to create sustainable value rather than paying lip service to it. This promise needs to be appreciated within the current context.

The twenty-first century features scholars, business leaders, investors and policy makers' calls for responsible capitalism (Mayer 2016, 2017; Mayer et al. 2017; Starbuck 2014), shareholder and stakeholder activism, the rise of social media (Joe et al. 2009; Liu and McConnell 2013) and a pervasive crisis of trust towards business (Edelmann 2020; Starbuck 2005). Climate change, social tensions and unrest, and governance issues pose real threats to companies but also represent opportunities for innovation, new products, new lines of business or business models, therefore these factors are relevant to companies (The Economist 2019a, b). Consequently, it can be expected that if corporations are to create sustainable value they need to be able to measure, decide upon and account for ESG factors that are material to their businesses. They need to do so because these factors encapsulate the contemporary business conditions for sustainable value creation.

ESG are prompting a corporate realization that value creation can only be achieved sustainably, in other words along environmental, social, financial and governance dimensions which exist both in the short and long term. The following examples illustrates how critical both the short and long dimensions of ESG factors are. In October 2018 Patisserie Valerie, a listed UK coffee chain, went into administration. The case revealed serious governance issues, while the company had been scrutinised for its sourcing of palm oil, in common with similar businesses. Long-term thinking had pushed Patisserie Valerie's investors and leaders to investigate the palm oil sources for its finest cakes and eclairs, without noticing that the business was close to bankruptcy (Financial Times 2019a; Montagnon 2019). The second example relates to the coronavirus pandemic upending the world economy in 2020. Since the beginning of this crisis, a growing list of companies from Rolls-Royce to Disney, Electrolux and H&M have slashed dividend payouts. Most importantly, some investors are not convinced that the measure is temporary or short-term

(Edgecliff-Johnson and Thomas 2020). The pandemic seems to have accelerated a significant shift in how companies dispose of their free cash flow, following increasing acceptance that they bear a responsibility towards all stakeholders (including investors) and the wider society. The above examples indicate how critical it is for businesses to attend to both short and long term dimensions of ESG if they are to "sustain" and create value next quarter as well as in the next 20, 40 or 100 years. Although there is a tendency to think of social responsibility, ESG and sustainability in terms of a distant future, their integration in business decisions needs to take the form of a "pattern of converging decisions"(Gray and Ariss 1985, p. 707), in other words an organized whole of "short-term steps for long-term change" (Kemp et al. 2007, p. 315). These steps are to be measured and reported upon, so that they can be fully integrated in decision making and corporate reporting, a matter that remains filled with tensions.

3 Integrated Reporting

Historically, the concept of measuring and reporting the social, environmental and economic impact of a corporation was popularized in the late 1990s and early 2000s, partly due to John Elkington's (1997) book *Cannibals with Forks: The Triple Bottom Line of 21st Century Business.* Elkington's work is credited with starting new, non-financial reporting frameworks from an environmental and social point of view (Dumay et al. 2016; Gray 2006).

Integrated reporting is a framework of reporting where non-financial and financial information are considered jointly, allowing companies to better detect risks and opportunities not only retrospectively, but also before they arise. In this way, ESG can provide managers, investors and other stakeholders (i.e. employees, customers and policy makers) comprehensive insights into how a business operates and its financial, social and environmental impact, including how sustainable the business is or is becoming. Integrated reporting requires globally accepted ESG standards of measurement and reporting, as is the case for the current financial standards and accounting practices. The meaning of *integration* is quite different from the current understanding, where integration refers to the inclusion of non-financial reporting in a company's annual report. According to the EU Non-financial Reporting Directive II (EU 2014) which came into effect in 2018, and the succeeding Non-binding Guidelines (EU 2017), environmental, social and governance matters are to be reported separately from, or in separate sections of, financial statements, rather than fully integrated (i.e. accounted for in the figures) into the same financial reports.

Since the 1980s, a plethora of ESG rating agencies have emerged (Avetisyan and Hockerts 2017; Brest et al. 2018), mainly focussed on retrospective ratings, examples of which include Morningstar, B Analytics, TSE4Good and the Dow Jones Sustainability Index. A variety of cross-industry initiatives on ESG reporting also exist, and include the Global Reporting Initiative, the Financial Stability Board Task Force on Climate Related Financial Disclosures, and the Sustainability Accounting

Standards Board. Corporations also face a multitude of single-issue niche groups, and 230 ESG standards for over 80 industry sectors in 180 countries (Solvang 2018). These represent several competing sets of standards by which to measure and report ESG. Furthermore, a common preference to categorise ESG as non-financial shapes the current policy debate on integrated reporting as a "1+1" discussion, promoting the illusion that these factors are anything other than financial. Consequently, recent developments in integrated reporting (EU 2014, 2017) contribute to relegate ESG to non-financial business dimensions and represent a tension corporations and their capital providers face in delivering sustainable value. How can a business be socially responsible if it cannot understand and measure the threats, opportunities and effects of climate change and incorporate them in its quarterly reporting, profit & loss statement or balance sheet, as a first step to contribute solutions to the climate crisis? After the 2019–2020 ravaging fires in California or Australia, how can corporations with facilities and operations in those areas account for their tangible and intangible losses and future exposure? How can corporations anticipate the social effect of the global pandemic on their employees or the effect of social unrest on their retail distribution and brand image? These are questions that global ESG standards and truly integrated reporting could contribute to address.

The multitude of non-globally accepted ESG standards, coupled with a myopic regard on ESG as non-core factors, impair a comprehensive consideration of their related risks and opportunities. This status quo also reinforces views of these factors as having little to no impact on the bottom-line (Klasa 2018; Van Duuren et al. 2016). The EU-recently mandated integrated reporting (in force since June 2019) (EU 2014, 2017), and the US Congress rejection of the adoption of ESG standards (Temple-West 2019) offer institutional support for views of value creation which are divorced from reality. They also represent a critical tension impairing the institutionalization of the corporate contribution to sustainable development. Through institutionalisation, CSR and sustainable development can become embedded in the social structure (Li 2017; Zucker 1977), in other words they can reach a point where nobody would dream of questioning what they are and why they matter, and therefore why ESG need to be integrated in decision-making and reported on.

In its current form, financial reporting is in itself a tool no longer fit for purpose. Financial reporting and accounting standards, as well as the network of professionals supporting them, developed from humble beginnings in the fifteenth century (Financial Times 2019b). In the twenty-first century, sources of, and impact on, value creation can take the shape of human, social or natural capital: hence, there is a need to go beyond a financial view of how a business is doing. ESG standardisation is an enabler of integrated reporting and it demands patience. Despite the current lack of global standards, recently mandated ESG disclosure (EU 2014, 2017) is an important, first step forward, but in its current form it is not enough to support the integration of ESG in business and investment decisions.

Although it has taken more than four hundred years to arrive at contemporary financial standards and accounting practices, the magnitude and urgency of the challenges of our time require immediate attention to make integrated reporting truly integrated, with "non-financial" factors regarded as financial and accounted for in numbers within financial reports. Society, policy makers, companies and their investors can ill-afford to wait another four hundred years for a globally accepted integrated reporting framework to emerge.

References

Armstrong, R. (2020). The dubious appeal of ESG investing is for dupes only. *Financial Times*. 23 August. Retrieved from https://www.ft.com

Avetisyan, E., & Hockerts, K. (2017). The consolidation within the ESG rating industry as an enactment of institutional retrogression. *Business Strategy and the Environment, 26*, 316–330.

Bansal, P. (2019). Sustainable development in an age of disruption. *Academy of Management Discoveries, 5*, 8–12.

Bansal, P., Gao, J., & Qureshi, I. (2014). The extensiveness of corporate social and environmental commitment across firms over time. *Organization Studies, 35*, 949–966.

Brest, P., Gilson, R. J., & Wolfson, M. A. (2018). How investors can (and can't) create social value. *ECGI Working Paper Series in Law. Working Paper N. 394*. Retrieved from https://papers.ssrn.com

Campbell, D., & Slack, R. (2011). Environmental disclosure and environmental risk: Sceptical attitudes of UK sell-side bank analysts. *The British Accounting Review, 43*, 54–64.

Cheng, B., Ioannou, I., & Serafeim, G. (2014). Corporate social responsibility and access to finance. *Strategic Management Journal, 35*, 1–23.

Cheng, M. M., Green, W. J., & Chi Wa Ko, J. (2015). The impact of strategic relevance and assurance of sustainability indicators on investors' decisions. *Auditing: A Journal of Practice & Theory, 34*, 131–162.

Chossudovsky, M. (1997). Economic reforms and social unrest in developing countries. *Economic & Political Weekly, 32*, 1786–1788.

Cornell, B. (2020). The ESG concept has been overhyped and oversold. *Financial Times*, 16 July. Retrieved from https://www.ft.com/content/719e6253-4d07-402a-9f36-d461671657a1?desktop=true&segmentId=d8d3e364-5197-20eb-17cf-2437841d178a#myft:notification:instant-email:content

Deegan, C., & Rankin, M. (1997). The materiality of environmental information to users of annual reports. *Accounting, Auditing & Accountability Journal, 10*, 562–583.

Dhaliwal, D. S., Li, O. Z., Tsang, A., & Yang, G. Y. (2011). Voluntary disclosure and the cost of equity capital: The initiation of corporate social responsibility reporting. *The Accounting Review, 86*, 59–100.

Dinesen, C. (2020). *Absent management in banking: How banks fail and cause financial crisis*. London: Palgrave MacMillan.

Dumay, J., Bernardi, C., Guthrie, J., & Demartini, P. (2016). Integrated reporting: A structured literature review. *Accounting Forum, 40*, 166–185.

Edelmann. (2020). *Edelman trust barometer*. 19 January. Retrieved from https://www.edelman.com/trustbarometer

Edgecliff-Johnson, A., & Thomas, D. (2020). Companies axe dividends in global push for cash. *Financial Times*. 23 March. Retrieved from https://www.ft.com/content/e9102d80-6d2c-11ea-89df-41bea055720b

Elkington, J. (1997). *Cannibals with Forks: The Triple Bottom Line of 21st Century Business*. Oxford: Capstone Publishers.

Espahbodi, L., Espahbodi, R., Juma, N., & Westbrook, A. (2019). Sustainability priorities, corporate strategy and investor behavior. *Journal of Financial Economics, 37*, 149–167.

European Commission. (2011). *A renewed European Union Strategy 2011–2014 for Corporate Social Responsibility*, 681, 3.1. Retrieved from https://www.europarl.europa.eu/meetdocs/2009_2014/documents/com/com_com(2011)0681_/com_com%282011%290681_en.pdf

European Union (EU). (2014). *Directive 2014/95/EU of the European Parliament and of the Council of 22 October 2014 amending Directive 2013/34/EU as regards disclosure of non-financial and diversity information by certain large undertakings and groups*. Retrieved from https://ec.europa.eu/info/business-economy-euro/company-reporting-and-auditing/company-reporting/non-financial-reporting_en

European Union (EU). (2017). *Guidelines on non-financial reporting (methodology for reporting non-financial information)*. Retrieved from https://ec.europa.eu/info/publications/non-financial-reporting-guidelines_en

Financial Times. (2019a). *Patisserie Valerie offers a bitter financial lesson*. 23 January. Retrieved from https://www.ft.com/content/6a8bacc0-1f0b-11e9-b126-46fc3ad87c65

Financial Times. (2019b). *Ethical investment needs more transparency*. 11 June. Retrieved from https://www.ft.com/content/e75917a0-86ef-11e9-a028-86cea8523dc2

Financial Times. (2020). *Walmart's sustainability chief: 'You can't separate environmental, social and economic success*. 17 July. Retrieved from https://www.ft.com/content

Fuks, A. (1984). *Social conflict in ancient Greece*. Jerusalem: The Magnes Press.

Glozer, S., & Morsing, M. (2020). Helpful hypocrisy? Investigating 'double-talk' and irony in CSR marketing communications. *Journal of Business Research, 114*, 363–375.

Goyer, M., & Jung, D. K. (2011). Diversity of institutional investors and foreign blockholdings in France: The evolution of an institutionally hybrid economy. *Corporate Governance: An International Review, 19*, 562–584.

Gray, B., & Ariss, S. S. (1985). Politics and strategic change across organizational life cycles. *Academy of Management Review, 10*, 707–723.

Gray, R. (2006). Social, environmental and sustainability reporting and organisational value creation? Whose value? Whose creation? *Accounting, Auditing & Accountability Journal, 19*, 793–819.

Grewal, J., Riedl, E. J., & Serafeim, G. (2019). Market reaction to mandatory nonfinancial disclosure. *Management Science, 65*, 3061–3084.

Hanson, D. (2013). ESG investing in Graham & Doddsville. *Journal of Applied Corporate Finance, 25*, 20–31.

Haskel, J., & Westlake, S. (2018). *Capitalism without capital: The rise of the intangible economy*. Woodstock: Princeton University Press.

Hollensbe, E., Wookey, C., Hickey, L., George, G., & Nichols, C. (2014). Organizations with purpose. *Academy of Management Journal, 57*, 1227–1234.

Hoskins, W. G. (1964). Harvest fluctuations and English economic history, 1480–1619. *The Agricultural History Review, 12*, 28–46.

Jackson, G., Bartosh, J., Avetisyan, E., Kinderman, D., & Knudsen, J. S. (2020). Mandatory non-financial disclosure and its influence on CSR: An international comparison. *Journal of Business Ethics, 162*, 323–342.

Joe, J. R., Louis, H., & Robinson, D. (2009). Managers' and investors' responses to media exposure of board ineffectiveness. *Journal of Financial and Quantitative Analysis, 44*, 579–605.

Kemp, R., Rotmans, J., & Loorbach, D. (2007). Assessing the Dutch energy transition policy: How does it deal with dilemmas of managing transitions? *Journal of Environmental Policy & Planning, 9*, 315–331.

Khan, M., Serafeim, G., & Yoon, A. (2016). Corporate sustainability: First evidence on materiality. *The Accounting Review, 91*, 1697–1724.

Kiernan, M. J. (2007). Universal owners and ESG: Leaving money on the table? *Corporate Governance: An International Review, 15*, 478–485.

Klasa, A. (2018). Sustainable finance: Integrated reporting offers a fix for 'insufficient' status quo. *Financial Times*. 5 December. Retrieved from https://www.ft.com/content

Korner, A. (2019). Hong Kong protests have united people of all ages. *Financial Times*. 6 September. Retrieved from https://www.ft.com/content

Li, Y. (2017). A semiotic theory of institutionalization. *Academy of Management Review, 42*, 520–547.

Liu, B., & McConnell, J. (2013). The role of the media in corporate governance: Do the media influence managers' capital allocation decisions? *Journal of Financial Economics, 110*, 1–17.

Mayer, C. (2016). Reinventing the corporation. *Journal of the British Academy, 4*, 53–72.

Mayer, C. (2017). Who is responsible for irresponsible business? An assessment. *Oxford Review of Economic Policy, 33*, 1–25.

Mayer, C., Wright, M., & Phan, P. (2017). Management research and the future of the corporation: A new agenda. *Academy of Management Perspectives, 31*, 179–182.

Montagnon, P. (2019). Patisserie Valerie offers a bitter financial lesson. *Financial Times*. Retrieved from https://www.ft.com/content

Mooney, A. (2020). Calpers, Schroders call for mandatory inclusion of climate risks in accounts. *Financial Times*, 15 August. Retrieved from https://www.ft.com

Nauman, B. (2020). ESG pressure on dividend payout to continue after crisis. *Financial Times*. 1 June. Retrieved from https://www.ft.com/content/da8b6f40-8afa-11ea-a109-483c62d17528

Neri, S. (2019). *Director engagement with corporate purpose: The contribution and potential of institutional investor stewardship.* Unpublished doctoral dissertation, The British University in Dubai, United Arab Emirates.

Neri, S. (2020a). Corporate purpose. In S. Idowu, R. Schmidpeter, N. Capaldi, L. Zu, M. Del Baldo, & R. Abreu (Eds.), *Encyclopedia of sustainable management*. Cham: Springer.

Neri, S. (2020b). Director engagement with corporate purpose: The contribution and potential of institutional investors. *Academy of Management Proceedings*. Retrieved from https://journals.aom.org/doi/abs/10.5465/AMBPP.2020.14538abstract

Neri, S., Pinnington, A. H., Lahrech, A., & Al-Malkawi, H.-A. (2019). Top executives' perceptions of the inclusion of corporate social responsibility in quality management. *Business Ethics: A European Review, 28*, 441–458. https://doi.org/10.1111/beer.12235. https://doi.org/10.1007/978-3-030-02006-4_1077-1.

Organisation for Economic Cooperation and Development (OECD). (2011a). *OECD Guidelines for Multinational Enterprises*. Retrieved from https://www.oecdwatch.org/oecd-ncps/the-oecd-guidelines-for-mnes/

Organisation for Economic Cooperation and Development (OECD). (2011b). *The role of institutional investors in promoting good corporate governance*. Retrieved from https://www.oecd.org/daf/ca/49081553.pdf

Organisation for Economic Cooperation and Development (OECD). (2015). *G20/OECD Principles of Corporate Governance*. Paris: OECD Publishing.

Organisation for Economic Cooperation and Development (OECD). (2018). *OECD due diligence guidance for responsible business conduct*. Retrieved from https://www.oecd.org/investment/due-diligence-guidance-for-responsible-business-conduct.htm

Powell, J. (2020). *Coronavirus is the acid test of ESG*. 7 April. Retrieved from https://ftalphaville.ft.com/2020/04/02/1585807115000/Coronavirus-as-the-ESG-acid-test/

Reghunandan, A., & Rajgopal, S. (2020). Do the socially responsible walk the talk? *SSRN*. 24 May. Retrieved from https://papers.ssrn.com/sol3/papers.cfm?abstract_id=3609056

Semenova, N., & Hassel, L. G. (2019). Private engagement by Nordic institutional investors on environmental, social, and governance risks in global companies. *Corporate Governance: An International Review, 27*, 144–161.

Shafer, W., & Lucianetti, L. (2018). Machiavellianism, stakeholder orientation, and support for sustainability reporting. *Business Ethics: A European Review, 27*, 271–285.

Singhapakdi, A., Vitell, S. J., Rallapalli, K. C., & Kraft, K. L. (1996). The perceived role of ethics and social responsibility: A scale development. *Journal of Business Ethics, 15*, 1131–1140.

Skouloudis, A., Avlonitis, G. J., Malesios, C., & Evangelinos, K. (2015). Priorities and perceptions of corporate social responsibility: Insights from the perspective of Greek business professionals. *Management Decision, 53*, 375–401.

Solvang, T. E. (2018). Sustainable finance: Danske Bank's shows need for bigger accounting picture. *Financial Times*. 5 December. Retrieved from https://www.ft.com/content

Starbuck, W. H. (2005). Four great conflicts of the twenty-first century. In C. L. Cooper (Ed.), *Management and leadership in the 21st century* (pp. 21–56). Oxford: Oxford University Press.

Starbuck, W. H. (2014). Why corporate governance deserves serious and creative thought. *Academy of Management Perspectives, 28*, 15–21.

Temple-West, P. (2019). US Congress rejects European-style ESG reporting standards. *Financial Times*. 12 July. Retrieved from https://www.ft.com/content

The Economist. (2019a). *Business and the effects of global warming*. 21 February. Retrieved from https://www.economist.com/business/2019/02/21/business-and-the-effects-of-global-warming

The Economist. (2019b). *Climate change and the threat to companies*. 21 February. Retrieved from https://www.economist.com/leaders/2019/02/21/climate-change-and-the-threat-to-companies

United Nations (UN). (2020). *Global Compact. See who is involved*. Retrieved from https://www.unglobalcompact.org/participation

Van Duuren, E., Platinga, A., & Scholtens, B. (2016). ESG integration and the investment management process: Fundamental investing reinvented. *Journal of Business Ethics, 138*, 525–533.

Vashchenko, M. (2017). An external perspective on CSR: What matters and what does not? *Business Ethics: A European Review, 26*, 396–397.

Vasuveda, G., Nachum, L., & Say, G. (2018). A signalling theory of institutional activism: How Norway's sovereign wealth fund investments affect firms' foreign acquisitions. *Academy of Management Journal, 61*, 1583–1611.

Vitell, S. J., & Paolillo, J. G. (2004). A cross-cultural study of the antecedents of the perceived role of ethics and social responsibility. *Business Ethics: A European Review, 13*, 185–199.

Wang, Z., Li, F., & Sun, Q. (2018). Confucian ethics, moral foundations, and shareholder value perspectives: An exploratory study. *Business Ethics: A European Review, 27*, 260–271.

World Economic Forum. (2019). *Davos Manifesto 2020: The universal purpose of a company in the fourth industrial revolution*. 2 December. Retrieved from https://www.weforum.org/agenda/2019/12/davos-manifesto-2020-the-universal-purpose-of-a-company-in-the-fourth-industrial-revolution/

Zucker, L. G. (1977). The role of institutionalization in cultural persistence. *American Sociological Review, 42*, 726–743.

Selina Neri is a Professor of Management and Corporate Governance at HULT International Business School (Dubai and London), where she designs and teaches post graduate and executive education courses in global leadership, talent management, business ethics, and luxury marketing. In 2020, Dr. Selina won the Best Professor of the Year (London campus) and the HULT Faculty of the Year award. Her most recent research focuses on board engagement and corporate purpose. Dr. Selina's research has been published in the Academy of Management Proceedings (2020), the Encyclopaedia of Sustainable Management (2020) edited by Springer, and in peer-reviewed journals including Business Ethics: A European Review (2019), the Journal of Studies in International Higher Education (2019), The Journal of Higher Education Policy and Management (2018), and The International Journal of Management Education (2018). For over 26 years, Dr. Selina has worked in executive management in the technology and luxury industries. She has served as a Non-Executive Director in the travel industry and is a senior industry advisor in the area of corporate governance. In recognition for her services in governance, she was included in "The Female FTSE Board Report 2016: 100 Women to Watch", published by Cranfield School of Management for the UK Government, Department for Business, Innovation & Skills (Lord Davies Review). Dr. Selina holds a Ph.D. in Business Management from The British University in Dubai (UAE), an MBA. from the University of Clemson (USA), and B.A. in Economics from the University of Parma (Italy). She is fluent in four languages and a resident of Dubai.

Index

A

Accountability, xiv, xxv, xxvi, 4, 15, 44, 48, 56, 107, 194, 199, 250, 251, 253, 254, 256, 258–264, 274
Adult development, 218
Africa, xxi, xxii, xxv, xxvi, 24, 52–54, 57, 69, 70, 77, 78, 80, 110, 203, 206, 257
African National Congress (the ANC), 44–46
Altruistic, 5, 12
American Accounting Association, 274
Apartheid, 40, 42–45, 48, 58
Awareness campaigns, 175, 176

B

Baselines, 276
Bell type containers, 182, 184, 185
Beyond philanthropy, 10
Biowaste, 181
Black economic empowerment (BEE), 45, 46
Bombay stock exchange of India (BSE), 113
Brexit, xxii, 99, 100
Brundtland Report, 272
Business and Human Rights, 205–206
Business and Human Rights framework, 205
Business ethics, xix, xxiii, xxv, 70, 106, 194
Business for society, 10, 88, 93, 98–100, 192, 195, 199, 200, 204, 207–210, 230
Business strategy, 5, 9, 107, 231, 234, 241, 294
Business sustainability, 229–242
Business sustainability strategy, 239, 242
Business with purpose, 4

C

Cause-related marketing, 10–12
Centre, xxii, xxvi, 15, 26, 29, 45, 49, 88–101, 126, 176, 193
Change management, 235
Circular economy, xii, xiii, 142–156, 164–167, 177, 178
  policy, 142, 145–154
  strategy, xiii, 141–157
Climate change, v, vii, x, xii, xxii, 14, 31, 40, 50, 58, 107, 108, 111, 123, 125, 143, 152, 163, 173, 197, 209, 216, 238, 261, 296, 298
Closed loop systems, 143, 146, 154, 156
Coaching, xxiv, 217
Coaching practice, 217
Colonialism, 68, 72, 73, 75–78, 90, 93
Colonisation, 42
Communications, xxi, xxiv–xxvi, 12, 77, 81, 146, 175, 179, 222, 224, 236, 288
Community, vii, xix, 4, 7, 10, 11, 22, 25–27, 29–32, 34, 40–42, 58, 68–71, 75–77, 79, 80, 89, 94, 96, 97, 100, 101, 108–110, 124, 146, 149, 154, 175, 176, 180, 182–185, 193, 196–202, 206–208, 210, 220, 222, 223, 230, 234, 237, 238, 251, 254–257, 263, 264, 284, 294
Companies Act of 2013, 107, 112–114, 123
Competence-oriented, 6, 7
Competency development, 235
Competitiveness, 5, 6, 152, 154, 230
Conflicts, xxiv, 31, 75, 89, 90, 97, 98, 110, 216–218, 221, 222, 258, 264

Congruence, 12
Consciousness, 219–221
Constricted, 7, 9
Contribution to theory, 240–241
Core business, xiii, 7–9, 12, 15, 100, 193, 196, 198, 199, 230, 231, 241
Coronavirus (COVID-19), x, xi, 8, 15, 21–35, 46, 50, 54, 147, 197, 296
Corporate citizens, 209, 255
Corporate Communications, xxii
Corporate corruption, 258, 261
Corporate culture, 11, 235, 237
Corporate disclosures, 157, 250, 253, 258
Corporate governance, xix, xx, xxiv–xxvi, 56, 194, 258, 260, 262, 264, 295
Corporate social data, 273
Corporate social indices, 271–289
Corporate social responsibility (CSR), vi–xv, xix–xxvii, 4–16, 25, 44, 56–59, 67–81, 88–101, 105–135, 192–210, 216, 219, 221, 229–242, 250–264, 277, 287, 288, 294, 295, 298
   as catalyst for development, 191–210, 214
   expenditure activity wise, 111, 112, 124
   in India, 107–108
   investments, 123, 145
   Policy and Program Development, 235
   pressing global issues, x
   pressing societal issues, xiii, 201–202, 207–208
   reporting, xiv, 199, 250–264, 272
Corporate social responsibility index (CSRI), 111, 113, 114, 117–119, 121, 123
Corporate strategy, xiii, 231–232, 242
Corporations, ix, x, xxii, 25, 42, 50, 51, 54–56, 70, 88, 91–97, 100, 101, 106, 119, 120, 123, 124, 145, 156, 194, 200, 205, 233, 239, 251, 257, 260, 263, 272, 280, 282, 287, 288, 293, 294, 296–298
Corruption, xxvi, 41, 46, 47, 50, 51, 55–56, 58, 59, 76, 97, 173, 252, 257–260, 264
Covid-19 pandemic, xi, xiv, xix, 22, 29, 30, 32, 33, 46, 47, 59, 99, 100, 197
Creating shared value (CSV), 5
Creativity, xxii, 15, 222, 241
Credibility, 12, 262
Cross-organizational comparability, 272, 280

D
Decent work, xii, 14, 30, 39–59, 208
Decision making, xiii, xiv, xxii, xxiv, 13, 30, 81, 108, 109, 197, 201, 209, 217, 218, 220, 223–226, 234, 235, 272, 275, 285, 288, 294, 297, 298
Dependency, 52, 73, 76, 148, 169
Developed countries, 23, 52, 167, 196
Developing countries, 24, 29, 35, 69, 75, 91, 92, 101, 110, 111, 145, 154, 196, 203, 204, 252, 253, 258, 264, 287
Developing economy, 58, 77
Developments, v, vii, xi, xii, xix, xxi–xxiv, 10, 22, 23, 25, 29, 30, 32, 35, 40, 42–44, 47–51, 53, 55, 56, 68–77, 79, 80, 88, 90, 91, 94–96, 98–101, 106–110, 119, 120, 124–127, 142, 144, 149, 156, 176, 178, 179, 193, 196, 198, 199, 208, 210, 217–220, 231–239, 263, 272, 278, 281, 282, 284, 287–289, 298
Disaster relief, 112, 114, 117–120, 123, 133
Diseases, xi, 23, 24, 29, 50, 76, 202
Dispersed, 7, 8
Drinking water and sanitation, 112, 117, 118, 120, 123

E
Eco designs, 142, 143, 146, 151, 154
Ecological, xxvii, 4, 107, 126, 144, 156, 181, 183
Economic growth, xii, 14, 30, 39–59, 106, 197, 208, 282
Economics, v–vii, ix–xiii, xix, xx, xxii, xxv–xxvii, 5, 6, 15, 22–30, 40–45, 49, 51, 53–56, 58, 69–74, 77–81, 88, 89, 91–93, 95, 98–101, 106, 107, 109, 110, 141, 142, 144–146, 155, 156, 195, 196, 199, 207, 230, 234, 236, 239, 254, 272–274, 281, 294–297
Eco systems, 154
Education, vii, x, xii, xxi, xxii, xxv–xxvii, 10–12, 14, 28, 29, 40, 51, 68, 73, 74, 77, 78, 80, 112, 114, 116–119, 123, 124, 126, 127, 166, 176, 177, 179, 180, 183, 185, 194, 196, 197, 200, 208, 226, 232, 281, 282, 284–286, 288
Emissions, v, xiii, 14, 26, 31, 107, 142, 143, 145, 146, 148, 152, 154, 155, 163, 164, 239, 258, 273, 275
Employee codes of conduct, 234
Employee Communications, 235
Empowerment and skill development, 112, 123
Energy, xii, xiii, 14, 29, 30, 74, 89, 123, 144–146, 148, 151, 156, 167, 201, 209, 221, 234, 288

Entrepreneurship, xxiv, 57, 59
Environment, xi, 5, 9, 22, 23, 26, 32, 47, 49, 54–57, 69, 70, 79, 80, 94, 101, 106–108, 110–112, 117–119, 123, 129, 142–145, 149–156, 164–168, 171–174, 176, 180, 181, 183, 185, 195, 196, 198, 204, 209, 225, 230–232, 234, 237, 239, 241, 251, 254, 256, 258, 259, 264, 286
Environmental impacts, ix, xxii, 79, 109, 146, 147, 156, 239, 276, 277, 279, 297
Environmental issues, 4, 167, 170, 174, 175, 177, 180, 184, 185, 197, 210, 255, 278
Environmental reporting, 272, 273
Environment protection, 14, 172, 175, 176, 180, 182, 185
Environment, social and governance (ESGs), xxii, 293–298
Ethical, xxiii, xxvii, 5, 6, 69, 81, 192, 194, 195, 205, 225, 230, 234, 239, 251, 254, 257, 262, 279–281, 287
Ethical trading, 205
Ethics, vi, xii, xix, xxvi, 11, 26, 56, 79, 192, 233, 235, 279, 294
European Directives on waste management, 178
European Union (EU), xii, xiii, xxvi, 52, 99, 142, 145–156, 163–167, 169, 170, 173, 175, 179, 182, 184, 185, 297, 298
 circular economy policy, xiii, 141–157
 policy, 142, 145–154, 163
 sustainability policy, 147
Explanation of Conceptual Framework, 240
Exploitation of natural resources, v, 123
Extent of responsiveness, 112–113, 117, 120

F
Failed state, 257, 264
Financial reporting, xiv, 209, 258, 260, 274, 277, 279–281, 296, 298
Fragile state, 263–264

G
General Agreement on Trade and Tariffs (GATT), 89–91, 95, 96
Generated waste, 151, 178, 181
Ghana, xx, xxi, 80, 87, 92, 96, 97, 99
Global Compact, 69, 199
Global Fund, 202
Global Fund for AIDS, malaria and tuberculosis, 202
Globalization, ix, 241
Global north, vii, x, xii, 74, 77, 80

Global Reporting Initiative (GRI), xiv, xxiii, 250, 253, 271–289, 297
Global south, vii, x, xii, 67–81
Governance, vi, xii, xiv, xix–xxi, xxv, 24, 33, 41, 42, 44, 48, 49, 51, 54, 56, 58, 80, 108, 145, 194, 209, 230, 232, 252, 257, 258, 262–264, 294, 296, 297
Government, v, ix, x, xxii, xxv, 7, 15, 22–24, 26–30, 33–35, 39, 44–46, 48–51, 53–59, 68, 70, 75, 78, 80, 88, 89, 92–94, 96–101, 109, 110, 114–116, 124–127, 142–145, 152–156, 166, 169, 170, 193, 196–200, 202–204, 207, 210, 234, 251, 254, 258, 261, 263, 282, 286, 288
GRI Standards, xiv, 272–281, 285, 287–289

H
Health, vi, vii, xi–xiii, xxv, 9, 12, 14, 15, 23, 24, 27–30, 33–35, 40, 43, 68, 71, 74, 77, 79, 80, 94, 99, 109, 112, 114, 117–119, 123, 124, 126, 128, 150, 154, 164–166, 169, 172, 180, 181, 185, 193–197, 200–203, 208, 209, 276, 282, 284–288, 294
Healthcare systems, 10, 23
High level of responsiveness to CSR implementation, 114, 117, 118
Historically disadvantaged persons, 45, 46
Holistic change, 219
Human Development Index (HDI), xiv, 272, 280–285, 287, 289
Human resource management (HRM), xiii, xxi, xxiii, xxv, xxvi, 201, 231, 233, 236, 239
Human resource management strategy, 232–233, 240
Human resources (HR), xxi, 11, 13, 231–236, 241
Human resource strategy, 242
Humanity, 4, 22–24, 32, 34, 35, 40, 78, 92, 295

I
Identity politics, 92, 95, 99
Incommensurability, 275, 281
Incomparability, 275
Independence, xii, 47, 53, 71, 72, 74, 78, 90, 199, 220
Industrial economics, 41, 42
Industries, xi, xiv, xx, xxi, xxv, 14, 22, 30, 33, 42, 43, 46, 50, 52, 53, 56, 68, 69, 80, 89, 98–100, 107–111, 116, 142, 144–150, 152–156, 164, 179, 192, 197, 204, 205, 208, 237–240, 250, 258, 260, 298

Inequalities, v, ix–xi, xiii, 15, 24, 29, 30, 40–46, 50, 56, 58, 71–73, 76, 79, 126, 196, 197, 208, 209, 216, 282, 295
Informal sectors, 24, 41, 42
Information campaigns, 175–183, 185
Informed decisions, 272, 274–278, 280, 281, 285, 288, 289
Infrastructure, xxi, 14, 30, 68, 95, 144, 146, 169, 172–178, 180, 182–185, 208, 288
Innovation, xi, xxv, xxvi, 9, 11, 14, 22, 30, 32–34, 40, 41, 53, 145, 146, 154–156, 167, 196, 197, 200, 208–210, 217, 235, 237–239, 241, 296
Innovation promoted by CSR, 11, 196, 200, 235, 241
Institutionalization, 298
Institutional theory, 250–252, 257, 259
Institutional voids, 259, 262
Institutions, vii, x, xiv, xxvii, 7, 15, 25, 31, 46, 51, 54, 55, 68, 73, 75, 77, 91, 127, 145, 239, 250–253, 256–264, 282, 286
Integral change, 217
Integral theory, 217
Integrated reporting, xiv, xxiii, 293–299
Intercommunity Development Associations (IDAs), 169, 174, 175, 180, 183–185
International Labour Organisation (ILO), 41, 47, 49, 52–54, 56
Investors, xxi, xxii, xxv, 53, 93, 156, 204, 207, 256, 277, 287, 288, 293–298
Isomorphism, 252, 257, 262

K
Knowledge, viii, xiii, xix, 11, 67–82, 152, 171, 173–175, 177, 178, 180, 182–185, 220, 237, 240, 242, 252

L
Labour standards, xxv, 49, 52
Landfilling, 165–167, 169, 170, 173, 178, 184
Languages, xii, xxii, xxv, 11, 71, 73, 78, 88, 94, 95, 101, 172, 222, 261, 279
Leadership, xx, xxi, xxiv–xxvi, 22, 32, 33, 74, 110, 148, 217, 231–233, 241, 262
Lesotho, xii, 39–59
Level of knowledge, 170, 172, 174, 175, 177, 178, 181, 183–185
Limitation of research, 171, 253
Livelihood, xiii, 30, 41, 74, 99, 112, 116–118, 120, 123, 125, 126, 196–198, 206–209, 216

Local content, vii, xii, xxi, 81, 87–101
Lockdown, 22, 23, 25, 26, 28, 31, 32, 46, 50
Low level of responsiveness to CSR implementation, 114, 117, 118, 123
Loyalty, ix, 13, 239

M
Managerial decisions, 294
Managerial implications, 142, 241, 242
Managers, xx, xxiii, 4, 6, 7, 9, 87, 232–235, 241, 254, 262, 297
Marikana, 43
Market-oriented, 6, 7
Materiality, 250, 255, 261
Measurement and reporting, 297
Millennium Development Goals (MDGs), 47, 50, 54, 58, 196
Mindset, vi
Mining operations, 108, 110
Ministry of Corporate Affairs, 107, 123
Missions, 8, 57, 71, 79, 125, 197, 231, 234, 235
Moderate level of responsiveness to CSR implementation, 114, 118, 123
Modern slavery, 194, 203–206, 208, 294
Moral philosophers, 279
Motivations, 9, 171, 239, 253–257, 264, 277, 294

N
National insecurity, 95
National stock exchange of India (NSE), 113, 258
National Strategic Development Plan (NSDP), 49–52, 58
Neo-liberalism, vii, 67
New circular economy plan, 154
New Product development, 236
Nigeria, xiv, xxiii, xxvi, 70, 76, 80, 250–264
Norway, 95, 96, 165–169, 179, 262

O
Oil industry/sector, xiv, 249–265
Operational efficiency, 142, 147, 152
Opportunities, vi, ix, x, xii, xiv, 4–7, 9, 11, 13, 15, 24, 27, 29, 31, 41, 43–53, 58, 70, 72, 75, 76, 91, 116, 146, 147, 193, 196, 197, 200, 201, 205, 208–210, 220, 223, 224, 226, 235, 238, 239, 241, 294–298
Ordinal ranking, 275
Organisational legitimacy, 254

Index   307

Organisation for Economic Cooperation and
    Development (OECD), 14, 142, 153, 293
Organizational experience, 240
Orientation training, 235
Others, x, xv, xx, xxiii, xxvi, 4, 7, 23, 24, 26,
    33, 53, 55, 75, 89, 93, 94, 112, 114, 117,
    118, 121, 123, 124, 135, 144–146, 150,
    155, 156, 164, 197, 207, 216, 220, 222,
    251, 253, 274, 294
Overall CSR expenditure, 111, 112, 118, 119,
    121, 122, 124

P

Packaging waste, 151, 164–166, 184
Pandemic, v, vi, x, xi, 15, 21–35, 46, 47, 99,
    193, 197, 295–298
Pandemic/COVID, xi, xiv, xix, 22, 29, 30, 32,
    33, 46, 47, 59, 99, 100, 197
Partnerships, 12, 13, 27, 29, 31, 40, 41, 45,
    57–59, 78, 81, 176, 197, 198, 200, 202,
    208, 209, 262, 263
Pay as you throw, 178, 181–185
Peripheral, 7, 8, 74, 90, 98
Periphery, 74, 87–102, 198, 199
Perspectives, 4–7, 9, 11, 39–58, 96, 99, 106,
    141–157, 170–185, 193, 230, 233, 238,
    253, 256, 275, 282, 284
Philanthropic, 5–7, 9–11, 34, 73, 80, 94, 193,
    194, 198, 199, 201, 253, 264, 276
Philanthropy, xiii, xiv, 4–10, 26, 70, 107,
    192–194, 198–201, 259, 264
Policy, viii, xii, xiii, xix, xxii, xxv, xxvi, 23, 42,
    44–49, 53–58, 68, 70, 71, 74, 76, 77, 81,
    88–90, 93, 95, 96, 98, 106, 109, 112,
    126, 142, 145–147, 149, 152, 154, 155,
    163, 175, 177, 178, 194, 195, 198, 200,
    206–210, 229, 233, 235, 236, 241, 259,
    278, 282, 296–298
Political leadership, 258, 262, 264
Pollution, x, 26, 27, 31, 107, 108, 143, 150,
    153, 154, 178, 238
Populism, 98, 99
Post covid-19, 30
Postcolonial, xii, xxvi, 67–81
Post-colonialism, 72, 96
Power, x, xii, 5, 44, 52, 57, 67–81, 88–101, 109,
    110, 116, 220, 232, 255, 260, 261, 264,
    282
Power shift, xxii
Priority, 5, 13–15, 30, 34, 47, 49, 70, 74, 89,
    109, 110, 142, 149, 153, 154, 193, 194,
    198, 200, 202, 210, 218, 221, 231, 264

Product development, 236–239
Product service systems, 145, 155
Progressive companies, 114
Promotional, 10–12
Protectionism, 92, 98, 99
Public authority, 170, 171, 174–185
Purpose, vi, ix, xxv, 4, 9, 15, 16, 39, 44, 48, 70,
    97, 107, 149, 152, 196, 199, 200, 223,
    257, 274, 275, 298

R

Rating agencies, 297
Recycling, xiii, 142, 143, 146–151, 154, 155,
    164–167, 169–185
    in European Union, xiii
    rate target, 180, 182, 183
    resources, 146
    in Romania, 167, 170
Reduce, 8, 12, 30, 110, 142, 143, 146, 147, 149,
    151–153, 163, 166, 170, 177, 178, 185,
    196, 197, 201, 202, 204, 205, 230, 232,
    239, 256, 264
Reduced consumption of resources, 153
Regional integration, 40, 53, 57–59
Regulation, ix, xi, xii, xxv, xxvi, 55, 67, 73, 75,
    79, 81, 87–89, 95–97, 99, 107, 123, 145,
    152, 166, 175, 238, 250, 253, 254, 259,
    260, 262, 263, 285, 294, 296
Regulatory regime, 258, 264
Reluctance to change, 163–185
Reporting, xiii, xiv, xxvi, 68, 107, 109, 117,
    123, 126, 145, 156, 193, 196, 198, 199,
    204, 206, 236, 250–252, 255–261, 263,
    264, 271, 273, 274, 278, 280, 282, 284,
    285, 288, 289, 297, 298
Research and Development (R&D), 30, 236,
    237, 242, 286
Resilience, xix, 22, 31–33, 197, 220
Resource efficiency, 146, 147
Resource nationalism, 88, 92, 93, 95, 99
Responsibility, vii, xi, xiv, xxiv, 4–6, 9, 16, 25,
    55, 67, 80, 81, 94, 110, 126, 152, 156,
    164, 170, 173–176, 179, 180, 182, 184,
    192, 195, 230, 233, 235, 251, 252, 254,
    256, 258–262, 264, 282, 284, 297
Responsible business, 108, 156, 195, 200, 230,
    256, 262–264, 288, 295
Responsible investing, 156
Responsible supply chains, 214
Restoring resources, 155
Reuse, xiii, 142, 143, 146, 149, 151, 152, 154,
    155, 167, 177, 185

Rhetoric, xxii, 88–101, 109, 273, 279
Risks, xiv, 12, 13, 28, 34, 68, 148, 149, 152, 205, 209, 241, 250, 254, 256, 264, 284, 294–298
Rural upliftment, 112, 114, 117–120, 123, 124, 134

S
Safety net, 24, 35
Sanitation company, 178, 181, 183
Schedule VII, Environmental and Social Impacts, 126
SD and pressing global issues, 295
SD and pressing societal issues, 192, 195, 272, 295
SDG8, xii, 30, 40–42, 46–50, 52–57, 208
Sen's capability approach, xiv, 272, 273, 276, 282, 283
Separate collection, 169–185
Shared value, 193, 195, 199, 200, 222
Shareholder, ix, 4, 5, 93, 192, 194–196, 198, 200–202, 207, 209, 230, 235, 250, 254, 256, 262, 264, 296
Skills development, 52, 68
Social accounting, 273
Social and environmental accounting, 273
Social and environmental performance, 250, 274
Social and environmental reporting, 273, 279
Social auditing, 205
Social dialogue, 41, 42, 53, 55
Social distancing, 8, 23
Social entrepreneurship, 57–59
Social indices, xiv, 272, 275, 276, 279, 281–289
Social performance, xiv, 106, 230, 236, 238, 251, 256, 264, 272–277, 282, 285, 287, 288
Social responsibility, xiii, 4, 6, 56, 81, 106, 108, 110, 111, 192, 195, 199, 229, 235, 237, 238, 241, 251, 294, 295, 297
Social setting, 264
Society, vi, viii, x, xxiii, xxiv, xxvi, xxvii, 4–6, 9, 12, 15, 16, 24, 25, 27, 29, 32, 34, 35, 40, 42, 43, 56, 57, 68, 88, 93, 94, 98–100, 106, 110, 124, 143, 191–210, 219, 226, 229, 230, 232, 250, 253, 255, 258, 261, 264, 287, 294, 295, 297, 298
Solidarity, 26, 34
Sorting plants, 181
South Africa, xii, xx, xxiv, xxv, 39–59, 206
Southern African Development Community (SADC), xii, 40–42, 49, 51–53, 56–59

Sponsorship, 10, 13
Stakeholder, xiv, xxiii, 5–9, 14, 25, 27, 42, 48, 49, 53, 69–71, 75, 80, 81, 93, 94, 101, 107, 110, 123, 144, 146–148, 153–156, 192–198, 200, 201, 207, 221, 229–231, 235, 237–239, 241, 250, 251, 253–259, 261–264, 272–281, 285, 287–289, 296, 297
Stakeholder theory, 93, 106, 278
Standards, xii–xiv, 40, 55, 56, 69, 77, 79, 89, 91, 125, 142, 145, 148, 195, 199, 200, 203, 205, 210, 235, 258, 274, 280–282, 284–287, 295–298
State, xxi, xxiii, xxvii, 8, 24, 30, 32, 40, 44, 46, 47, 52, 56–59, 72, 73, 89–93, 95, 96, 98, 101, 114–116, 124, 126, 127, 146, 147, 149–151, 154, 155, 164, 165, 169, 175, 192, 198, 217, 218, 225, 230, 231, 236–238, 261, 279, 280, 282, 286, 288
Strategic CSR, xiii, 5, 6, 70, 199, 230, 235
Strategy, vii, xiii, xx, xxi, xxvi, xxvii, 8, 9, 14, 23, 25, 28, 46–50, 52, 53, 57, 58, 69, 70, 75, 79–81, 88, 93, 97, 142, 146, 147, 151, 154, 180, 183, 194, 195, 197, 200, 203, 230–242, 250, 255, 259, 262–264, 272, 278, 288
control, 97
development, 231, 232, 234
implementation, 231–233
Strict monitoring and accountability, 125
Sub Saharan Africa, 52, 55, 202
Supply chain audits, 195, 205
Supply chains, xiii, xxi, 93, 146, 147, 156, 192–195, 198, 201, 203–206, 208, 209
Sustainability, v, vi, xi, xiii, xix–xxiv, xxvi, xxvii, 3–16, 22, 33, 57, 97, 101, 106, 126, 147–152, 154, 193–207, 209, 210, 216, 234, 239, 240, 250, 251, 253, 254, 271–274, 277, 287, 297
Sustainability Accounting Standards, 297
Sustainability reporting, 199, 273, 274
Sustainable consumption, xxvii, 31, 125, 142, 146
Sustainable Development, vii–xv, xxi, xxiii, xxv, xxvi, 21, 27, 35, 41, 48, 56, 69, 74, 88, 96, 106, 108, 166, 177, 180, 192, 195, 199–201, 263, 272–274, 294, 295, 298
Sustainable Development Goals (SDGs), v, vi, xi–xiii, 13, 14, 16, 22, 27–32, 35, 39–59, 145, 147, 193–198, 200, 206–210, 222, 225, 226, 263

Sustainable production, xiii, 142, 143, 145, 147, 154–156
Swachh Bharat Abhiyan, 125
System door-to-door, 175, 182, 185
Systemic change, 219, 225–226

T
Tax payments, 274, 286
Teal Mindset, 215
Trans-national corporations (TNCs), ix–xi, 68–70, 73–76, 78–81
Transparency, xiii, xiv, xxv, 15, 55, 56, 107, 148, 173, 204, 239, 251, 253, 258–261, 263, 264, 272, 274–279
Triple bottom line (TBL), 69, 193, 196, 198, 200, 272, 273, 279, 297
Triple Win CSR, 199
2030 agenda for sustainable development, 57, 147, 163

U
UK, xix, xx, xxii, xxiv, xxv, 26, 34, 99, 204, 206, 255, 296
Uncertainty, vi, 12, 22, 23, 25, 32, 51, 68, 92, 99, 220
Unemployment, 24, 40–43, 45, 46, 48–50, 52–54, 58, 295
United Nations (UN), 13, 27, 40, 49, 53, 163, 193, 196, 199, 205–206, 216, 281, 287
  Framework on Business and Human Rights, 205–206
  Global Compact, 192, 199, 210, 239, 250, 277, 281, 293

UN Sustainable Development Goals (UNDP), xxvi, 4, 13–16, 50, 53, 216, 282
USA, xxi, xxiv, xxv, 99, 192

V
Value creating, 4, 10–12
Value creation, 5, 7, 10–13, 231, 294, 296, 298
Values, xiii, xxiv, xxv, 4, 7, 11, 12, 14, 15, 25, 44, 46, 56, 57, 69, 70, 73–75, 77, 80, 91, 94, 95, 98, 106, 141–156, 194, 196, 200, 201, 205, 207, 209, 215–227, 229, 231, 233–237, 239, 250, 254, 256, 264, 274, 279–281, 283, 288, 294, 296–298
Values-based business, 231, 234
Values-based change, 218
Values-based decision making, xiii, 209, 218
Vision, xxiv, 45, 47, 51, 53, 197, 223–225, 231, 234, 235, 262
Vulnerabilities, vii, 24, 27, 28, 35, 50

W
Waste management, 143, 146, 149, 154, 164, 167, 169–175, 178–180, 182–185
WEEE collection, 175, 176
Workers' rights, 43, 53, 80, 275
Workforce planning and recruitment, 234
World Trade Organisation, 90, 91

Z
Zimbabwe, xii, 39–59
Zimbabwe Congress of Trade Union (ZCTU), 54

Lightning Source UK Ltd.
Milton Keynes UK
UKHW022046170821
388990UK00001B/9